Reproductive Physiology
of Mammals and Birds

A Series of Books in Agricultural Science

Animal Science
Editor: G. W. Salisbury

THIRD EDITION

Reproductive Physiology of Mammals and Birds

The Comparative Physiology of Domestic and Laboratory Animals and Man

A. V. Nalbandov
UNIVERSITY OF ILLINOIS

With a chapter by
Brian Cook
ROYAL INFIRMARY, GLASGOW

W. H. Freeman and Company
San Francisco

Library of Congress Cataloging in Publication Data

Nalbandov, Andrew Vladimir, 1912–
    Reproductive physiology of mammals and birds.

    Previously published under title: Reproductive
physiology.
    Includes bibliographies and index.
    1. Reproduction.   2.   Physiology, Comparative.
I. Cook, Brian.   II.   Title.   [DNLM: 1.   Physiology,
Comparative.   2.   Reproduction.   QT4 N165r]
QP251.N32   1975   599'.01'6   75-25890
ISBN 0-7167-0843-4

Printed in the United States of America

10   9   8   7   6   5   4   3   2   1

To the memory of
A. N. N. and V. S. N.

Contents

Preface

The physiology of reproduction is a composite science. For an understanding of the subject it is necessary to draw upon the accumulated knowledge of endocrinology, histology, embryology, anatomy, physiology, and genetics. To write a truly comprehensive treatise on reproductive physiology would require many volumes; such a work would be excellent for purposes of reference but unusable as a textbook. Accordingly, the present edition, like preceding editions, is limited to essential aspects of the subject. It was thought unnecessary to submerge the reader in the minutiae that are important to the mature specialist but distracting and puzzling to the beginner. This book allows the young scientist to become acquainted with the entire field and to decide for himself what is worth pursuing in detail.

The book emphasizes the diversity of the reproductive process by comparing the reproductive systems of different species. Quite often it is not possible to extrapolate from our knowledge of one species to another. It is becoming more and more apparent that much of the experimental work done on rats—hitherto the most widely used experimental animal—will have to be done over in order to be applicable to domestic animals or man.

Dr. Brian Cook of the Royal Infirmary, Glasgow, has written a new chapter on the chemistry of hormones involved in reproduction. This chapter includes a discussion of the theory and use of radioimmunoassay, a technique that is widely used at present.

I have followed the practice established in the previous editions of not documenting all statements (about 1,500 citations would be required to provide complete documentation). Although some readers have deplored the paucity of references, others feel that frequent citations of the literature are distracting. One instructor who uses this book requires his students to find documentation for statements that he considers important, and thinks that this practice is an excellent pedagogic device. I frequently cite the excellent articles on reproductive endocrinology published by the American Physiological Society in their *Handbook of Physiology* (Section 7). This work and the annual series *Recent Progress of Hormone Research* contain important new contributions as well as reviews by outstanding authorities in the field of reproductive endocrinology. Both are also an invaluable source of references to original research papers. A recent and unfortunate trend among younger colleagues is the tendency to ignore the older literature. As a result, their research frequently "rediscovers" phenomena that were recorded years ago. For this reason I have in many instances retained older data and references, in the hope that students will delve into the older literature.

Although comparisons between primates and nonprimates are frequently made in this book, experimental data for human reproduction are almost totally lacking. The reasons for this are the sanctions against purely experimental surgical incursions in humans, which makes them poor experimental subjects. Ovulation in ewes injected with an LH-releasing substance can be verified by actually counting the ova after laparotomy. On the other hand, results in women treated with an LH-releasing hormone must be inferred, and it is assumed that if blood progestin rises, ovulation has occurred. However, it is equally possible that the follicles may have luteinized without shedding the ovum.

Similarly, it is impossible to include large numbers of women or men of similar ages in single endocrine experiments, although this is routinely done with laboratory and domestic animals. For these reasons I have relied heavily on data from domestic and laboratory animals, without neglecting what I consider to be good data on human subjects.

As in previous editions I want to acknowledge the enormous debt I owe to the graduate students and postdoctoral fellows with whom I have worked, and who continue to provide stimulation both in the laboratory and in discussions. I gratefully acknowledge the help of Elizabeth Waldron in the preparation of this book.

July 1975   A. V. Nalbandov

Reproductive Physiology
of Mammals and Birds

Introduction

The biological problem of the perpetuation of the species has been solved in a bewildering variety of ways in nature. Comparing across species one is struck by the diversity of reproductive systems and the comparative uniformity of other physiologic systems, such as respiration, digestion, and circulation. Evolution seems to have arrived at good solutions for some physiologic problems, modifying the basic principles only slightly or not at all in the species; by contrast, the reproductive process has been drastically modified in some species. As we shall see, these modifications extend from the anatomy of the female reproductive system to the neuroendocrine mechanism controlling reproduction. (Curiously, the evolutionary modification of reproductive processes is largely restricted to females; reproduction in males is remarkably uniform across species.)

The assumption that reproduction is the most important physiologic function and that diversity maximizes the chance of success in the competition between species is not tenable. No one mechanism of reproduction has an advantage over another simply because *all* species, including the primitive platypus and the didelphians, have been more or less equally successful in maintaining themselves over the ages. Moreover, other physiologic systems are equally vital to the survival of the species; inadequate circulatory or digestive systems, for instance, would not even permit young members of a species to reach reproductive age.

Some scientists resist accepting diversity in reproductive systems, insisting instead on finding a unified scheme that would fit all mammals. The danger of this thinking is that one is tempted to generalize from studies of single species, a temptation to which many scientists have succumbed in the past. That this cannot be done will be frequently illustrated in this book. Readers should be alert to this fact and beware of making unwarranted generalizations.*

Because this book deals with mammals and birds, we shall be concerned mostly with sexual reproduction of the dioecious type. Nevertheless, asexual reproduction, in the general sense of the term, occurs even in mammals and birds; all processes of growth are basically due to the asexual fission of somatic cells.

The two sexes involved in dioecious reproduction are highly specialized. The degree of specialization is the result of the complex synchronization of reproductive events, which, as a rule, follow predictable, well-regulated patterns. The details of these patterns differ in various species. Reproductive events are regulated by a complex of interlocking hormone systems, which are themselves locked into a neural control system. The neurohormonal control system provides a maximum of checks that ensure a normal and well-balanced functioning of the end organs affected by hormones, and this in turn results in synchronization of the functions of the sex mechanism. The neuroendocrine system and the anatomy and physiology of the sex mechanisms, both of which are largely controlled by the neuroendocrine system, are the main subjects of this book.

The processes of reproduction occasionally decline in efficiency because one of the hormonal, neural, or humoral links becomes impaired, or because they function out of phase, disrupting the sensitive mechanism of synchronization that underlies most reproductive events. When this happens, partial or complete sterility results. Because of its importance in domestic animals and humans, sterility is discussed later in the book. First, however, we shall consider some of the basic concepts of sex, the anatomy of the reproductive system, and the hormonal and other signaling systems that have evolved in the different species. The

*For a more detailed discussion of diversity in reproduction see Nalbandov, "The Fourth Hammond Memorial Lecture. Puzzles of reproductive physiology." *J. Repro. Fert.,* 34:1, 1973.

rat, the elephant, and the chicken all face basically the same problems in reproducing the species. Eggs must be matured and shed, and a signal must be given to the male that the latter event is about to happen in order to ensure the presence of viable sperm. These and subsequent events, such as pregnancy, parturition, and lactation, or the incubation of eggs by birds, depend on humoral and neural feedback mechanisms that form chains of events. It is essential to remember that no reproductive event ever stands alone, that any one event is only one link in a chain. Although thinking in terms of chains of events rather than isolated phenomena may be difficult at first, such an approach is necessary to understanding reproductive phenomena.

1

The Biology
of Sex

This chapter briefly summarizes present thinking on the subject of sex ratios. (Students interested in details should consult the works of Lawrence and Crew, in that order.) Sex ratios deviate from equality frequently and significantly; this deviation may be caused by a variety of factors, such as genetic selection, season, age of dam, parity of dam, and frequency of ejaculation by the male. Because the internal and external environment influences sex ratios, it is incorrect to assume that the "normal" sex ratio must be 50/50. The sex ratios of dogs, pigs, cattle, and rodents are normally high; those of horses, sheep, and chickens are normally low. The origin of sex chromosome aneuploidy is discussed and it is pointed out that reproductive or endocrine disorders are very common in both men and women with the genotypes XXY or XO. Whether sex chromosome aneuploidy occurs in other animals as frequently as in man is not known. The sex chromatin serves as a diagnostic feature of the genetic sex of embryos before sex differentiation as well as of normal and intersexual adult mammals.

INTRODUCTION

The word "sex" comes from the Latin word *sexus*, which means division (from *secare*, to cut or separate). Sex is not a biological entity, but "the sum of the peculiarities of structure and function that distinguish a male from a female" (*Webster's New International Dictionary*, second edition). Not all populations can be divided into two distinct groups of males and females, especially if one uses only one or two external sex characters as the basis of division. Populations are actually composed of individuals differing to a greater or lesser degree in the sum of their peculiarities of structure and function. In some instances the sexes overlap completely, making it impossible to distinguish them externally. For a clearer understanding of the continuous variation between the sexes, we shall distinguish between *genotypic* and *phenotypic* sex.

GENOTYPIC SEX

Genotypic sex is determined by the sex chromosomes received from the parents (Table 1–1). In man much work has been done on the genotypic (chromosomal) sex of individuals who show endocrine (including reproductive) and mental disorders. In individuals showing the Klinefelter syndrome, which is characterized by gonadal hypofunction, the genotype is usually XXY (instead of XY); in those showing Turner's syndrome, which is characterized by gonadal agenesis or aplasia, the genotype is frequently XO rather than XY or XX.

These unusual genotypes of the sex chromosome complex are called *aneuploidy;* they are ascribed to errors in chromosome division occurring during meiosis. If as the result of nondisjunction both X chromosomes go to the ovum, fertilization by a sperm will lead to the formation of either an XXX or an XXY zygote (maternal nondisjunction). In paternal nondisjunction some sperm will be of the genotype XY and others will have no sex chromosome (O). Fertilization of eggs of a normal genotype will result in zygotes of either the XXY or XO genotypes. As already noted, such aneuploid individuals very frequently show gonadal or endocrine defects.

Similar studies in other animals are lacking, and it is not known how frequently aneuploidy of sex chromosomes occurs in them. It appears that cases of gonadal disfunction or gonadal aplasia seen in domestic animals may possibly be due to aneuploidy, unless the profit motive in

Table 1-1.   Sex chromosomes in male and female mammals and birds

	TYPE OF CHROMOSOME IN GAMETES	PROPORTION OF CHROMOSOME IN GAMETES
Female mammal	X only	All eggs carry X
Male mammal	X and Y	Half of sperm carry X, half carry Y
Female bird	X and Y	Half of eggs carry X, half carry Y
Male bird	X only	All sperm carry X

animal breeding has resulted in the elimination from the breeding population of families prone to aneuploidy. Research along these lines would be of great theoretical and perhaps practical significance and might explain why intersexuality is so frequently seen in some species (goats and pigs), but rare in others.

PHENOTYPIC SEX

To some extent genotypic sex may be compared to a genetic character controlled by a single gene—it is an all-or-none character. With respect to sexuality, the "sum total of peculiarities of structure and function" is either male or female. This implies, and to some extent correctly, that the characteristics of "maleness" or "femaleness" are unalterably fixed when the sperm fertilizes the egg and are not subject to environmental modification, just as the genetic whiteness of an albino animal is not subject to environmental modification. Actually, the manifestation of sexuality is as variable as the somatic characters that are determined by multiple genes. This variability can be seen in the different degrees of masculinity or femininity normally found in a population. The degree of expression of one or the other sex is called phenotypic sex.

To explain how such an all-or-none character as sex can be subject to variability, C. B. Bridges advanced the *genic balance* theory of sex determination. This theory accepts the basic assumption that sex is determined at the time of fertilization by a combination of the sex chromosomes. In addition, the theory assumes that the autosomes carry genes for maleness and the sex chromosomes carry genes for femaleness. If the number of genes for maleness on the autosomes is large, then in the case of an XY male these genes overbalance the genes for femaleness carried on the sex chromosomes. Such a male may be phenotypically more masculine than a male that has received fewer maleness genes on its autosomes. It is postulated that though the XY type of sex determination is permanently fixed and is not subject to genetic variability the presence of maleness genes on the autosomes makes divergence in the phenotypic expression of sexuality possible. The sex chromosomes establish the blueprint for genotypic sex; autosomal genes cannot alter this blueprint drastically in most higher vertebrates, but they may be responsible for the variations in the expression of phenotypic sex observed in most higher animals, including man.

It is not known how the autosomal sex genes accomplish phenotypic, or external, sex modification. They may act by controlling the glandular mechanisms of both sexes and by governing the rates at which sex hormones (which are responsible for phenotypic sex traits) are secreted. It may be that large amounts of one or the other of the sex hormones produce "masculine" individuals, smaller amounts produce "average" males or females, and still smaller quantities result in "effeminate" individuals.

SEX RATIOS

In view of the preceding discussion it may be assumed correctly, that the combination of sex chromosomes effectively determines the proportions of the sexes in a population and holds them near equality. A casual inspection of available statistics dealing with sex ratios bears out this assumption. Nevertheless, sex ratios sometimes deviate very significantly from the expected equality. These exceptions deserve close scrutiny; but before we take them up, a few general statements concerning terminology and the factors affecting sex ratios will be useful.

It is conventional to express sex ratios as the number of males per 100 females. If we say that the sex ratio of a certain population is 108, we mean that there are 108 males per 100 females at the time the count was made. Occasionally, for greater ease of statistical manipulation, it is desirable to express sex ratios as the percentage of males (as in Table 1-2). Three different figures expressing sex ratios are generally found in the literature: the primary, secondary, and tertiary sex ratios.

The *primary sex ratio* is the proportion of males to females at the time of fertilization or conception. This is only a theoretical concept, for it is virtually impossible to determine the sex of the zygote or of the very early embryo. The concept is useful, however, and we shall return to its meaning later. If a way were found to ascertain routinely the sex of zygotes or very early embryos, significant contributions could be made to the understanding of the problem of sex ratios at the time of fertilization.

The *secondary sex ratio* expresses the proportion of the sexes at the time of birth. Strictly speaking, this term is used only with reference to sex at birth, but frequently it is made to include the proportion of the sexes during intrauterine development, from the time the sex can first be known to the time of birth. For reasons that will become obvious shortly, this use should be avoided.

The *tertiary sex ratio* is of no great concern here; it is used in vital statistics to express the proportion of the sexes in the postpubescent population.

We have agreed that if the theory of the chromosomal type of sex determination is correct we should find the two sexes about equal in the primary and secondary ratios. Table 1–2 summarizes the sex ratios at birth of a variety of mammals and birds. One is struck by the great variation in the sex ratios of the animals listed and particularly by the deviations from equality even within the same species. Before we attempt to interpret these observations, let us see what factors have been found to modify sex ratios experimentally.

In a classic study Helen Dean King found that sex ratios may be modified by genetic selection. She started with a strain of rats that had a secondary sex ratio of about 110. By combining inbreeding with selection, she was able to split this strain into two substrains, which differed significantly in their sex ratios. By the twenty-fifth generation of inbreeding and selection, the sex ratio of the high strain had risen to 124; that of the low strain had dropped to 82. This and other experiments show that the sex ratios of animals can be modified by genetic selection. The mechanism through which this is accomplished remains unknown. It is improbable that genetic selection modifies the role of the entire sex chromosome in sex determination. It is more likely that modification is produced by selection for a greater or lesser number of the sex genes on the autosomes and the X chromosomes. Selection for a larger number may shift the sex ratio in the direction of an increasing proportion of males; selection for fewer maleness genes would cause the ratio to go in the direction of a preponderance of females.

The frequency of ejaculations by the male also seems to have an effect on the secondary ratio. Stallions that were used frequently for mating produced a high sex ratio of 101, whereas those used one-third as often produced a ratio of 97. Seasonal factors seem to influence the ratio in mice. A. S. Parkes found a sex ratio in mice 98.9 during March through June, and 133.2 during July through October. I. Johansson, making a similar comparison in dairy cattle, found no such difference.

The ratio may change with increasing parity of the female. The ratio in rats drops from 122 for the first litter to 103.1 for the fourth. However, Parkes found the opposite effect in mice: the first birth had a sex ratio of 124.3 but subsequent births rose to 150.0.

These findings strongly suggest that both genetic and environmental (internal and external) factors are capable of modifying the secondary

Table 1-2.   Secondary sex ratios in some birds and mammals

	SAMPLE POPULATION	MALES (HIGH RATIO)	SAMPLE POPULATION	MALES (LOW RATIO)
Canary	68	77.9%	200	43.5%
Dog	1,400	55.4	6,878	52.4
Pig	2,357	52.8	16,233	48.8
Mouse	2,903	52.6	1,464	44.4
Rat	1,001	51.9	1,862	46.2
Cattle	4,900	51.8	982	48.6
Guinea pig	7,989	51.7	2,014	49.4
Horse	25,560	49.9	135,826	49.1
Sheep	50,685	49.5	8,965	49.2
Domestic chicken	20,037	48.6	2,501	46.8

SOURCE: Data modified from various sources, especially from P. S. Lawrence, *Quart. Rev. Biol.*, 16:35, 1941.

sex ratio. It is reasonable to suspect that the so-called "normal" sex ratios (Table 1-2) were subjected to modifying influences of one kind or another and that therein may lie the explanation for the great variability in ratios found both between and within species. In studying the sex ratios of a population, one should find the "normal" sex ratios that are typical of the population, rather than assume that they should be near equality and agree with the ratios found by other investigators. One should also remember that some species have high "normal" ratios (dog: 110-124), others low "normal" ratios (horse, sheep, and chicken all have ratios of less than 100).

One disturbing finding has yet to be explained satisfactorily. It has been found repeatedly in humans that the sex ratio of abortions (both spontaneous and induced) is extremely high and that at certain stages of pregnancy three or four times as many male as female fetuses may be aborted. Studies of fetal ratios in domestic animals have shown that during the early stages of gestation there are significantly more males than females among the conceptuses. A University of Wisconsin study, for instance, showed that the ratio of fetal calves 5-10 cm in length was 193, or 66 percent males. It decreased to about 100 near term. Similar observations have been made on the ratios of sheep and pigs.

There are two plausible explanations for the high ratio of males in human abortions and the preponderance of males during early phases of gestation in domestic animals. It is possible that the primary sex ratio is not equality, and that more males than females are conceived. If the intrauterine mortality is primarily at the expense of the male, the excess of males is eliminated and the sex ratio approaches equality by the time of parturition. If this assumption is made, it does not invalidate the chromosome theory of sex determination, but it calls for the further assumption that in mammals the two types of sperm are not produced in equal numbers, or that, if they are produced in equal numbers, the Y-bearing sperm has a much greater chance of fertilizing the egg than the X-bearing sperm. Although either or both of these suppositions are possible, there is no experimental evidence for them.

The second explanation, which fits the facts of the case and which seems probable, assumes that the primary ratio is equality but that during the early embryonal stages, during which the sex of the conceptuses cannot be readily determined, embryonal mortality is primarily at the expense of the female. By the time the sex can be determined, the ratio is in favor of the male. From then on until parturition, fetal mortality occurs mostly at the expense of the male; so, by the time of birth, equality has been reestablished. There is need for much additional work on this interesting subject, but what is needed most is a method that would permit the routine determination of the primary sex ratio.

Many attempts have been made to control sex. None of the attempts—electrophoretic, mechanical, or chemical separation of the X- and Y-bearing sperm—has succeeded. A solution of this problem would have an enormous significance to the field of animal breeding, for in many branches of livestock production males are unwanted. A fortune awaits the scientist who finds a practical method of regulating the proportion of the sexes without resorting to the introduction of lethal genes and thus reducing the productivity of the mother. It should be remembered that any method that succeeds in domestic animals will also work in man. Whether we are ethically, morally, and politically advanced enough to use such a method wisely, once it becomes available, remains to be seen.

SEX CHROMATIN

We have noted that the primary sex ratio is only a theoretical concept because it is virtually impossible to diagnose sex at the time of conception. This statement is essentially correct, but it has to be modified to some extent, for it is now possible to determine the genetic sex of very young embryos by use of the "sex chromatin." This mass of chromatin is present in the somatic nuclei of all the mammals studied with the possible exception of rodents. The sex chromatin is about 1 micron in diameter, is feulgen-positive, and is usually found lying against the inner surface of the nuclear membrane, although in nerve cells it may lie adjacent to the nucleolus (Figure 1-1). The sex chromatin is present only in genetic females and is thought to be derived from the fused heteropyknotic portions of the two X-chromosomes.

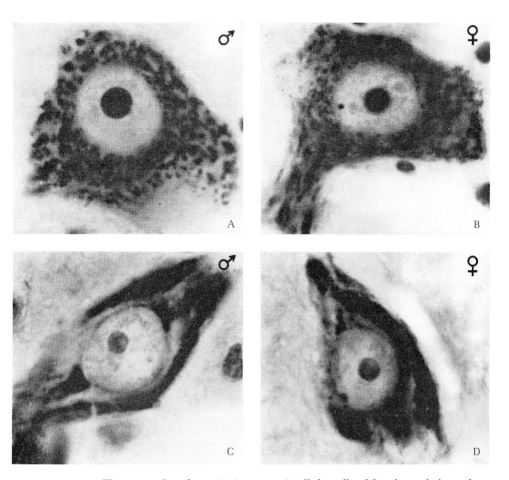

Figure 1-1 Sex chromatin is present in all the cells of females and absent from those of males. **A.** A cell from the spinal cord of a normal bull calf. **B.** A similar cell from a heifer calf; note the chromatin mass at nine o'clock. **C.** A nerve cell from a male monkey. **D.** A nerve cell from a female monkey; the sex chromatin lies against the cell membrane. [From Moore *et al.*, 1957, *J. Exp. Zool.*, 135:101; and Prince *et al.*, 1955, *Anat. Rec.*, 122:153.]

The sex chromatin can be used for diagnostic purposes in a variety of ways. The genetic sex of embryos as small as 1 mm in length can be determined before sexual differentiation and long before the gonads have differentiated to the point where sexing can be done histologically. The sex chromatin is even present in primordial germ cells before they have arrived at the genital ridge. Now it is possible to predict the sex of human babies from biopsy specimens of embryonal tissue without obvious damage to the fetus or to the normal course of gestation.

In adult human intersexes, in whom both phenotypic sexes may be equally well developed, it is possible to determine the genetic sex from skin biopsies and oral or nasal smears, and to decide accordingly whether hormonal or surgical therapy should favor maleness or femaleness.

The discovery of the sex chromatin is still too new to permit an evaluation of the contributions that such studies can make to our understanding of early sex differentiations. Probably most important is that the sex chromatin, as a diagnostic criterion of genetic sex, permits much greater assurance of accuracy of sexing in studies concerned with the embryology of sex. Only additional studies will show whether the sex chromatin is unalterably fixed at the time of conception or can be modified by the action of sex hormones. Apparently, as in freemartins, the sex chromatin is not modified because genetic females are exposed to the prolonged action of male sex hormones. Similar studies in human intersexes are in progress. The sex chromatin now appears to be a useful tool for the study of sex embryology, sex differentiation, primary sex ratios, and intersexuality.

SELECTED READING

Bridges, C. B. 1939. The genetics of sex in drosophila. In *Sex and Internal Secretions,* 2nd ed., Williams & Wilkins.

Crew, F. A. E. 1952. The factors which determine sex. In *Marshall's Physiology of Reproduction,* 3d ed., Vol. 2, Longman.

Lawrence, P. S. 1941. The sex ratio, fertility, and ancestral longevity. *Quart. Rev. Biol.,* **16**:35.

Lennox, B. 1956. Nuclear sexing: A review incorporating some personal observations. *Scottish Med. J.,* **1**:97.

Nalbandov, A. V. 1973. The Fourth Hammond Memorial Lecture. Puzzles of reproductive physiology. *J. Reprod. Fert.,* **34**:1.

Nowakowski, H., and W. Lenz. 1961. Genetic aspects of male hypogonadism. In *Recent Progress in Hormone Research,* Vol. 17. Academic Press.

Pearl, R. 1939. *The Natural History of Populations.* Oxford University Press.

Tietze, C. 1948. A note on the sex ratio of abortions. *Human Biol.,* **20**:156.

2

The Structure of the
Male and Female
Reproductive Systems

This chapter lays the groundwork for the understanding of much of what is to follow. It discusses the basic anatomy and histology of the male and female reproductive systems. It contains enough information to refresh the memories of students who have had courses in histology, embryology, and anatomy. Because the endocrine system controls the development of the end organs discussed here, it is necessary to have a thorough knowledge of the basic structure and histology of the reproductive systems as well as of their embryonic origins.

THE INDIFFERENT STAGE

Though the genotypic sex of an individual is fixed at the time of fertilization, the reproductive systems of both sexes go through a stage of embryonal development during which it is difficult, if not impossible, to tell the sexes apart by either gross or microscopic examination. This period is known as the indifferent stage. In this stage all anlagen for sub-sequent differentiation into complete male and female systems are present in a rudimentary form; the blueprint, as it were, is finished. All the materials for the later elaboration of the fittings and furnishings of the structures are present, but no attempt is made to arrange the internal furnishings permanently until final orders are received for the emphasis of either male or female aspects of the different structures. This state of affairs makes it possible to discuss the early development of the repro-ductive systems of both sexes together without distinguishing between them.

The question whether hormones secreted by embryonal glands cause differentiation of sex during embryonal development is of great interest, but cannot be answered with any finality at the present time. It has been shown that embryonal testes of both bulls and rats contain the male sex hormone, androgen; but no androgen is present in the urine of postnatal calves. The thyroid gland of larval amphibians, if macerated in situ, will induce metamorphosis of the larva into an adult animal. If the gland is left undisturbed, it does not cause metamorphosis until a considerably later time in the life of the larva. These and other facts argue for the assumption that embryonal glands are capable of produc-ing their specific hormones, but that these hormones are not released into the circulation of the embryo, or at least that they are not released in sufficient quantities to have any discernible effect on their specific target organs.

Other experiments have shown that tissues in the adult animal that depend on hormones for maintenance and normal function may differ-entiate in the embryo without the aid of hormones. This is suggested by experiments carried out by C. R. Moore, in which pouch young of the opossum were castrated. The accessory glands and structures developed and differentiated normally in spite of the absence of the gonads and hence of androgen. This led Moore to the idea that during the early stages of differentiation the sex apparatus does not depend on hormonal stimulation, but grows under the impetus of the "genetic potential" (whatever that may mean). A. Jost's objection to this interpretation is that although embryonal hormones may initiate differentiation they may not be necessary to its maintenance once it has begun.

In general, maternal steroid hormones cross the placental barrier quite easily, whereas the protein and polypeptide hormones do not. (One exception to this rule, to be discussed later, is the pregnant mare serum produced in the endometrial cups of the uterus and which drastically affects the genitalia of the fetus.) The interesting work of Jost and his collaborators has shown that the glands of the fetal rat and rabbit do depend on hypophyseal trophic substances of fetal origin.

Thus, the fetal thyroids and adrenals both atrophy after in utero de-capitation (that is, "hypophysectomy") of the fetus. These experiments show that the fetal hypophysis does produce trophic hormones that stimulate their respective end organs and that the appropriate trophic hormones of maternal origin are either unable to cross the placental barrier or are of the wrong configuration (so far as the fetus is concerned) to be stimulatory. The latter possibility is partly ruled out because Jost's group established that the glycogen content of the liver increases spec-tacularly during the last third of pregnancy in rats and rabbit fetuses. In decapitated fetuses this increase in liver glycogen is absent, but it can be caused to occur if the in utero decapitated young are injected with a combination of adrenal steroid hormones and growth hormone. (Prolactin can be substituted for growth hormone, but much larger quantities are required, and it remains unclear whether its ability to cause glycogenesis is due to its possible contamination with growth hor-mone or to its conjectured similarity in physiologic action to somato-trophin.) These findings show that not only does the fetal hypophysis secrete and release trophic hormones but that the fetal end organs are capable of responding.

A graphic summary of the role of genetic and endocrine factors in prenatal and postnatal differentiation of the reproductive systems is shown in Figure 2-1.

DIFFERENTIATION OF THE SYSTEMS
Origin of Germ Cells

It is now well established that germ cells originate extragonadally in the yolk-sac endoderm and that they migrate from there by ameboid move-ments (in mammals) or via the circulatory system (in the chick) to the genital ridge, where the first microscopically demonstrable aggregation of germ cells occurs. Embryological analysis of several species shows quite clearly that the formation of a gonad depends on the arrival of germ cells and that the genital ridge alone is incapable of developing into anything resembling a gonad. If the germ cells are prevented from migrating to the genital ridge, either experimentally or by accidents of development, the afflicted animal will be born without one or both of the gonads. Unilateral and bilateral agonadism are not rare and have been seen in several species (for one such case, in the female pig, see Figure 12-1e, p. 308).

Primordial germs cells are able to cause endodermal and mesodermal cells to enclose them. This ability is of interest because it permits germ cells, after they arrive at the genital ridge, to form follicle cells. The

Prenatal period   Postnatal period

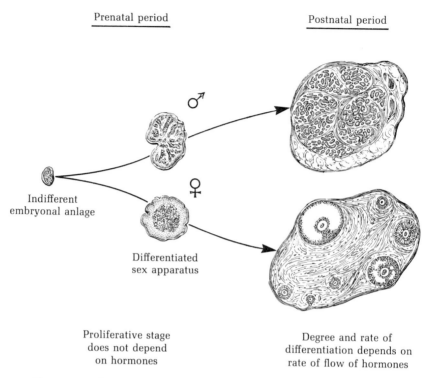

Indifferent
embryonal anlage

Differentiated
sex apparatus

Proliferative stage   Degree and rate of
does not depend   differentiation depends on
on hormones   rate of flow of hormones

Figure 2-1 The role of genetic and endocrine factors in sexual differentiation.

genital ridge itself consists of a mesenchymal thickening covered by
mesothelium. The latter gradually thickens and becomes the germinal
epithelium. The cells within this epithelium differentiate, the larger
ones becoming the primordia of the gonads. If a gonad is to become a
testis, the cells in the germinal epithelium grow into the underlying
mesenchymal tissue and there form chordlike masses, which eventually
become the seminiferous tubules of the testis. If a gonad is to become
an ovary, the primordial germ cells grow into the mesenchyme, where
they become differentiated into ovarian follicles containing ova.

Origin of the Duct System

It is significant that the genital ridge lies close to the mesonephric kidney,
which at this stage of embryonal development is still the functional
kidney. The significance of this proximity is that the differentiating
gonad utilizes parts of the duct system of the embryonal kidney for its

own excurrent ducts. The mesonephric tubules and the mesonephric ducts are used very extensively by the male, in fact, for its permanent duct system; in the female, however, all these ducts are seen only as rudiments. Because the structures of the reproductive systems in the two sexes arise from indifferent rudiments in the embryo, we speak of the structures in the two sexes as being homologous. It would lead us into too much detail to discuss the differentiation of the male and female duct systems from the indifferent stage. The male and female systems are compared in Table 2-1 and Figure 2-2.

Origin of the External Genitalia

Like the gonads, the external genitalia develop from an indifferent structure possessing all the rudimentary anatomic features of both sexes. As development progresses, some of these features become accentuated while others remain undeveloped, according to the genotypic sex of the embryo (Table 2-1). The earliest sign is the genital swelling, which later differentiates into the genital tubercle and the genital folds. If the embryo is genetically destined to become a male, the genital tubercle greatly elongates to become the penis. The penis is partly enclosed by the prepuce, which develops from the genital folds. The genital swellings enlarge to form the scrotal pouches. If the embryo is destined to become a female, the genital tubercle becomes the clitoris, the homolog of the penis. The genital folds turn into the labia minora and the genital swellings into the labia majora. The opening of the urogenital sinus in the embryo does not undergo any changes in the female (in contrast to the male), but persists almost in its original position. The prepuce is rudimentary in females.

Aside from illustrating the opportunism of organisms in utilizing parts of embryonal organ systems for subsequent use in postnatal life, this discussion should serve as a reminder that both sexes develop from an indifferent stage and that each sex carries, as rudiments, the genital organ systems of the opposite sex. As a rule, these rudiments remain such and do not impede the normal function of the prevalent organ system. Occasionally, however, mistakes do happen; the system that is intended to be rudimentary may differentiate and, under the influence of hormones, gain considerable prominence. This may lead to aberrations or obstructions of the duct system or, if the gonads are involved, to intersexuality, which is not rare in animals. These aberrations may be slight and therefore harmless, but they may become great enough to lead to complete sterility. We shall return to this problem later in greater detail.

18

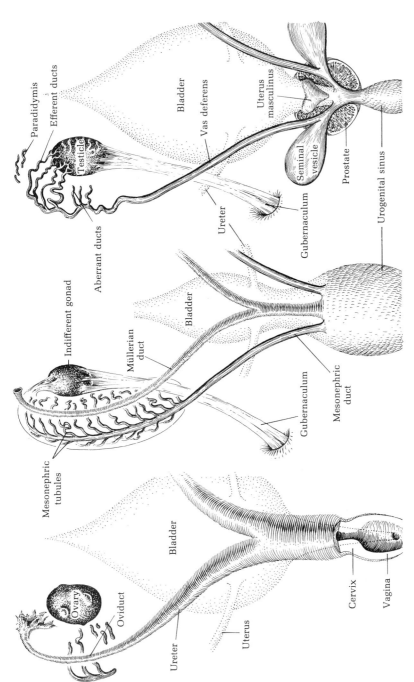

Figure 2–2 The indifferent reproductive system (center) and its modification into the female (left) and male (right) systems. In the indifferent stage the duct systems of both sexes are present. Müller's duct (except for the uterus masculinus) disappears in phenotypic males; the mesonephric duct (except for oviducal appendages) disappears in phenotypic females. See Table 2–1 for details.

Table 2–1. Homology of the male and female reproductive systems

FEMALE ◄—————	INDIFFERENT EMBRYO STAGE —————►	MALE
INTERNAL GENITALIA		
Ovary	Gonad	Testis
Rete ovarii*		Rete testis
Epoophoron*	Mesonephric tubules	Vas efferens
Paroophoron*		Paradidymis*
		Vas aberrans*
Duct part of epoophoron	Mesonephric duct	Epididymis
(Gartner's duct)*		Vas deferens
		Ejaculatory duct
		Seminal vesicle
		Appendix of the epididymis*
Appendage of the ovary	Müllerian duct	Appendage of testis*
(hydatid)*		
Fimbria of oviduct		
Oviduct		
		Prostatic utricle (uterus
Uterus		masculinus)*
Vagina [all or part?]		
	Urogenital sinus	Prostatic, membranous, and
Urethra		cavernous urethra
Vestibule		
Vagina [in part?]		
Vestibular glands		Bulbo-urethral glands
(Bartholin's glands)		(Cowper's glands)
Para-urethral glands*		Prostate
EXTERNAL GENITALIA		
Glans clitoris	Genital tubercle	Glans penis
Corpus clitoridis		Corpus penis
Labia minora	Urethral folds	Raphe of scrotum and penis
Labia majora	Labioscrotal swellings	Scrotum

*Rudimentary.

THE FEMALE REPRODUCTIVE SYSTEM

The reproductive system of the female consists of the ovaries and the duct system. The latter not only receives the eggs ovulated by the ovary and conveys them to the site of ultimate implantation, the uterus, but also receives the sperm and conveys it to the site of fertilization, the oviduct. The duct system of birds differs in all essentials from that of mammals and will be discussed separately.

In mammals the ovaries and the duct portion of the reproductive system are connected with each other and are attached to the body wall by a series of ligaments. The ovary receives its blood and nerve supply through the hilus, which is also attached to the uterus. The oviduct lies in a fold of the mesosalpinx, which in turn is attached to the mesovarian ligament. This ligament is continuous with the inguinal ligament, which is a homolog of the gubernaculum of the testes. Another part of this ligament forms the round ligament of the uterus, which extends from the uterus to the inguinal region.

Mammalian Ovaries

In all mammals the ovaries are paired. They remain near the kidneys, where they first differentiate, and they do not undergo the elaborate descent that is typical of the testes. The free surface of the ovary (which is not taken up by the hilus or the mesovarium) bulges into the body cavity.

The size of the ovary depends largely on the age and reproductive state of the female. The growth of the ovary and the development of its histological components are controlled by hormones from the pituitary gland; much can be learned therefore about the quality and quantity of these hormones from the size and histological appearance of the ovary. For larger animals, microscopic and macroscopic examination of the ovaries is especially useful in diagnosis of abnormalities of the ovary-pituitary relationship. In laboratory animals this relationship is the key to the bioassay of pituitary hormones.

The shape of the ovary varies greatly with the species and depends largely on whether the females are habitually litter-bearing (polytocous) or single-bearing (monotocous). In the former the ovary is usually berry-shaped; in the latter it is ovoid. In the mare the ovary is kidney-shaped, with a distinct depression (the ovulation fossa) on one surface. It was once thought that all ovulations occurred from this depression (hence the name), but now it is known that ovulations may occur in other areas as well. For the purposes of this book a brief summary of the histology of the ovary will suffice. The two most important components of the ovary are the follicle and the corpus luteum.

THE OVARIAN FOLLICLE. Ovarian follicles go through three stages of growth. In the embryo, as well as in the postnatal female, the great majority of them are *primary follicles*. They form a thick layer under the tunica albuginea and are distinguished by the fact that the ova contained in them have no vitelline membrane. The ova are surrounded by many layers of follicular cells, which form the granulosa layer of the more mature follicle. When an ovum acquires a membrane (the zona pellucida) and when the follicle has grown, it becomes a *secondary follicle*. At this stage the follicle has also assumed a more oval shape and has moved away from the cortex toward the medullary portion of the ovary. Eventually a clear, fluid-filled space (the antrum) forms around the ovum and the granulosa cell layers surrounding it. The fluid is called follicular fluid or liquor folliculi. Follicles with antra are called *tertiary follicles;* the major difference between them and mature Graafian follicles is size. As a follicle grows, the antrum enlarges until it extends through the whole thickness of the ovarian cortex. The mature follicle, distended by accumulated follicular fluid, protrudes like a blister above the free surface of the ovary (Figure 2-3a).

At this stage the egg is embedded in a solid mass of follicular cells forming the cumulus oopherus (or discus proligerus), which protrudes into the fluid-filled antrum. In most follicles the cumulus is located on the surface opposite the side that will rupture at ovulation. Occasionally, however, in histological preparations, one finds a cumulus on the side that will rupture. Whether such follicles ovulate normally is not known.

The other important cellular components of the follicle are the granulosa cells, which line the antrum and form the cumulus oophorus, and the corona radiata immediately around the egg (Figure 2-3b). In some animals the corona cells remain attached to the ovum after ovulation; in others the eggs are ovulated naked. The significance of this distinction is not known. The granulosa cell layer is separated from the theca folliculi (which consists of the theca interna and the theca externa) by the basement membrane and connective tissue cells. The granulosa cells and theca cells play an important role in the formation of the corpus luteum, and we shall return to them in that connection.

In addition to the normally developing follicles, an ovary always has a certain number of degenerating follicles and follicles undergoing atresia. Follicular atresia normally accompanies the formation and maturation of follicles; it becomes abnormal only when large numbers of follicles become atretic. The endocrine causes and the significance of atresia will be discussed later. We simply note here that it always begins with the ovum. In an otherwise normal-appearing follicle, impending atresia is signaled by the staining reaction of the membrane around the eggs. The membrane takes on a deep blue or purple stain, and stands out very prominently in the otherwise normally staining follicle. In later stages of atresia other histological symptoms of disintegration appear:

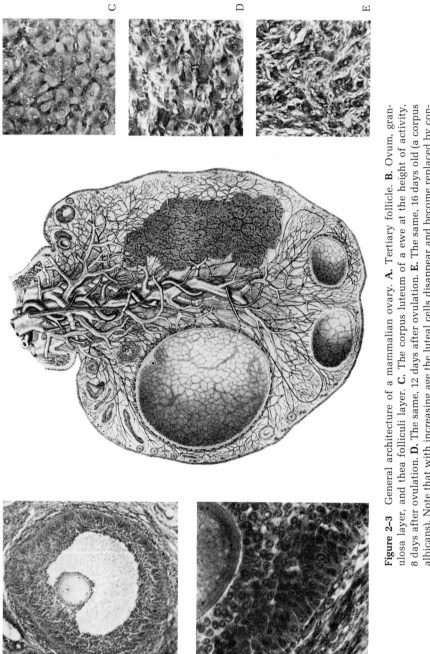

Figure 2–3 General architecture of a mammalian ovary. **A.** Tertiary follicle. **B.** Ovum, granulosa layer, and thea folliculi layer. **C.** The corpus luteum of a ewe at the height of activity, 8 days after ovulation. **D.** The same, 12 days after ovulation. **E.** The same, 16 days old (a corpus albicans). Note that with increasing age the luteal cells disappear and become replaced by connective tissue. [Drawing of ovary by permission of Dr. Eugen Seiferle.]

accumulation of fat droplets and coarse granules inside the ovum; shrinkage or collapse of the ovum itself; detachment of the ovum from the surrounding granulosa cells; and eventual degeneration of the granulosa cells. Atresia may affect follicles in all three stages of development, but in domestic animals it is most commonly seen in tertiary follicles.

CORPORA LUTEA AND ALBICANTIA. After ovulation the follicular cavity is filled with blood and lymph. In some species, e.g., swine, these fluids distend the ovulated follicle so much that for from five to seven days after rupture it is larger than at any time before. In other species, usually in sheep and cattle, the accumulation of fluids is insignificant, and the follicle is smaller than it was before ovulation. Gradually the blood clot is resorbed as luteinization progresses, and eventually the space is filled by the corpus luteum. Histologically, the corpus luteum is composed mostly of granulosa cells, but theca cells also may contribute to its formation. The increase in the size of the corpus luteum occurs because of the hypertrophy and hyperplasia of the granulosa and theca cells. Except in women, these two cell types generally lose their identity in the formed corpus luteum.

As a corpus luteum passes the peak of its functional activity, more and more connective tissue, fat, and hyaline-like substances appear among the luteal cells. The whole corpus luteum gradually decreases in size, eventually becoming a barely visible scar on the surface of the ovary. It also loses its initial red-brown color and becomes white or pale brown. It is then called the corpus albicans (Figure 2–3e).

OTHER STRUCTURES. In the ovaries of pigs, sheep, cattle, and humans one may occasionally see a more or less elaborate duct system, usually confined to the medulla of the ovary. This system has been traced out in the ovaries from a sheep and a pig and was found to end blindly. The ducts are remnants of the epoopheron, which may be lined by either tall-columnar or cuboidal epithelia. They do not interfere with normal reproduction.

In the granulosa layers of the follicles of many species one often sees small, dark-staining granules that form the center of a stellate arrangement of granulosa cells. The granules (the Call-Exner bodies) may be very numerous in the ovaries of some females and completely absent from other females of the same species. Their significance is unknown. An intensive study of the granules was made in female swine; there was no apparent correlation between their presence and reproductive performance.

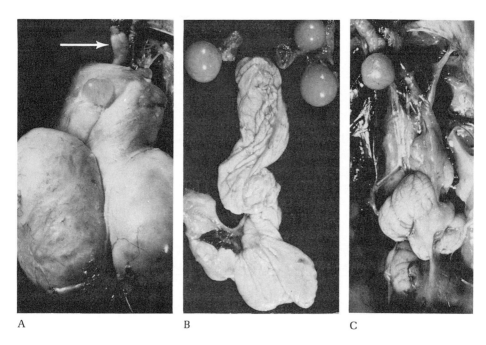

A   B   C

Figure 2-4 Rudimentary gonads in a chicken with the left ovary removed.
A. Testis (arrow). The oviduct is filled with coagulated albumen (the large blister
below the testicle is not a follicle, but a cyst of the oviduct). **B.** Right ovary (left
side of the picture) and regenerated normal left ovary, both containing mature
follicles. In addition to a normal (large) oviduct, there is a small rudimentary
oviduct on the right side. **C.** Ovotestis on right side (testis on top, follicle below).
[From W. Kornfeld, 1953, dissertation, University of Illinois.]

Avian Ovaries

THE RUDIMENTARY RIGHT OVARY. In the great majority of Aves only
the left ovary is functional. The right ovary is present in embryos and is
macroscopically visible for a few days after hatching. In adult females
it persists only as a microscopic vestige. If the functional ovary is removed surgically or, as frequently happens, is destroyed by disease, the
right rudiment enlarges and becomes functional.

The age at which the removal of the functional ovary occurs determines the future development of the rudiment. If the ovary is removed
from chickens less than twenty days old, the rudiment hypertrophies into
a structure resembling a testis (Figure 2-4*a*) and capable of spermatogenesis. However, because the Wolffian duct system does not develop
in genetic females, there is no duct connection between the testis and

the copulatory organ in the cloaca. The celebrated case of the hen that laid eggs and became the mother of chicks, then lost her ovary, differentiated the rudiment into a testis, and fathered a brood of chicks, is probably a canard. Such versatility is, of course, theoretically possible, but highly improbable.

If the left ovary is removed at a greater age, the rudiment develops into a functional ovary (Figure 2-4b), complete with ova capable of being ovulated. However, the ovulated egg cannot be conducted to the outside since the Müllerian duct system in the female is developed unilaterally— there is no oviduct on the side of the rudimentary gonad. Operations at intermediate ages (as well as in a certain proportion at the other two ages) cause the rudiment to form an ovotestis, in which the germ cells of both sexes are present but usually in an undifferentiated condition (Figure 2-4c). If either a testis or an ovotestis is formed after the loss of the functional ovary, the genetic female develops all the sex characteristics of the male, including a large red comb, male plumage, crowing, and male copulatory behavior toward the normal female.

This ambisexual versatility of birds (it is not restricted to the Gallinaceae but has been recorded in ducks, songbirds, and many other kinds) has excited the imaginations of naturalists and folklorists for many centuries. Sex changes in birds were described and speculated on by Aristotle and were noted even before his time. As late as the eighteenth century "hens that crow" and "cocks that lay eggs" were associated with the supernatural, either as indispensable appurtenances of sorcery and witchcraft or as portents of evil. The cockatrice is a legendary serpent or lizard thought to have hatched from cock's eggs, and whose breath or glance was said to be fatal to men.

A wealth of biologically interesting secrets remains hidden in avian sex changelings. It is not known why the age at operation plays a role in determining the fate of the rudimentary gonad. Nor is the chromosome constitution of the differentiated gonad of these genetically female intersexes known. It has been found that in the normal female with a functional ovary the rudiment is prevented from developing by the estrogen secreted by the ovary. Immature females, even shortly after hatching, secrete enough estrogen to suppress the development of the rudiment. This is why the right gonad remains rudimentary as long as the ovary functions.

THE FUNCTIONAL LEFT OVARY. The avian ovary differs morphologically from the mammalian ovary in that it consists of two major lobes. Within each lobe, large numbers of follicles are carried on follicular stalks (Figure 2-5a). A more important difference is that the avian ovum contains a large quantity of yolk, compared to which the germinal

26

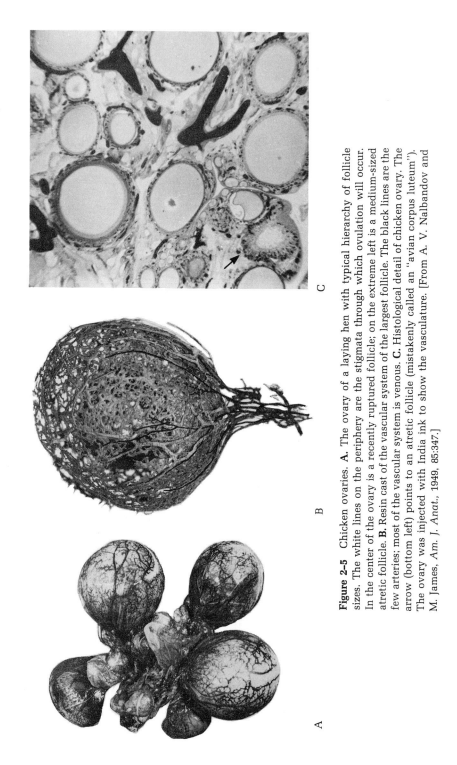

Figure 2-5 Chicken ovaries. **A.** The ovary of a laying hen with typical hierarchy of follicle sizes. The white lines on the periphery are the stigmata through which ovulation will occur. In the center of the ovary is a recently ruptured follicle; on the extreme left is a medium-sized atretic follicle. **B.** Resin cast of the vascular system of the largest follicle. The black lines are the few arteries; most of the vascular system is venous. **C.** Histological detail of chicken ovary. The arrow (bottom left) points to an atretic follicle (mistakenly called an "avian corpus luteum"). The ovary was injected with India ink to show the vasculature. [From A. V. Nalbandov and M. James, Am. J. Anat., 1949, 85:347.]

portion of the egg is negligible in size. The avian follicle has no antrum and no follicular fluid, the ovum filling the follicular sac completely. The avian follicle is lined by granulosa cells, and the arrangement of the theca interna and externa is very similar to the arrangement found in mammals. In their histological appearance these cells are also very similar to their mammalian counterparts.

Avian follicles are probably the fastest-growing structures found in the higher vertebrates. Starting from a diameter of less than 1 mm and a weight of less than 100 mg, the ovum reaches a mature size and a weight of 18–20 grams in nine days. To accomplish this prodigious task of transportation and deposition of yolk material into the ovum, a very complex circulatory system has been developed. The main feature of this system is the enormous development of the venous supply, which is arranged in three concentric layers around the follicle and which terminates in an extremely fine venous capillary network enveloping the growing ovum. By contrast, the arterial supply is poorly developed. The system apparently depends on getting the blood into the venous network, where it stays long enough to permit transfer of the yolk antecedents to the ovum (Figure 2-5b). The mechanics of the transfer of lipoproteins through the membranes of the capillaries, the cells lining the follicle, and the vitelline membrane of the ovum have not been thoroughly investigated in chickens. It has been possible, with vital stain techniques, to demonstrate yolk precursors in the follicular circulatory system, in the granulosa cells, and, of course, inside the ovum. How these relatively enormous globules pass, apparently intact, from the capillaries into the granulosa cells, through the vitelline membrane, and into the ovum remains a mystery.

The intricate follicular circulatory system described serves primarily one of the larger follicles, but it is also utilized by smaller follicles, which grow in large numbers on the same follicular stalk on which the large follicle is growing. When the largest follicle ovulates, spiral arteries in its wall become constricted, largely because of the release of tension in the collapsed follicle wall. The blood flow to the now empty follicular sac is thus greatly reduced. This may be one reason why there is rarely if ever any bleeding into the lumen of the ovulated follicle, or even in the torn stigma through which the follicle is ovulated. The stigma itself is less vascular than the adjacent follicle wall but is by no means devoid of blood vessels. In spite of the reduced blood flow to the ovulated follicle sac, vascularization remains sufficient to permit rapid growth of the smaller follicles clustered along the follicle stalk. One of these is destined to enlarge more rapidly than the rest; as it enlarges, its blood system rapidly increases in complexity until it reaches ovulatory size and ruptures; then the whole process is repeated with the second largest follicle. Thus, the major part of the intricate vascular system is used for a succession of follicles, which are growing on the follicular stalk and awaiting their turn until the circulatory system provides for their own rapid growth.

After ovulation an empty follicle shrinks but remains visible for a considerable time. No structure comparable to the mammalian corpus luteum is formed in birds, but the empty follicle seems to play a role in the reproductive cycle of chickens. More will be said about this in connection with the endocrinology of the chicken ovary (p. 144). In at least two species of wild birds, the mallard and the pheasant, ovulated follicles persist in microscopically and macroscopically recognizable form for as long as 90 days after ovulation. One can therefore roughly estimate the seasonal egg production of these species by counting the number of ovulated follicles present in the ovary.

In chicken ovaries large, clear interstitial cells are found in clusters close to the theca. Similar cells are found in the three types of rudimentary gonads. These cells degenerate after hypophysectomy. Because the hen secretes considerable amounts of androgen, there is the temptation to ascribe androgen secretion to these cells. But experimental data on this point are lacking, and it is equally plausible to assume that these clear cells secrete progesterone, which has been found in the blood and the follicle wall of chickens. It is also possible that the clear cells have no secretory function at all.

The Mammalian Duct System

The duct system in female mammals consists of the oviducts, or Fallopian tubes; the uterus, including the uterine horns and the uterine body; the cervix; the vagina; and the external genitalia (Figure 2-6).

OVIDUCTS.   These paired tubes, the connecting tubes between ovaries and uterus, are long and convoluted derivatives of the Müllerian ducts. The ovarian end of an oviduct is flared out into a fimbria, or funnel, with fringed edges. In their natural position the fimbriae either envelop the ovaries or are very close to them. At the time of ovulation the fimbriated ends of the oviducts show great motility, which probably helps the ovum to find its way into the oviduct. Although the fimbria may not actually "massage" the ovary to aid the process of ovulation, as some earlier anatomists believed, they can pick up ova that have been lost in the body cavity or that have been ovulated from the opposite ovary. This has been demonstrated experimentally by unilateral castration of females and occlusion of the oviduct on the opposite side. Animals prepared in this way showed some fertility, thus indicating that the patent oviduct has the ability to pick up ova ovulated by the contralateral ovary.

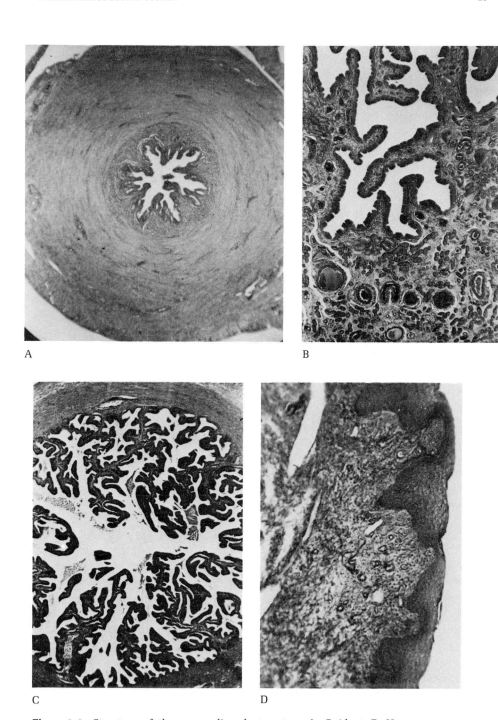

Figure 2-6 Structure of the mammalian duct system. **A.** Oviduct. **B.** Uterus. **C.** Cervix. **D.** Vagina.

In some species the ovarian end of the oviduct forms a complete capsule, which encloses the ovary as in a sac. This sac (as opposed to the open, funnel-shaped fimbria) is called the bursa ovarii. In the rat and the mouse the bursa is complete except for a small perforation in one wall. In the dog, the fox, and the mink the bursa has a slit that is big enough to permit the ovary to pass through with ease. Some litter-bearing animals (such as those named above) have a bursa, but others (such as the pig and the rabbit) do not. The bursa undoubtedly improves an egg's chance of arriving in the uterus; but most animals have no bursa, and the reproductive waste due to loss of ova between the ovary and the uterus is no greater among such animals than among animals that have a bursa.

The lumen of the oviduct is lined by a much-folded mucous membrane (Figure 2-6a). The epithelium lining the lumen is simple-columnar and ciliated. The cilia beat away from the ovary, creating a current within the oviduct toward the uterus. In all mammals except primates the cilia are constantly present and functional throughout the reproductive life. In primate females oviducal cilia grow during the follicular phase, when estrogen predominates, and regress in the luteal phase, when progesterone is the major hormone. Girls are born with fully functional oviducal cilia (probably grown under the influence of maternal placental estrogen), but shortly after birth the cilia disappear and do not reappear until the girl reaches puberty and begins to synthesize her own estrogen. For the same reason cilia are absent in the oviducts of postmenopausal women. Some of the nonciliated cells are transformed into goblet cells, and there are no other glands in the oviduct. In rabbits the ovum acquires a layer of secretions during the trip through the oviduct. This layer was once thought to be albuminous but is now known to be a polysaccharide. The ova of most other common mammals remain uncoated.

The musculature of the oviduct consists of the inner circular layer and the outer longitudinal layer. Both peristalsis and antiperistalsis take place; the latter occurs less commonly and may be an artifact. Eggs spend most of the time required for their passage through the oviduct in the ovarian half of the duct, and it is there that fertilization takes place. Fertilization does not take place in the uterus, as is sometimes stated.

UTERUS.   The uterus usually consists of two horns and a body. The whole organ is attached to the pelvic and abdominal walls by the broad ligament of the uterus. Through this ligament the uterus receives its blood and nerve supplies. The outer layer of the broad ligament forms the round ligament of the uterus. The types of uterine anatomy most commonly encountered are compared and described in Figure 2-7.

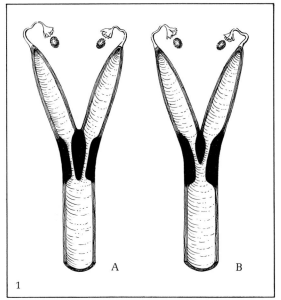

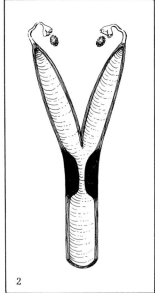

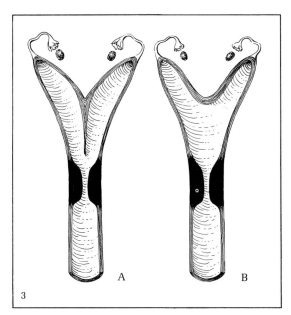

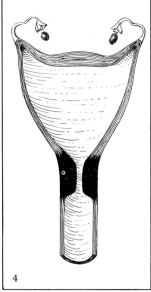

Figure 2–7 Comparison of the four basic types of uteri.

    1. *Duplex uterus.* Two cervixes, no uterine body, horns completely separated. (*A*) Rat, mouse, rabbit. (*B*) Guinea pig.

    2. *Bicornuate uterus.* One cervix, uterine body very small. Pig, insectivores.

    3. *Bipartite uterus.* One cervix, uterine body prominent. (*A*) A septum separates the horns. Cat, dog, cow, ewe. (*B*) Uterine body especially prominent. Mare.

    4. *Simplex uterus.* One cervix, body very prominent, horns absent. Primates.

In the Didelphia the two halves of the duct system have separate external openings (two vaginae, hence the name "didelphia"), two cervices, and two separate and unconnected uterine horns. This arrangement is found in all pouch-bearing animals, such as the platypus and the opossum. Accordingly, the male has a forked penis capable of entering the two vaginae simultaneously, giving rise to the well-known superstition that copulation in the opossum is accomplished through the nostrils. None of the common laboratory and domestic animals has two vaginae, but there is a great variety of peculiarities in the uterine anatomy of these mammals.

Abnormalities in uterine anatomy are common, and reversions of complex types to the simpler and more primitive conditions from which they evolved have been recorded in all species. In primates, for example, bipartite, bicornuate, and even duplex uteri occur occasionally. Women with two cervices and duplex uteri have been known to conceive in the two horns separately to ovulations in two different intermenstrual periods twenty-eight days apart. Cows, sheep, and pigs with two cervices and two completely separate horns are also occasionally seen. These reversions to the more primitive types of duct systems do not seem to interfere with fertility.

The uterine wall consists of the following layers: (1) the serous membrane, which covers the hole viscus; (2) the myometrium, which itself consists of three layers—the internal circular muscle, the external longitudinal muscle, and, separating the muscle layers, the vascular layer; and (3) the endometrium, which consists of the epithelial lining of the lumen, the glandular layer, and the connective tissue.

The myometrium is usually the thickest of the three layers. Its vascular layer carries the extremely important blood vessels supplying the uterus. The smooth muscle cells in the myometrium are capable of increasing greatly in length during pregnancy; they are also responsive to hormones, which, at parturition and occasionally at other times, cause uterine contractions.

The epithelium lining the lumen of the uterus is simple-columnar. In women some patches of the epithelium may bear cilia, which are said to beat toward the cervix, but no cilia have been seen in the uteri of other mammals.

The uterine glands are the most important component of the endometrium. The glands are tubular invaginations of the epithelium, and they too are lined with simple-columnar epithelium. The distal ends of the glands may be either straight or convoluted, depending on the stage of the estrous cycle at which they are observed. (The changes occurring in the glands and their function are discussed in detail in Chapter 4.) In ruminants the uterine surface has special areas for the attachment of the placenta, called the cotyledonary areas; no glands are found underneath the cotyledons.

CERVIX. The cervix is a sphincter muscle that lies between the uterus and the vagina. The anatomy of the cervix varies among mammals, but in most the lumen is interrupted by transverse interlocking ridges called annular rings, which are developed to different degrees of prominence. Annular rings are very prominent in the cow, less so in the pig, and least so in the mare.

The lumen of the cervix is lined by tall-columnar epithelium. Goblet cells are present in the mucosa, which is so intricately folded and branched that it has an enormous secretory surface. The secretion is a mucus, which changes in amount and viscosity with the stages of the cycle. The intricate folds look like a fern leaf and give the cervix its characteristic appearance under the microscope. The myometrium of the cervix is very rich in dense fibrous tissue, has a large number of smooth muscle cells, and includes much collagenous and elastic tissue.

The main function of the cervix is to close the uterine lumen against microscopic and macroscopic intruders. The cervical canal is closed at all times except during parturition. In pigs castration and progesterone cause considerable relaxation of the cervix, whereas estrogen causes the greatest contraction of the sphincter muscle and thus a tightening of the cervical lumen (Figure 2–8). In pregnant animals the cervical mucus hardens and seals off the canal by forming the cervical plug. Immediately before parturition this plug liquefies, probably under hormonal influence, and shortly thereafter the cervix as a whole relaxes (for the hormonal control of this effect see Chapter 11). Breaking the cervical seal in pregnant cows usually leads to abortion or mummification of the fetus. This effect is apparently due to bacterial invasion of the uterine lumen, for it can be prevented by antibiotics given when the cervical seal is broken. Nevertheless, it is important to bear in mind that intrauterine artificial insemination of cows that show heat after conception may lead to interruption of pregnancy.

VAGINA AND EXTERNAL GENITALIA. The vagina is divided into two parts: the vestibule (the outermost part of the vagina) and the posterior vagina (extending from the urethral opening to the cervix). The muscular coat is less well developed in the vagina than in the other portions of the duct system. It consists of a thin layer of longitudinal fibers and a thicker layer of circular fibers. Much loose and dense connective tissue, well supplied with a venous plexus, nerve bundles, and small groups of nerve cells, is characteristic of the vagina.

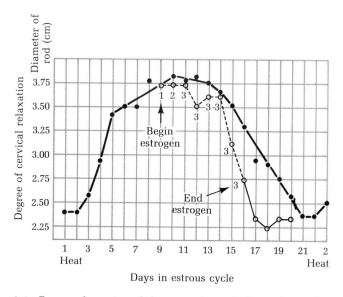

Figure 2–8 Degree of opening of the cervical canal of pigs during the estrous cycle, as measured by the diameter of the rod that can be inserted into the lumen. The dotted line shows that injection of estrogen (1, 2, and 3 mg daily) hastens closure of the lumen somewhat. [From J. C. Smith, 1955, MS thesis, University of Illinois.]

In normally cycling females the epithelial lining of the vagina undergoes periodic changes, which are controlled by hormones secreted by the ovaries. The epithelium may be low-cuboidal or stratified-squamous, depending on the stage of the estrous cycle at which it is examined. The histology of the vagina in most mammals, especially in those with short cycles, has been found to reflect quite faithfully the events that take place in the ovaries. (This important relationship—between ovarian cycles and changes in vaginal histology—is discussed in greater detail in Chapter 4.)

There are no glands in the vagina. The mucus normally found in its lumen, which becomes especially copious in females in heat, originates largely in the cervix, whence it flows into the vaginal lumen.

In virginal females there is a transverse fold, the hymen, which forms the border between the anterior part of the vagina and the vestibule. The hymen has a core of thin connective tissue, which is covered with stratified-squamous epithelium. Normally it is torn and disappears by the time reproductive age is reached, but occasionally it may persist in mature females because of unusual toughness and may prevent normal copulation. In such cases it can be cut.

The external genitalia comprise the clitoris, the labia majora and minora, and certain glands that open into the vaginal vestibule. The clitoris is the embryological homolog of the penis (Table 2-1) and consists of two small erectile cavernous bodies ending in the rudimentary glans clitoris. The paired glands of Bartholin, the vestibular glands, open on the inner surface of the labia minora. They are tubo-alveolar glands and closely resemble the bulbo-urethral glands of the male. They secrete a lubricating mucus.

The labia minora have a core of spongy connective tissue and are covered with stratified-squamous epithelium. Many large sebaceous glands are found on their surface. The labia minora are most prominent in human females, quite small in domestic animals, and almost completely missing in laboratory mammals.

The labia majora are folds of skin with a large amount of fatty tissue and a thin layer of smooth muscle. The outer surface is covered with hair, and the inner surface is smooth and hairless. On both surfaces there are many sebaceous glands. The outer genitalia are well supplied with sensory nerve endings, which play an important role during sexual excitation of the female. The clitoris is capable of limited erection during the sex act, and the labia, because of an increased flow of blood, become extremely turgid.

The Avian Duct System

As we have seen, the mammalian duct system consists of several morphologically separate parts, which differ in circumference and in external and internal architecture. The duct system of birds is a tube of almost uniform diameter with a single unilateral distention near the cloaca (Figure 2-9). Furthermore, whereas in mammals the duct system is paired, in the great majority of birds only the left half of the Müllerian duct system persists; the right half degenerates completely or persists only in rudimentary, usually nonpatent form. There are, however, exceptions to this rule; in certain Raptores both sides of the duct system persist. In chickens and in other birds one occasionally finds individuals (or even genetic strains of chickens) in which both sides of the duct system are developed and capable of normal physiological function.

ANATOMY.   On the basis of the physiological function and the microscopic anatomy of the Müllerian duct (which in birds is usually called the oviduct), it is possible to distinguish the following parts: infundibulum, magnum, isthmus, shell gland, and vagina.

The *infundibulum* consists of the funnel, or fimbria, which receives the ovulated ovum, and the chalaziferous region, in which the chalazae—two coiled, springlike cords that extend from the yolk to the poles of the egg—are formed. Though speculations concerning their function began in antiquity, the reason for their being remains unknown.

The infundibular region merges, externally imperceptibly, into the *magnum,* which is the longest portion of the oviduct. The magnum is also called the albumen-secreting region, because here the white of the egg is deposited round the yolk as it passes through.

Between the magnum and the next region, the *isthmus,* there is a distinct and externally visible line of demarcation, called the magnum-isthmus junction (Figure 2–9), which girdles the duct. The yolk reaches the isthmus already surrounded by albumen and acquires the soft shell membranes in this part of the oviduct.

From the isthmus the egg passes into the expanded caudal portion of the oviduct, the *shell gland,* where it acquires the hard calciferous shell. The term "uterus" should not be used for the shell gland, for this region of the oviduct is in no sense comparable or homologous to the mammalian uterus.

After formation of the egg is completed in the shell gland, the egg is expelled through the rather short *vagina,* in which it remains only a short time and in which it becomes coated with mucus (the "bloom" of the egg). The mucus seals the shell pores and thus, presumably hinders bacterial invasion.

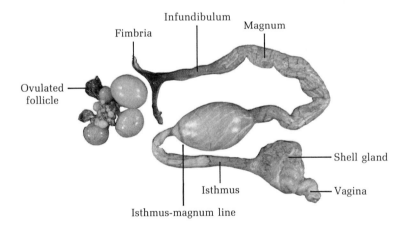

Figure 2–9 The oviduct and ovary of a hen. Note the hierarchy of sizes of follicles in the ovary. Below the ovulated follicle is another one, one day older. The egg is just emerging from the magnum and passing into the isthmus, which is separated from the magnum by the isthmus-magnum line. [Photo by K. Imai.]

Table 2–2. The parts of the oviduct of hens: average lengths, contributions to the egg, and time spent by the egg in each

| | AVERAGE LENGTH (cm) | CONTRIBUTION | | | TIME SPENT BY EGG (hours) |
		KIND	AMOUNT (grams)	PERCENT SOLIDS	
Infundibulum	11.0	Chalaza	. . .	. . .	1/4
Magnum	33.6	Albumen	32.9	12.2	3
Isthmus	10.6	Shell membrane	0.3	80.0	1 1/4
Shell gland	10.1	Calciferous shell	6.1	98.4	18–22
Vagina	6.9	Mucus	0.1	. . .	1/60

The oviduct needs little time to perform its various tasks. The time spent by the egg in the different parts of the duct, in relation to their length, are summarized in Table 2–2. Each 5 cm of the magnum secrete 5–6 grams of albumen in the 30 minutes that it takes the egg to move through that distance. This is a truly prodigious accomplishment even though most (88 percent) of that secretion is water. The egg spends the longest time in the shell gland, where the secretion is least in weight but where 98 percent of the secretions are solids. The calcium carbonate and traces of other minerals are deposited at the rate of 0.3 gram per hour.

HISTOLOGY.   To accomplish the phenomenal task of transporting and secreting about 40 grams of material, consisting of 10 grams of solids and 30 grams of water, in about 26 hours, the chicken oviduct must be complex. Only an abridged account of its histology can be given here. Those interested in details should consult the unsurpassed 1935 monograph by K. C. Richardson (see end of chapter).

The gross histological structure of the oviduct consists of the external peritoneal coat (serosa), the outer longitudinal and inner circular muscle coats, the connective tissue layer carrying the blood vessels and nerves, and, finally, the mucosa lining the whole duct. In the young chick the mucosa is a simple, ungrooved, and unfolded lining. As sexual maturity approaches, and as the oviduct begins to be stimulated by estrogen and progesterone, the mucosa become intricately grooved and folded into primary, secondary, and tertiary folds. Numerous glands are found throughout the length of the whole duct. At the height of secretory activity the cells vary in type from simple tall-columnar cells to transitional columnar cells, the great majority being ciliated. The glands secrete the substances peculiar to the region of the duct in which they are found, albumen aggregating in large coarse globules that fill the cells and the gland lumina. The endocrinology of the secretion of these substances is known (see p. 153), but the mechanism by which they are excreted to the surface of the duct, where they are deposited round the forming egg, remains unknown. Avian oviducts cannot distinguish between an ovum and any other foreign body, and they secrete albumen, soft membranes, and a hard shell round any oviducal inclusion. Nothing is known about the nature of the physical forces responsible for the shape of the egg (sharper end posterior): The shape is apparent long before the egg acquires the rigidity imparted to it by the hard shell; in fact it is easily discernible in the upper portions of the magnum, where only a very thin layer of albumen has been deposited round the yolk.

THE MALE REPRODUCTIVE SYSTEM

The reproductive system of the male consists of paired testes, paired accessory glands, and the duct system, including the copulatory organ.

The Testes

The indifferent gonads of early embryos differentiate in females into ovaries and in males into testes. In all species the testes develop in the vicinity of the kidneys, in the region of the primitive ridge. In mammals the testes undergo an elaborate descent, ending, for most species, in the scrotum. In birds the testes do not descend but remain approximately in the position in which they originate. The function of testes is twofold: they produce the male sex hormone, androgen, and they form the male gametes, the sperm.

The sperm are produced in the seminiferous tubules, which make up over 90 percent of the testicular mass. The tubules are extremely convoluted; each testis contains tubules that would be several miles long if straightened out. The histology of the tubules changes progressively with age. In young males the tubules are simple; the germinal epithelium consists only of spermatogonia and Sertoli cells. In older males the spermatogonia give rise to primary spermatocytes, which, after the first miotic division, give rise to the haploid secondary spermatocytes. These, in turn, become spermatids, which, after a series of transformations called spermiogenesis, give rise to sperm cells consisting of a head, a middle piece, and a tail. The Sertoli cells, which are found along the basement membrane of the tubules, are called "sperm mother cells" because it is thought that the sperm heads, which become embedded in these cells, undergo a ripening process in them. The processes of spermatogenesis and spermiogenesis, which are controlled by hormones, will be discussed in another section. These processes are fairly uniform in all animals studied. The major differences arise from the seasonal spermatogenesis of many species: during the nonbreeding season the testes are completely inactive (juvenile) and recrudesce during the breeding season or shortly before its onset. Testicular regression and recrudescence are both caused by the ebb and flow of pituitary hormones.

The other major function of the testes is the secretion of the male sex hormone. The best available evidence indicates that only the Leydig cells of the interstitial tissue secrete androgen, but it has not been possible to rule out completely the slight possibility that components of the seminiferous tubules may participate in this function. There is considerable difference between and within species in the degree of development of the Leydig cells. In young cocks the Leydig cells are much more numerous than in older birds; it is difficult, in fact, to find Leydig cells in the interstitial tissue of mature cocks. However, there is no evidence that adult cocks secrete less androgen than growing males, and the assumption is valid that the few Leydig cells found in adults produce androgen as efficiently as the more numerous cells of younger birds. Leydig cells are seen in all mammals at all ages when the males are in breeding condition, but again there are differences between species in the number of Leydig cells present. In boars and rats the interstitial tissue is very well developed, and large aggregates of Leydig cells occupy a good share of the total testicular volume. In men and bulls Leydig cells are much less prominent and do not form large nests as they do in the other species (Figure 2–10a, b). Secretion of androgen by Leydig cells is controlled by pituitary hormones, and the rate of the secretion depends on the rate of the pituitary function.

The capsule enveloping the testicles contains muscle cells capable of contraction. In experiments conducted in vivo on male rabbits it was demonstrated that the capsule contracts spontaneously and massages the testicle contained in it with a frequency of about four contractions per minute. It is reasonable to speculate that this massaging action of the testicular capsule causes the sperm, which are nonmotile in the seminiferous tubules, to be moved from the tubules to the excurrent testicular ducts and into the vas deferens. Whether these contractions are under hormonal control is not known (Davis et al., 1970).

The Scrotum

In all mammals, except those living in the sea and the pachyderms, the testes descend into the scrotum. In continuously breeding species they remain in the scrotum; in seasonally breeding species they ascend through the inguinal canal and remain in the body cavity during the

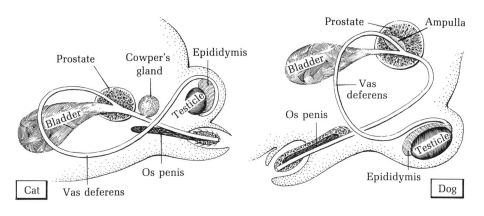

Cat

Prostate — Cowper's gland — Epididymis — Testicle — Bladder — Os penis — Vas deferens

Dog

Prostate — Ampulla — Bladder — Vas deferens — Os penis — Testicle — Epididymis

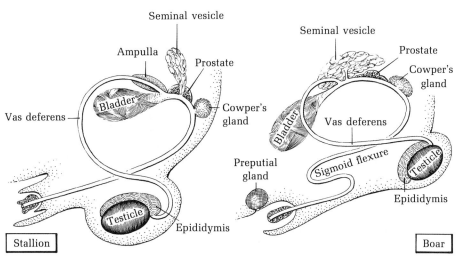

Stallion

Seminal vesicle — Ampulla — Prostate — Bladder — Cowper's gland — Vas deferens — Testicle — Epididymis

Boar

Seminal vesicle — Prostate — Cowper's gland — Bladder — Vas deferens — Preputial gland — Sigmoid flexure — Testicle — Epididymis

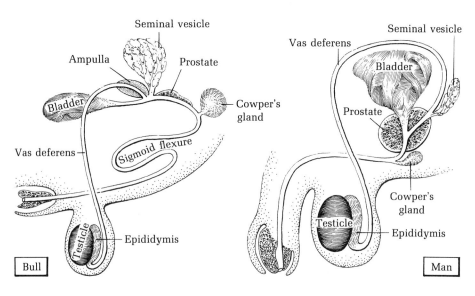

Bull

Seminal vesicle — Ampulla — Prostate — Bladder — Cowper's gland — Vas deferens — Sigmoid flexure — Testicle — Epididymis

Man

Vas deferens — Seminal vesicle — Bladder — Prostate — Cowper's gland — Testicle — Epididymis

40

nonbreeding season. In birds the testes do not descend but remain in the vicinity of the kidneys. The scrotal pouch is formed as the result of the double action of the physical pressure exerted by the testes and the stimulatory effect of androgen. Neither of these factors alone is sufficient to induce the formation of the scrotum. The main function of the scrotum is to provide the testes with an environment that is 1–8°F cooler than the body cavity.

Though the mere presence of the testes in the scrotum assures them of an environment cooler than the body cavity, the scrotum is capable of further regulating their temperature. It accomplishes this by a double muscle system that draws the testes close to the body wall for warmth or lets them fall away from the body wall for cooling. The two muscles involved are the external cremaster and the tunica dartos. The former passes through the inguinal canal and is attached to the tunica vaginalis enveloping each testis; when it contracts, it raises the tunica vaginalis and with it the testis. The tunica dartos adheres to the scrotal skin very closely and forms the septum that separates the scrotum into the two pouches. Exposure to cold makes the dartos contract, thus causing the scrotal skin to pucker and forcing the testes closer to the body wall. Both muscles atrophy and lose their ability to contract after castration, but they regain it after treatment with androgen.

There is a second mechanism, originally suggested by R. G. Harrison, which is equal in importance to the scrotum in providing the temperature regulation (cooling) that is so vital for the normal physiological function of the testes. This mechanism involves the so-called pampiniform plexus, which is the intricately and prodigiously looped system of veins and arteries that lies on the surface of the epididymis and which, gradually becoming less looped, follows the spermatic cord into the inguinal canal (Figure 2–11). The arterial branch of this plexus is supplied by the spermatic artery and the venous part empties into the spermatic vein. Actual measurement in rams has established that the blood enters the plexus at a temperature of 39.0°C; by the time it enters the testicle it is at or near the temperature (34.8°C) typical of that organ. Similarly, the cool venous blood leaving the testicle and entering the pampiniform plexus becomes gradually warmer as it approaches and joins the main peripheral circulation. The cellular and biochemical

Figure 2-10 Reproductive systems of the male cat, the dog, the stallion, the boar, the bull, and man. Compare the relative sizes of the various accessory glands, and note that all these species have the prostate; that the dog and the cat have no seminal vesicles; that the dog has no Cowper's gland; that the cat, the boar, and man have no ampullar swelling; that the bull and the boar have the sigmoid flexure of the penis; that the dog and the cat have the os penis; that only the boar has the preputial pouch.

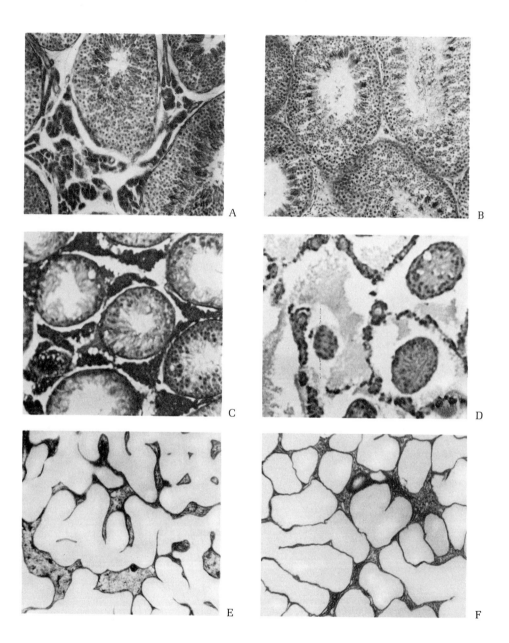

Figure 2-11 Histology of normal and abnormal testes. **A.** Normal testis of a boar; note the massive development of interstitial tissue. **B.** Normal testis of a bull; note the scant intertubular tissue. **C.** Testis confined to the body cavity for 20 days; note that here, as in *D* and *E,* the interstitial tissue is nearly intact, whereas the tubules show progressive degeneration. **D.** Testis confined to the body cavity for 80 days. **E.** Testis from a spontaneous bilateral cryptorchid goat that was probably two years old. **F.** Testis of a rat injected with massive doses of desoxycorticosterone. The same degree of destruction is noted in males on diets deficient in vitamin E.

mechanisms by which the descending blood loses heat are completely unknown. It is probable that in such males as stallions, boars, rats, and others in which the scrotum is tight and not pendulous—as it is in men, rams, or bulls—the pampiniform plexus plays a major role in cooling the testes (see Figure 2-12).

It is now known that the subcutaneous injection of minute amounts of cadmium—for instance, in the neck region of a male rat—rapidly causes sterility, owing to a complete eventual destruction of the seminiferous tubules and, to a lesser extent, of the interstitial tissue. It is interesting to note that cadmium exerts its initial effects on the lining of the blood vessels of the pampiniform plexus. The ultimate destruction of the tubules is due to the inability of the impaired plexus to cool the arterial blood. The female, which has no such plexus, is not affected by cadmium. It is not known what it is about the blood vessels of the pampiniform plexus that makes only them vulnerable to the cadmium effect, since none of the other vessels show detectable damage. (Parenthetically, it may be mentioned that if rats are pretreated with zinc, or if zinc is given simultaneously with cadmium, the animals are completely protected against the deleterious effects of the latter.) It is interesting that in shrews (as well as cocks), which have no scrotum but have abdominal testes, cadmium injections have no significant detrimental effect on fertility.

Domestic animals with scrotal testes injected either intratesticularly or systemically with cadmium show extensive destruction of testes. In goats, bull calves, and boars the intratesticular injection of cadmium is more effective than the systemic or intravenous administration. However, the treatment of these animals causes undesirable side effects such as loss of weight which precludes the possibility of using cadmium as a means of chemical castration (Pate *et al.*, 1970).

Sheepherders learned long ago that it is possible to cause temporary sterility in rams by tying the testes close to the body wall, thus preventing them from cooling off when the environmental temperature rises. It has also been known for a long time that cryptorchids—animals whose testes remain inside the body cavity instead of descending into the scrotum—are sterile. Cryptorchism occurs spontaneously in practically all mammalian species. It is common in man and most domestic animals, and in pigs and goats it is frequently hereditary. It may be caused by hormonal deficiencies and may then be alleviated by the injection of androgen, which causes testicular descent when given in proper doses. If cryptorchism is caused by an anatomical obstruction of the inguinal canal that prevents the testes from entering the scrotum, hormonal therapy is useless, although surgical correction of the obstruction may permit testicular descent.

In most mammals, especially in laboratory mammals, it is easy to produce experimental cryptorchism by pushing the testes through the inguinal canal into the body cavity and preventing their descent by surgical closure of the canal. Such cryptorchids continue to mate, but the semen they produce contains increasing amounts of abnormal sperm. Eventually, spermatogenesis stops completely. The seminiferous tubules become more and more disorganized; the germinal epithelium sloughs off and the debris fills the lumen of the tubules. After prolonged confinement to the body cavity, the germinal epithelium disappears more or less completely; the spermatogonia are the last affected. If the testes are returned to the scrotum at this stage, complete recovery may occur and spermatogenesis may be resumed. After prolonged exposure to body temperature, the damage becomes irreversible. The testes of natural and experimental cryptorchids are smaller than normal testes; the seminiferous tubules are shrunk considerably and the intertubular spaces become proportionately larger than in normal testes (Figure 2–10c, d).

The effect of body temperature on the ability of testes to secrete androgen is not quite clear. There is probably no effect on the Leydig cells or on androgen secretion when the testes of adult males are confined to the body cavity, but cryptorchism in young males seems to reduce the rate of androgen secretion somewhat. There is probably considerable difference between species in this respect. Unilaterally cryptorchid stallions (called ridglings) are said to be sexually more vigorous and aggressive than normal males.

Even in males whose testes have descended, the scrotum cannot always compensate for the effect of high temperatures on spermatogenesis. Notable examples of this are man, who may become temporarily or permanently sterile after prolonged fever, and rams, which are subject to the so-called summer sterility. In rams the higher temperature of summer causes the production of increasing numbers of defective spermatozoa and eventually brings about impaired fertility or complete sterility for the duration of hot weather. That these deleterious effects are due to heat can be inferred from the fact that the onset of lower fall temperatures alleviates summer sterility and the fact that rams placed in cool environments during summer are significantly more fertile than those exposed to normal summer temperature. Finally, when the scrotum and the body of rams are sheared (thus lowering body temperature and scrotal temperature), the fertility of the rams remains high in spite of higher environmental temperatures.

Why high temperatures are injurious to spermatogenesis is not known. According to some reports, summer sterility in rams can be overcome by treatment with thyroid preparations if the treatment is begun before spermatogenesis begins to deteriorate. The feeding of androgen to rams does prevent summer sterility, but the results obtained from such treatments have varied considerably in different experiments.

The Duct System

The duct system of the male is derived, to a large extent, from the Wolffian duct system of the mesonephric kidney (Figure 2-2). A part of the Müllerian system, which the female utilizes for most of its ducts, persists in the male as a rudiment in the prostate, where it forms the prostatic utricle (uterus masculinus). This structure retains its ability to respond to female sex hormones and is frequently responsible for the prostatic enlargement that commonly occurs in men and dogs past the prime of life.

The mesonephric tubules develop into the vas efferens, the mesonephric duct to the epididymis, the vas deferens, and the seminal vesicles, the last being formed from evaginations of the duct. The remainder of the duct system (the prostatic, membranous, and cavernous urethra) comes from the urogenital sinus, as do the other two male accessory glands—the prostate and Cowper's glands (bulbo-urethral glands) (Table 2-1). The accessory glands are developed to different degrees in different mammals (Figure 2-12). Seminal vesicles are absent from the dog, the fox, and the wolf. The boar's preputial pouch, near the tip of the penis (Figure 2-12), cannot be properly classified as an accessory gland, but it regresses in size very significantly after castration. Its function is not known. Urine accumulates in this pouch and is responsible for the strong characteristic odor of boars, which permeates even their flesh and accounts for its disagreeable taste.

The ductuli efferentia become convoluted where they originate from the rete testes. These ducts are lined by alternate clusters of tall and low epithelial cells, which bear presumably nonmotile cilia. They gradually fuse into a single, intricately convoluted duct to form the head, body, and tail of the epididymis, which is lined by tall, pseudostratified epithelial cells bearing nonmotile stereocilia. Gradually becoming less convoluted, the duct acquires a bigger lumen and a thicker wall; as it leaves the tail of the epididymis it becomes the vas deferens. Here the pseudostratified, stereociliated epithelium is lower than in the epididymis. The vas is surrounded by well-developed muscle layers—inner and outer longitudinal layers with a circular layer between them. The contraction of these muscle layers may be partially responsible for the movement of the sperm through the duct system.

The two vasa deferentia lie side by side, without fusing internally in the region of the bladder, and form the ampulla of the vasa deferentia.

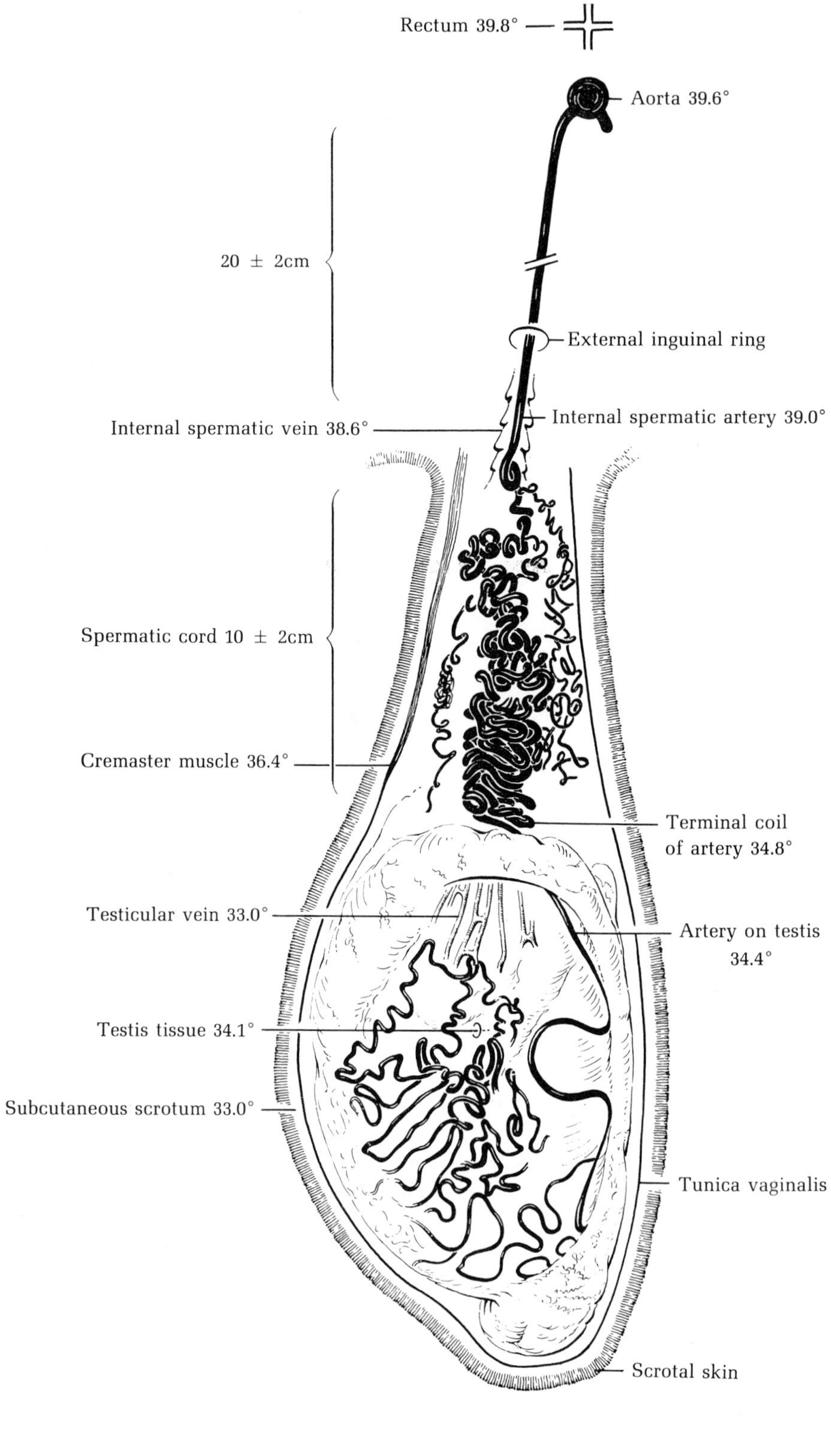

Rectum 39.8°

Aorta 39.6°

20 ± 2cm

External inguinal ring

Internal spermatic vein 38.6°

Internal spermatic artery 39.0°

Spermatic cord 10 ± 2cm

Cremaster muscle 36.4°

Terminal coil
of artery 34.8°

Testicular vein 33.0°

Artery on testis
34.4°

Testis tissue 34.1°

Subcutaneous scrotum 33.0°

Tunica vaginalis

Scrotal skin

In most species there is a thickening of the wall but very little increase in the lumen in the region of the ampulla. Here the urethra, the ducts from the paired seminal vesicles, and the multiple prostatic ducts all empty into the duct system. A short distance further the multiple ducts from the bulbo-urethral (Cowper's) glands also empty into the duct system, which at this point is called the prostatic urethra. The epithelium in this area is simple, pseudostratified, and columnar. (For further details of the gross and microscopic anatomy of the male urethra consult any textbook of histology.)

The Penis

In some mammals, especially in the bull and the boar, the male urethra forms a loop called the sigmoid flexure. In other mammals the flexure is very slight (as in man) or completely missing. In man the penis consists of three (and in all domestic and laboratory mammals of two) cylindrical bodies called the corpora cavernosa penis. These become filled with blood during sexual excitation; as the penis becomes more turgid and erect and is pulled against the bony rim of the pelvis, escape of the blood becomes increasingly (and eventually completely) blocked. Erection is thus largely due to this inability of the blood to drain.

In many species a bony structure is formed by ossification of a corpus cavernosum. This structure, called the os penis (or the baculum), is found in all Canidae, sea lions, Mustellidae, and in a variety of other animals.

The end of the penis, capping the corpora cavernosa, is called the glans penis. It is mushroom-shaped in man but assumes a variety of shapes among mammals. In the opossum it is forked, the fork including the terminal 2 cm of the shaft of the penis. In the boar it is corkscrew-shaped; in the bull there is a less pronounced twist. In the goat and the ram a thin projection extends 3–4 cm beyond the tip of the glans; this is called the urethral or vermiform process, a very apt descriptive term.

The Accessory Glands

The accessory glands include the paired seminal vesicles, the prostate (which in rats consists of three lobes, in other mammals of one single structure), and the paired bulbo-urethral or Cowper's glands. There is considerable variation in the relative size and anatomy of the glands in

Figure 2–12   The effect of the pampiniform plexus on the modification of temperature of blood entering and leaving the scrotum and testes of rams. Temperatures are °C; air temperature is 21 ± 4°C. [From Waites and Moule, 1961, *J. Repro. Fertil.*, 2:213.]

different species. In the cat and dog the seminal vesicles are absent and the prostate is relatively much better developed than it is in other mammals. In the bull and especially in the boar the seminal vesicles are greatly enlarged, but the prostate is relatively small and poorly developed (Figure 2–11). The anatomic structure is also extremely divergent. In man the seminal vesicles are hollow sacs with numerous outpocketings; in rats they are frond-shaped and elongated and consist of many compartments. The lumina of all three of these glands are lined with tall-columnar epithelium.

Sperm cells contained in the lumina of the seminiferous tubules and the proximal excurrent ducts are nonmotile. They become motile and presumably metabolically active when they come in contact with the so-called seminal plasma. There is considerable variation among the species in the amount of seminal fluid contributed by the various accessory glands and the testes. In man the prostate contributes 15–30 percent and the seminal vesicles 40–80 percent of the total ejaculate. In the boar 15–20 percent of the semen plasma is derived from the seminal vesicles, 2–5 percent from the testes and the epididymis, 10–25 percent from Cowper's glands, and the remainder from the urethral glands. In any species not more than 2–5 percent of the total ejaculate is made up of the secretion of the testes and the epididymis. The seminal plasma is added to the sperm during ejaculation.

In some animals (boars and man), in which the process of ejaculation lasts a considerable time, it is possible, by use of the fractional-ejaculate method, to determine the sequence in which the different accessory glands make their contribution to the total ejaculate. It is probable that in all animals (but certainly in boars and man) the urethrae are first "flushed" by a sperm-free ejaculate, probably originating from the urethral glands and the bulbo-urethral glands. This is followed by a second ejaculate, rich in sperm, which contains contributions from the seminal vesicles and the prostate. This sperm-rich fraction is followed by a sperm-free ejaculate, which is probably predominantly of vesicular origin.

The seminal plasma has two main functions: it serves as a suspending and activating medium for the previously nonmotile sperm cells, and it furnishes the cells with a substrate that is rich in electrolytes (sodium and potassium chloride), nitrogen, citric acid, fructose, ascorbic acid inositol, phosphatase, and ergothionene, and contains traces of vitamins and enzymes. Neither of these functions is essential to sperm fertility, for sperm become motile in any physiological liquid, and epididymal and even testicular sperm are capable of fertilizing eggs (though their fertilizing ability is less than that of more mature sperm). We have seen that cats and dogs have no seminal vesicles, and one can remove the seminal vesicles and the prostate from boars and rats without impairing their fertility. Nevertheless, the secretions of the male accessory glands should be considered an essential and integral part of semen, without which sperm cells could not attain their greatest fertilizing ability.

The semen of several species coagulates upon ejaculation. Human semen coagulates shortly after ejaculation but liquefies again, presumably under the action of proteolytic enzymes contributed by the prostate. The semen of rats coagulates because of the action of the secretion of the coagulating gland of the prostate on the seminal fluid. The resulting coagulum forms the vaginal plug in the vagina of the female. It was earlier thought that the formation of the vaginal plug was essential to the maximal fertility of the rat, but evidence shows that removal of the plug immediately after copulation does not reduce fertility; in fact, the plug (which can be used as an indication that a female has mated) very frequently drops out shortly after mating. The semen of guinea pigs also forms a jellylike vaginal plug, the coagulation being due to a substance contributed by a portion of the prostate gland. A vaginal plug is also formed in baboons and squirrels but not in other animals.

The pH of fresh semen is usually about 7.0. It has been found to be on the acid side in some species (bull, dog, fox, rabbit) and distinctly alkaline in others.

We shall not discuss in detail the chemical composition of seminal plasma, but we shall take note of a few peculiarities of various species. Fructose occurs in the semen of virtually all mammals; it is especially high in the semen of bulls and goats and considerably lower in that of stallions and boars (both of which ejaculate large volumes of semen). Cock semen contains no fructose, and rabbit semen contains fructose and glucose, the latter being absent from bull, ram, and human semen. In most mammals seminal fructose comes from the seminal vesicles, but in the rat it comes from the dorsal prostate. Citric acid in the semen of boars and bulls is produced in the seminal vesicles, but in men it is secreted by the prostate. The seminal vesicles of boars and rats secrete very large amounts of inositol (these vesicles, in fact, are the richest known source of inositol), but bull and human semen contain virtually no inositol. Though these differences are perplexing and their significance in the metabolic activity of the sperm of the various species remains unexplained, they point up the conclusion that different species may solve the same fundamental problem in different ways. It appears probable that inositol plays a role in the sperm metabolism of the boar, for it is secreted by that animal in large quantities, but that other mammalian sperm do not need this vitamin, for it is absent from their seminal plasma.

SELECTED READING

Amann, R. P. 1970. Sperm production rates. In *The Testis*, A. D. Johnson *et al.*, eds., Vol. 1. Academic Press.

Asdell, S. A. 1946. *Patterns of Mammalian Reproduction*. Comstock.

Brambell, F. W. R. 1956. Ovarian changes. In *Marshall's Physiology of Reproduction*, Vol. 1, Part 1, 3rd ed., Longmans.

Courot, M., M. T. Hochereau, and R. Ortovant. 1970. Spermatogenesis. In *The Testis*, Vol. 1. Academic Press.

Davis, J. R., G. A. Langford, and P. J. Kirby. 1970. The testicular capsule. In *The Testis*, Vol. 1. Academic Press.

Eckstein, P., and S. Zuckerman. 1956. Morphology of the reproductive tract. In *Marshall's Physiology of Reproduction*, Vol. 1, Part 1, 3rd ed., Longmans.

Forbes, T. R. 1947. The crowing hen: early observations on spontaneous sex reversal in birds. *Yale J. Biol. Med.*, **19**:955.

Gier, H. T., and G. B. Marion. 1970. Development of the mammalian testis. In *The Testis*, Vol. 1. Academic Press.

Gunn, S. A., and T. C. Gould. 1970. Cadmium and other mineral elements. In *The Testis*, Vol. 1. Academic Press.

Harrison, R. J. 1948. The changes occurring in the ovary of the goat during the estrous cycle and early pregnancy. *J. Anat.*, **82**:21.

Mann, T. 1964. *The Biochemistry of Semen and of the Male Reproductive Tract*. Methuen.

Pate, F. M., A. D. Johnson, and W. J. Miller. 1970. Testicular changes in calves following injection of cadmium chloride. *J. Animal Sci.*, **31**:559.

Richardson, K. C. 1935. The secretory phenomena in the oviduct of the fowl. *Philos. Trans. Roy. Soc. London (B)*, **225**:149.

Seiferle, E. 1933. Art- und Altersmerkmale der weiblichen Geschlechts-organe unserer Haussäugetiere, Pferd, Rind, Kalb, Schaf, Ziege, Kaninchen, Meerschweinchen, Schwein und Katze. *Z. Anatomie Entwicklungsgeschichte* **101**:1.

Selye, H. 1943. Factors influencing developing of scrotum. *Anat. Rec.*, **85**:377.

Sisson, S., and J. D. Grossman. 1940. *The Anatomy of the Domestic Animals*. Saunders.

Waites, G. M. H. 1970. Temperature regulation and the testis. In *The Testis*, Vol. I. Academic Press.

Winters, L. M., W. W. Green, and R. E. Comstock. 1942. *Prenatal Development of the Bovine*. University of Minnesota Technical Bulletin 151.

Witschi, E. 1951. Embryogenesis of the adrenal and the reproductive glands. In *Recent Progress of Hormone Research*, Vol. 6. Academic Press.

3

The Endocrinology
of Reproduction

This chapter deals with the fundamentals of reproductive endocrinology. It establishes the fact that both the anterior and the posterior lobe of the pituitary gland are controlled, to a considerable extent, by the hypothalamus. Evidence of this control is given. The interested reader might take each of the effects listed and diagram all the events that contribute to it. For instance, compare induced and spontaneous ovulation; diagram all the events, the end organs they affect, the hormones that participate, and the relations that exist between the duct system, the ovaries, the hypothalamus, and the pituitary gland. Similar diagrams for all the other events listed would clarify the various interrelations and reveal holes in the theories presented in the text.

INTRODUCTION

Sexual reproduction, even in the lower vertebrates, depends on synchronizing systems. The shedding of eggs by the female frog, for example, must be closely followed by the shedding of sperm by the male. Here the problem is rather simple; all that is needed is a system that will assure the simultaneous shedding of gametes by two individuals of opposite sex in the same vicinity of the pond at the most propitious season. In more complex animals a whole series of interlocking and synchronized events must follow one another if reproductive efficiency is to be attained. The shedding of gametes, fertilization, gestation, parturition, and lactation are all events that require accurate timing, and all of them must occur in such a way that the young life depending on them has a good chance of surviving.

Without synchronization and cyclic repeatability, reproduction would become a chaotic and completely inefficient game of chance. Although the problem has been solved in various ways in different species, all animals have evolved an intricate and interrelated system of checks and balances, a time clock consisting of two interlocking systems—the endocrine and nervous systems. These two systems usually function as a unit, although, as we shall see, one system may on occasion override the other, depending on the task to be performed. For ease of discussion we shall consider each system separately before we ask the question of why *two* interlocking systems have evolved to control the cycle of reproductive events.

Most individual reproductive events are themselves chains of interdependent, mutually controlling, and cyclic reactions. A single event in the reproductive cycle is the product of a whole series of phenomena, all of which must occur in a properly timed sequence if they are to culminate in a normal and predictable effect on an end organ. In order to understand the intricate hormonal interplay that produces an end result, one must think in terms of the *chain* of events that produces the result rather than the single events that make up the chain. The basic pattern of hormonal interaction is shown below. Although some interactions are simpler, and most are more complex, they all follow basically the same scheme.

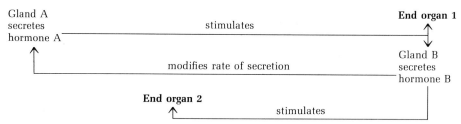

The governing parts of the reproductive system are the endocrine glands, the hormones secreted by them, the end organs (such as the ovaries and the uterus) that are acted on by the hormones, and, under certain circumstances, the nervous system.

The endocrine (or ductless) glands can be defined as specialized groups of cells that have the primary function of forming and elaborating chemical substances called hormones, which they secrete directly into the blood stream or the lymph. Hormones, which have also been called distance activators, can be defined as substances made by endocrine glands in one part of the body and carried by the blood or the lymph to other parts of the body, where they modify the activity of certain genetically conditioned end organs. The specifications "genetically conditioned" and "end organ" are important because all hormones are highly specific and selective in their action. Many hormones cause growth; but estrogen causes growth of the uterus and not of the comb, androgen causes growth of the comb and not of the ovaries, and growth hormone causes general body growth, not growth of the uterus, comb, or ovaries. In Chapter 6 we shall discuss some ideas of the possible mechanism of hormone action and the probable reasons why hormones show tissue specificity, and how tissues recognize their own trophic hormones and ignore others. Note especially that hormones *modify* the activity of genetically conditioned end organs. They do not initiate reactions that are not normally performed by the cells of those organs. Such cells are endowed with the ability to perform certain highly specialized chemical tasks, which, in the absence of hormones, may be accomplished only slowly and inadequately, but, in the presence of hormones, are speeded up or performed more efficiently.

It is now clear that optimal reproductive performance is possible only in euhormonal organisms, and that the entire endocrine system participates, to a greater or lesser extent, in bringing about maximal reproductive efficiency. Some glands are directly concerned with reproduction, and it is these that we shall examine in detail. (The role of the adrenal glands in reproduction will be ignored almost completely, not because it is insignificant, but because it is less precisely defined than the role of glands that are directly involved in this process. Nevertheless, students are urged to consult textbooks of endocrinology to acquaint themselves with all the endocrine glands and their important roles in the reproductive processes.)

THE PITUITARY GLAND

The pituitary gland, or hypophysis, is located in a bony depression at the base of the brain called the sella turcica. It consists of the adenohypophysis and the neurohypophysis; each of these can be subdivided into anatomically distinct parts.

The adenohypophysis, or anterior lobe, consists of the pars tuberalis, the pars intermedia, and the pars distalis, the latter forming the major part of the anterior lobe. A part of Rathke's pouch, which is an entodermal outpocketing of the stomadeum, forms the pars intermedia, and the floor of this pouch gives rise to the pars distalis. The pars tuberalis is the most vascular portion of the pituitary gland. It is found in all vertebrates, but its endocrine function is not known.

The anterior lobe is not innervated. Evidence presented later in the book indicates that many reproductive phenomena may be controlled by nerve impulses that signal the adenohypophysis that this or that hormone should be released. If the adenohypophysis were innervated such messages could reach the anterior lobe directly and speedily. However, it is now known that the nerve impulses terminate in the hypothalamic region, where the command for the release of hypophyseal hormones is translated into humoral agents—of which more will be said later—which in turn reach the adenohypophysis via the hypothalamo-pituitary portal system. It would seem that this system of relaying commands is inefficient because it appears to be slower than the direct and instant relay possible via the nervous system alone. Of course, this concern may be simply a reflection of our present ignorance of the rapidity with which hypothalamic humoral agents may be released into the portal system and the rapidity with which they may cause release of hypophyseal hormones.

The neurohypophysis, also known as the posterior lobe, is derived from the diencephalon and is thus ectodermal in origin. The posterior lobe consists of the pars nervosa and the median eminence; it is through the latter that both the anterior and posterior lobe of the hypophysis derive their blood supply. According to the best available evidence, only the pars nervosa is innervated, the nerves coming to it via the pituitary (hypophyseal) stalk from the hypothalamic region. The hormones usually associated with the posterior lobe are produced in the hypothalamus and find their way to the posterior lobe through nerve tracts connecting that lobe and the hypothalamus. If these tracts are blocked, secretory granules (presumed to be hormone precursors or even the hormones themselves) accumulate on the hypothalamic side of the block and gradually disappear on the pituitary side. This type of evidence makes it seem likely that the posterior lobe itself does not secrete the hormones found in it, but only stores and releases them.

Histology of the Anterior Lobe

Pituitary cytology is too involved and unsettled at the present time, in my opinion, to warrant a detailed discussion here. There are at least three distinct and important types of cells in the anterior lobe of the pituitary gland. The chromophobes, the largest of the three, are so named because they do not stain as readily as the other types, of which they are generally assumed to be the progenitors. They are not associated with the secretion of any hormone. The acidophilic chromophiles are smaller than the basophilic chromophiles. Both stain readily with the appropriate biological stains, and both are secretory.

The proportions of the three types differ with species, age, sex, and reproductive stage. Castration, pregnancy, and various thyroid and other endocrine states have a profound effect on the appearance and number of the acidophiles and basophiles. These observations provide a basis for associating the secretion of certain hormones with one or the other of the two secretory cell types, as will be shown later in this chapter.

Hormones from the Anterior Lobe

Before discussing the different hormones that are known to be produced by the anterior lobe of the pituitary gland, let us see how the information about the identity of the hormones secreted by that gland has been obtained. Endocrine experimentation usually involves the removal of the gland, followed by replacement therapy. If, for instance, we want to find the endocrine function of testes in the cock, we first remove them surgically. We find that the comb shrinks and eventually becomes very small and dry. We conclude that the growth of the comb in some way depends on the testes. But we do not know (1) whether there is a neural connection between the testes and the comb; (2) whether the action of the testes on the comb is direct; or (3) whether it is indirect, via some other gland that secretes a hormone responsible for growth and maintenance of the comb.

To resolve these uncertainties, we use replacement therapy. We can either inject an extract of testes into a castrated cock, or we can implant the testes into such an animal. In either case growth of the comb is resumed. We have thus confirmed the initial observation of a relation between testes and comb, and, because testicular extract was effective, we have eliminated the possibility that a neural connection is essential to growth of the comb. We still do not know whether the action of the testes on the comb is direct or indirect, via some intermediate gland and hormone. At this point it becomes necessary to use the technique of hypophysectomy, or the surgical removal of the pituitary gland. In 1927 P. E. Smith perfected the parapharyngeal approach of that gland in the rat. The technique became a laboratory routine, and as a result endocrinology became a much more exact branch of physiology and made impressive advances. Since the original operation on rats, the procedure has been used on all laboratory animals—cats, dogs, chickens, pigeons, frogs, pigs, and goats.

By removing the gland that is now known to control the function of every other endocrine gland in the body, hypophysectomy creates an endocrine vacuum, for all the other glands depend on pituitary gland secretions for normal functioning.

We now return to the relation between the testes and the comb of the cock. If a cock is hypophysectomized, both testes and comb shrink. We shall temporarily ignore the information that the testes seem to depend on pituitary gland secretions not only for size but also for functioning (the comb decreased in size!) and concentrate on the original problem of the testes-comb relation. If we inject a testicular extract into a hypophysectomized cock, growth of the comb is resumed. Since the pituitary gland has been removed, we know that it cannot be intermediary between the testes and the comb. It is conceivable that the extract may have produced growth of the comb by acting on one of the remaining glands and causing it to secrete the hormone that is responsible for such growth. To test this possibility, we can remove, in turn, the thyroids, the adrenals, etc., from hypophysectomized cocks and inject testicular extract. Or we can inject hormones from these other glands, but not testicular extract, into hypophysectomized cocks. In this way we can eliminate the contributions of all other glands and conclude that a hormone produced by the testes is directly responsible for growth of the comb (Figure 3–1).

This example illustrates two types of endocrine events: (1) the simple one-step reaction from the testes to their end organ, the comb; (2) the more complex two-step reaction from the pituitary gland to one of its end organs, the testes, and from the testes to their end organ, the comb.

If we removed the entire pituitary gland from the cock, we would have no way of knowing whether the hormone acting on the testes was secreted by the anterior or the posterior lobe. In some laboratory animals and in the chicken it is possible to remove only one lobe and leave the other in place to continue secretion of its peculiar hormones. It is also quite easy to separate the two lobes in glands obtained from slaughtered animals and to assay each of the lobes separately for the hormones it contains. If we were to do this, or if we were to remove only one of the lobes, leaving the other one in place in the cock, we should find that only hormones from the anterior lobe are capable of stimulating the testes to secrete the hormone that causes growth of the comb.

It is also possible to subject the two lobes of the pituitary gland to chemical fractionation and to extract from them several distinct hormones, which have different physiological effects.

All pituitary hormones are either polypeptides or proteins (see Chapter 6). The chemistry of some is known, and experiments of the type discussed and chemical fractionation have provided evidence of the secretion by the anterior lobe of at least six distinct and physiologically different hormones. Since there are only three types of cells in that lobe,

and only two of them are considered to be secretory, we can only conclude that the acidophiles and basophiles each secrete several of the
pituitary hormones. Investigators have obtained evidence on this point
by correlating pituitary cytology with certain physiological states of
animals: in dwarf mice, for example, no acidophiles are found in the

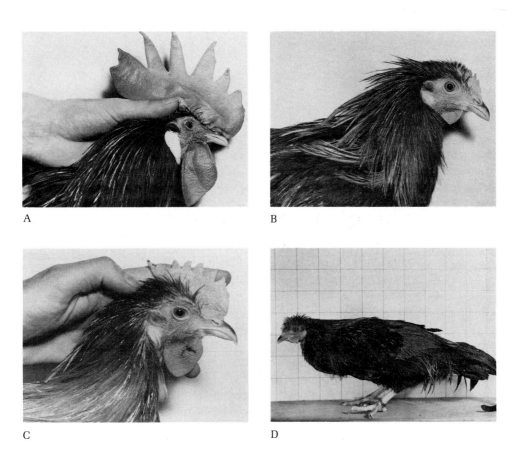

Figure 3–1 Effect of hypophysectomy and replacement therapy on male chickens.
Compare the size of combs and wattles and the size and color of earlobes in the
four pictures, and note that in D the feathers are long and narrow (thyroid deficiency) and the legs short, and that the bird appears juvenile. The fact that
chickens have externally visible and measurable secondary sex characters (comb,
wattles), and the fact that even traces of pituitary tissue will cause partial or
normal testicular or ovarian function, make chickens good experimental subjects
for endocrine research. **A.** Normal control. **B.** His hypophysectomized brother,
30 days after the operation. **C.** Another hypophysectomized brother, injected
with LH for 8 days. **D.** Hypophysectomized male that remained untreated for
14 months after the operation.

anterior lobe; and in certain types of gigantism and acromegaly, relatively common in men, acidophiles frequently form tumors and become very numerous. These observations lead to the assumption that the acidophiles secrete the growth (somatotrophic) hormone.

Castration causes vacuolation of basophiles, which are then called signet-ring cells because of the acentric nucleus and the large vacuole in the cytoplasm. In some cases of precocious sexual maturity, basophilic tumors occur in the anterior lobe. These findings suggest that the gonadotrophic complex (there are two different hormones) is produced by basophiles. Thyroidectomy also modifies the basophiles, suggesting that the thyrotrophic hormone also is secreted by them (then called the "thyrotrophs"). Similar evidence implicates the basophiles as the probable source of the adrenocorticotrophic hormone. Prolactin is assumed to come from the acidophiles. Much of the evidence presented here has been contradicted or corroborated by various staining techniques that are too complex to be discussed in detail (see Cowie and Folley, 1955).

INTERRELATION OF THE PITUITARY AND OVARY

When hypophysectomy was first performed on immature rats, the gonads failed to become functional; when the operation was performed on adult rats, the gonads shrank and eventually atrophied. These important observations, together with the fact that the gonads of hypophysectomized animals could be restored or maintained by the injection or implantation of pituitary tissue, conclusively established the dependence of the gonads on pituitary hormones. There remained the intricate job of finding out how the pituitary gland causes the cyclic behavior of the ovary. This was accomplished in a series of beautiful experiments by early endocrinologists, who laid the foundation for the currently accepted interpretation.

According to this interpretation, the pituitary gland secretes the gonadotrophic complex, which consists of two distinct substances. The identity of these two gonadotrophic hormones was first established by the use of female rats as test animals. The two hormones were therefore named, with reference to their effects on the ovary, the follicle-stimulating hormone (abbreviated as FSH) and the luteinizing hormone (or LH). Later it was shown that both hormones are also secreted by the pituitary glands of males. Occasionally the term "interstitial-cell-stimulating hormone" (ICSH) is used instead of LH when the discussion is restricted to the effects of LH in males. There seems little justification for the term ICSH, for it is identical with LH, and no effort has been made to rename FSH when its effects in males are discussed. In this book we shall use the terms FSH and LH whether the discussion pertains to males or to females.

Some endocrinologists also consider prolactin a gonadotrophic hormone, largely because of the luteotrophic effect of this hormone on the corpora lutea of rats and mice. But, as is pointed out elsewhere (pp. 68-79), prolactin is not the luteotrophic hormone of other mammals and for this reason should not be classified as a gonadotrophin. In fact, in some mammals and birds prolactin has an antigonadotrophic action. If it is injected into mature males and females with functional gonads, the gonads can be made to regress completely if the dose of prolactin is sufficiently high and if the injections are continued for several days. These antigonadotrophic effects of prolactin can be counteracted by the simultaneous injection with prolactin and FSH and/or LH, suggesting that exogenous prolactin inhibits the synthesis or the release of FSH and of LH.

Stage 1 of the Estrous Cycle—Follicular Phase

In the last ten years the development of two important analytical methods has made it possible to be more precise about changes in hormone levels during the estrous cycle. These are the radioimmunoassay (RIA) and the protein-binding assay, methods which have the distinction of being exquisitely sensitive for the detection of minute quantities of protein or steroid hormones.* (Nevertheless, earlier conclusions about hormone levels are not false, even though they were based on insensitive bioassay methods or chemical determinations.) Although reliable radioimmunoassays are available for LH in many species, at present complete confidence can be placed only in the RIA for FSH in women and rats. For this reason, part of the argument presented here is based on assays obtained in humans in which simultaneous assays for both FSH and LH have been made. The fragmentary information available for other species makes it seem probable that most, if not all, animals follow the same pattern.

We shall examine the hormone profile found in women with a normal menstrual cycle, shown in Figure 3-2. For the time being we shall concentrate on the FSH and LH values found in peripheral plasma. Two interesting features should be carefully noted because they are important and we will come back to them in other connections. First, note that there are concomitant peaks of LH and FSH release. This phenomenon is also found in the rat and there are some indications that double release is typical of all mammals, with the possible exception of the rabbit. Next,

---

*See p. 190 for a description of radioimmunoassay.

note that the post peak LH levels are very low and steady and that the
same appears to be true of the FSH levels, which in women actually
drop prior to the ovulatory peak. Another important feature is the extreme sharpness of the two hormone peaks, a feature that should be kept
in mind because if one wishes to detect these peaks frequent and closely-
spaced bleeding is mandatory. Because it is well established that both
FSH and LH are needed to cause follicular growth, it is somewhat dis-
concerting to find that the rates of secretion prior to the surge are so
uniformly flat. One wonders then how to explain the surge in follicular
growth and the concomitant peak of plasma estradiol that begin a few
days prior to gonadotrophin release and subsequent ovulation.

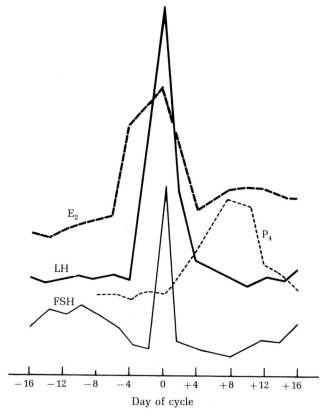

Figure 3–2 Relative interrelations between plasma FSH, LH, estradiol, and progesterone during the menstrual cycle of women. Note particularly that (a) both
FSH and LH peak just prior to ovulation; that there is a decrease in FSH and no
significant rise in LH prior to their peaks; (b) plasma estradiol rises prior to the
peaks of the gonadotrophins; (c) plasma progesterone rises significantly after
ovulation. [For details see *Recent Progress of Hormone Research,* 1970, 26:1–62,
63–103.]

Preliminary studies on sheep have produced essentially the same picture. In sheep a comparison can be made between the levels of FSH and LH (Figure 3–3), and the rate of growth of the largest follicle during the estrous cycle (Figure 3–4). (Since the two types of data came from different experiments, the curves cannot be superimposed.) Figure 3–4 shows that there is an increase in follicle size between days 0 and 5 of the cycle, followed by a plateau. The so-called follicular growth spurt occurs between days 15 and ovulation, which in these sheep can be assumed to have taken place on days 17 or 18. Figure 3–3 shows that there is no obvious change in the amounts of FSH and LH in the peripheral plasma that might account for the growth phase (with the possible exception of the preovulatory spurt).

An explanation may be that the sensitivity of the follicle to hormone stimulation changes drastically so that even small quantities of hormone are utilized much more effectively. One possible mechanism of increased sensitivity may be that the number of binding sites for the gonadotrophins increases in the follicle cell layers as the follicle approaches ovulation. Thus, more of whatever little hormone there is may be bound and cause more rapid follicle growth. If this interpretation is correct, the question remains what causes this increase in binding sites. We shall return to the problem of the physiological significance of low levels of hormone in the blood stream in the discussion of the hormonal control mechanism of the corpus luteum (p. 70).

The level of estrogen in the peripheral blood rises steeply *prior* to the peaks of LH and FSH (Figure 3–2). It has been known for many years that estrogen may be the trigger for LH release, since it has been shown in such diverse species as rats and sheep that the injection of small doses of estrogen could induce ovulation. Recently, attention has again been focused on the relation between estrogen and gonadotrophin release. One of the best demonstrations of estrogen-induced LH release has been obtained in anestrous sheep in which no spontaneous LH releases occur during the nonbreeding season (Figure 3–5). That the estrogen-controlled LH-releasing mechanism is physiological is suggested by the fact that a plasma estrogen peak is normally detectable in ewes just prior to the onset of heat (Figure 3–5) and such a transient peak is also seen in women (Figure 3–2).

Finally, in women no major secretion of progesterone takes place prior to release of the gonadotrophins (Figure 3–2), implying that progesterone is not involved in this release even though under certain experimental conditions progesterone is known to induce ovulation—notably in the chicken, but occasionally also in cows, rabbits, rats, and sheep. In another connection it will be pointed out that in several species there is good evidence for the fact that follicles do produce progesterone prior to ovulation (p. 174). In the absence of a careful analysis of the role

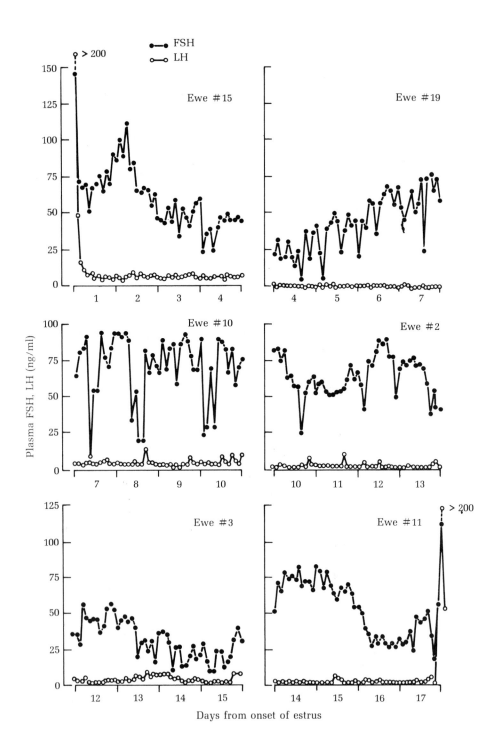

Days from onset of estrus

Figure 3-3 Plasma FSH and LH levels during the cycle of six ewes. Note that ewe No. 15 shows simultaneous FSH and LH on day 0 (day 1 is day of heat) and ewe No. 11 shows the same peaks at the end of her cycle. Also note that during increasing follicular growth (Figure 3-4) there are no apparent increases in plasma gonadotrophins. Ewe No. 11 shows a decrease in plasma FSH, similar to that seen in women (Figure 3-2) just prior to the gonadotrophin peaks. [From Salamonsen *et al.*, 1973, *Endocrinology*, 93:610.]

of progesterone in LH-release a conclusion as to its physiological role in ovulation in the different species must be withheld.

Note in Figure 3-5 that LH-release in sheep can be induced by very small doses of estradiol (10-50 µg/sheep) injected systemically. Considering the size of a sheep (about 50 kg), the amount of estrogen that can be expected to reach the hypothalamo-hypophyseal control center is indeed minute, probably in the picogram range. By contrast, if large estrogen doses are injected, secretion of gonadotrophins is totally blocked. (This of course is the basis for the effectiveness of contraceptive pills, some of which contain only estrogen, whereas others consist of a combination of estrogen and progesterone.) The fact that the same hormone, depending on the dosage, may either cause or block the release of a pituitary hormone should caution against those generalizations,

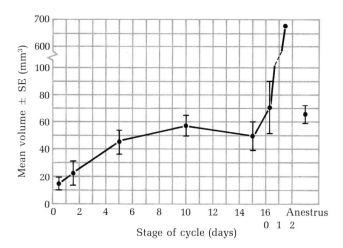

Figure 3-4 Changes in the volume of the largest ovarian follicle during the estrous cycle of sheep. (For comparison, note that the follicle of ewes during the nonbreeding season is as large as the follicle on day 16 of the cycle.) Significant follicular growth occurs throughout the cycle and culminates in an "ovulatory spurt" immediately prior to ovulation. Compare this with Figure 3-3 and note that during the follicular growth phase between days 0 and 15 there are no corresponding increases in plasma gonadotrophins. [From Hutchinson and Robertson, 1966, *Research in Vet. Sci*, 7:17.]

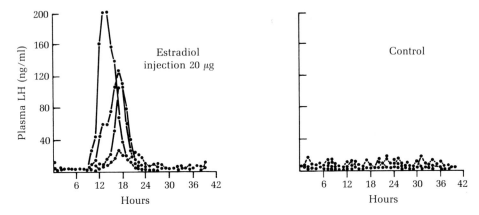

Figure 3–5 Example of effectiveness of a single injection of estradiol on the release of LH in anestrous ewes. Note the absence of such releases in uninjected control animals. [From Goding et al., 1969, Endocrinology, 85:139.]

frequently found in the literature, in which effects are ascribed to a hormone without stipulating the dose. If sufficiently large doses of estrogen are injected or implanted into male or female mammals or birds, a total blockade of the gonadotrophic secretion of the pituitary gland can be brought about. In heavily estrogenized males the testes and all the accessory organs can be caused to atrophy to a degree comparable to that caused by hypophysectomy. In females a similar pituitary blockade can be produced by large doses of estrogen but in them only the ovaries atrophy while the Müllerian duct system greatly hypertrophies because it is the target tissue for estrogen.

We now return to the curious fact that simultaneous preovulatory release of both FSH and LH is demonstrable in rats, primates, sheep, cattle, and other animals. The data in Table 3–1 demonstrate that even though FSH alone can cause follicular growth (ovarian weight is increased by FSH), the follicles are unable to produce estrogen, as shown by the fact that the uteri do not increase in size. Only after combined treatment with FSH and LH, which does not result in luteinization, is uterine weight increased—testifying to the fact that estrogen is being secreted. Thus, it appears probable that all or most physiological responses ascribed to gonadotrophins require the cooperation of both FSH and LH. In fact, in sheep both hormones are present in the peripheral blood throughout the cycle (see Figures 3–2 and 3–4). It also appears probable that they are both required for the induction of ovulation.

Shortly after the historical separation of the gonadotrophic complex into its two components, FSH and LH, it was established that LH is the ovulation-inducing hormone. Although LH is still considered the main

Table 3-1. Effect of FSH and LH on ovarian and uterine weights of
            immature hypophysectomized rats

FSH (days)	LH (days)		OVARIAN WEIGHT (mg)	UTERINE WEIGHT (mg)
4			16.3	15.7
7			17.9	14.1
8			18.2	16.7
9			14.0	12.5
10			15.1	12.9
7	plus	2	23.4	49.5
8	plus	3	29.4	61.1
9	plus	4	26.2	78.2
10	plus	5	22.8	101.4
Hypophysectomized controls			6.7	12.4

SOURCE: Greep, Van Dyke, and Chow, Endocrinology, 30:635, 1942.

ovulatory hormone, it has been found that in hypophysectomized rats
and intact rabbits ovulation can be induced with FSH and that in rats
there is a simultaneous depletion of both hypophyseal FSH and LH
on the morning or afternoon of proestrus. Moreover, work in the Physiology Research Laboratory at the University of Illinois shows that
although intrafollicular injection of about 10 ng of LH causes ovulation
in rabbits, so does injection of 100 ng of FSH (Jones, 1972; Goldman and
Mahesh, 1969). To complicate matters still further, systemic injection of
LH in laying hens causes ovulation, but intrafollicular injection does
not. Yet if a mixture of FSH and LH is injected into the wall of chicken
follicles, ovulation follows. It is obvious that much additional work is
needed before the interrelation between FSH and LH in the induction
of ovulation is understood. (In none of the cases in which FSH alone was
used can ovulation be ascribed to contamination of the FSH with LH.)

The simultaneous ovulatory release of FSH and LH raises one more
question that deserves thought. Ever since the fractionation of the
"gonadotrophic complex" into two components, it has been taken for
granted that the pituitary does indeed secrete two separate hormones,
FSH and LH. There is much evidence for this contention, such as the
fact that two separate hormones that have distinctly different physiological effects can be prepared from whole pituitary glands. Also,
expert cytologists can recognize two different pituitary cell types believed to secrete the two different hormones. Finally, the literature is
full of data that support the claim that there are two separable hypothalamic releasing factors (see Chapter 6 for details). More recently,
however, both published and unpublished results show that the case
made for two releasing factors may be weak and that there may be only

one factor, which would explain why both LH and FSH are released together prior to ovulation (see, for instance, Meites, 1970).

Although it is possible to obtain LH preparations that show no FSH activity, no one has ever succeeded in producing a "pure" FSH; all FSH preparations show evidence of LH contamination. This fact is commonly ascribed to the so-called common-core phenomenon, which is due to the possibility that both hormones have a certain number of amino acid sequences in common. This commonality may be responsible for the residual LH activity of so-called pure FSH preparations. If the common sequence is removed, the preparation loses not only its LH but also its FSH activity.

Thus we again come back to the question of the cause of follicular growth in view of the seemingly unchanging levels of FSH and LH during the rapid growth phase of follicles. We have already noted the possibility that even small doses of the hormones may cause follicular growth, provided that the physiological sensitivity of follicles changes late in the cycle. However, there is also the possibility that the RIA does not accurately mirror the true hormonal status of the animals because the assay is designed to measure two antigenically different biochemical substances (FSH and LH) instead of a single entity—the gonadotrophic complex. That some thought should be given to this possibility is illustrated by the data in Table 3–2. These data predate the presently available sensitive assays and are based on the ability of the gonadotrophic complex contained in the unextracted pituitary gland to cause an increase in the testes weight of day-old chicks. Data of this type have been deprecated on the grounds that pituitary content is not indicative of the hormone that actually reaches the target tissue and acts on it. Nevertheless, these data deserve some attention, especially in view of the dilemma posed above. We can see that in pigs from days 4 to 7 of the cycle, the total gonadotrophic content of the pituitary gland is low and that 100 percent of the follicles in the ovaries of the donors are less than 7 mm in diameter. On day 8 the concentration of hypophyseal gonadotrophic hormone increases significantly and is coincident with a very significant jump in the total number of follicles. From days 16 to 20 of the cycle there is no significant change in gonadotrophic hormone content, but an increasing proportion of the follicles begin to appear in the category of follicles of ovulatory size (8–10 mm). Finally, in pigs actually showing estrous behavior all the follicles are of ovulatory size; the average number of all follicles has decreased sharply from the pre-estrus period and so has the pituitary content of gonadotrophic hormone. The depletion of the pituitary gland on the days of heat presumably can be ascribed to the release of FSH and LH prior to ovulation.

Table 3–2. Changes in ovarian morphology and gonadotrophic potency of pituitaries of pigs

| DAY OF CYCLE | CLASSES OF FOLLICLES | | | | | AVERAGE NUMBER FOLLICLES IN BOTH OVARIES | GTH IN AP* | STAGE OF CYCLE |
| | FOLLICLE SIZE (mm) | | | | | | | |
	5	5–7	8–10	10	CYSTS			
4	100%	. . .	. . .	. . .	. . .	13	14.7	
5	. . .	100%	. . .	. . .	. . .	19	13.5	
6	56	40	. . .	. . .	4%	24	17.2	
7	47	53	. . .	. . .	. . .	19	14.1	
8	75	25	. . .	. . .	. . .	49	23.5	Luteal phase
9	. . .	89	. . .	. . .	11	49	29.6	
10	62	38	. . .	. . .	. . .	42	26.1	
13	68	30	. . .	. . .	2	51	28.1	
14	51	46	3%	. . .	. . .	37	22.5	
16	40	59	1	. . .	. . .	40	23.1	
17	47	52	1	. . .	. . .	45	26.4	
18	56	33	10	. . .	. . .	39	33.5	Follicular phase
19	44	9	47	. . .	. . .	57	20.5	
20	84	16	. . .	. . .	. . .	62	22.5	
1	5	9	76	9	1	16	13.7	Heat
1	. . .	. . .	69	31	. . .	13	17.1	

*Total gonadotrophic hormone in the anterior lobe of the pituitary gland measured by weight (mg) of testes of day-old chicks.

Source: Robinson and Nalbandov, *J. Animal Science,* 10:469, 1951.

Thus, while the correlation between the hormone content of the pituitary gland and morphological changes in the ovary is not perfect, there is an indication that throughout the cycle pituitary gonadotrophins change in relation to the events in the ovaries. By contrast, the RIA of FSH and LH shows only that a significant peak of these hormones occurs at the time of ovulation, and the question arises whether this is an accurate reflection of the hormonal status in the peripheral blood stream.

In view of the ambiguity of the gonadotrophic hormone titers during the cycle, it is now more difficult to discuss the hormonal interrelationship between the pituitary gland and the ovaries than it was in the days when the only available bioassays were those of the hormone content of the pituitary gland. Nevertheless, a few tenets still hold. We have already noted that the evidence is good that estrogen is responsible for LH release, although the possible implication of progesterone in this function cannot be totally dismissed at this time. The dilemma lies in the fact that, according to RIA, there is no obvious change in the blood FSH-LH levels during the time of rapid follicular growth and the increase in the estrogen level. It is possible that we should not be looking for a change in hormone titers in the blood; the effect of a sustained level of hormone stimulation may be cumulative and just as important in producing significant changes in the rate of follicle growth as a measurable increase in the level of the hormone. We have already noted the possibility that the sensitivity of follicles to hormonal stimulation may be increased, permitting accelerated growth of the follicles despite a constant hormone level. All of these problems remain unresolved.

Even though the interrelationship of the hypophyseal hormones and their role in ovarian stimulation remains uncertain, it is known that both estrogen and, as we shall see later, progesterone are powerful inhibitors of pituitary function. Thus, even very small doses of estrogen (0.015 to 0.075 μg of estradiol daily) injected into rats cause a decrease in the pituitary FSH levels. At these levels the estradiols do not cause stimulation of uterine growth of the treated rats. If the estrogen dose is high, a condition similar to "physiological hypophysectomy" is produced in that no gonadotrophic hormone is released, but large quantities of FSH and LH are accumulated in the hypophysis. This method of total blockade of the pituitary gland is practical in males but not in females because the Müllerian duct system and the breasts would be excessively stimulated by high estrogen levels. Sufficiently high doses of estrogen can block not only the release of the gonadotrophins but also thyrotrophin and, if the estrogen dose is very high, even somatotrophin, albeit not completely.

Stage 2—Luteal Phase

At one time it was believed that information obtained on the endocrinology of the luteal phase in the laboratory rat or mouse held for other mammals. However, work done on other mammals suggests a different endocrine mechanism in the formation and maintenance of corpora lutea. In fact, the rat must be considered an exception to the rule. Prolactin, which is generally assumed to be responsible for the maintenance of the formed corpus luteum, is luteotrophic only in the rat and not in other mammals in which it has been tried. Even though it can be argued that in other mammals the role of prolactin may be assumed by another as yet unidentified luteotrophic hormone (LTH), there are other problems (to be mentioned later) that make it advisable to discuss the luteal phase in rats separately.

LUTEAL PHASE IN RATS. During the last ten years there has developed a drastic revision of our understanding of the endocrinology of the luteal phase in the rat. In fact, our understanding of the entire reproductive cycle of the rat has changed.* Until recently it was thought that LH was the primary ovulation-inducing hormone, but, as we have already noted, it appears more probable that a mixture of FSH and LH is the ovulation-inducing complex. Furthermore, under certain experimental conditions FSH alone is capable of inducing ovulation. The rat is one of the species in which simultaneous preovulatory release of FSH and LH has been demonstrated. LH is commonly held responsible for mobilizing cholesterol, the essential precursor of progesterone, and moving it into the forming corpus luteum. If no mating occurs, the formed corpora lutea of the cycle produce very little progesterone, so little, in fact, that the uterus of the unmated rat does not undergo the typical progestational proliferation usually associated with progesterone action. However, if the rat mates, the nerve endings in the vagina and the cervix are known to be stimulated by the male's penis, transmitting a neural signal to the hypothalamus, where the prolactin-inhibiting factor is blocked by the neural signal and prolactin is released by the adenohypophysis. Prolactin maintains the cholesterol ester synthetase, which is essential in the conversion of cholesterol to cholesterol ester; it also maintains the esterase that plays a cardinal role in the conversion of cholesterol ester to free cholesterol.

Until recently it was assumed that the chain of events described above demonstrated that in the mated rat prolactin was responsible for progesterone synthesis, which continued through the first 13 days of pregnancy. However, reports began to appear in the literature that during in vitro incubation of rat corpora lutea the addition of prolactin to the medium did not cause progesterone synthesis, whereas the addition of LH did. Even more significant is the recent finding by N. R. Moudgal and his co-workers that the injection of a highly specific LH antiserum into pregnant rats between days 1 and 12 of gestation invariably leads to spontaneous abortion. Since the LH antiserum neutralizes the action of the endogenous LH, the logical conclusion was that LH normally participates in some manner in the maintenance of pregnancy. Since the effect of the LH antiserum could be offset in the preimplantation stages of pregnancy by estrogen and in the postimplantation stages by progesterone, the conclusion was made that LH normally acts on corpora lutea and causes progesterone synthesis. This conclusion was substantiated by the finding that the rate of progesterone synthesis in antiserum-treated rats was significantly depressed compared to that in untreated

---

*This should prove instructive to all researchers. The laboratory rat had been the most intensively researched of all species for at least 40 years and we had begun to think we had a complete understanding of the rat's reproductive endocrinology. That this proved to be wrong should keep those who think they have conclusively solved a problem alert and open-minded.

pregnant control animals. Moudgal and his colleagues (1974) postulate, with good reason, that the role of LH is activation of the esterase involved in the conversion of the free cholesterol to progesterone. The reason LH is needed only during the first 12 days of gestation is that the rat depends on the pituitary gland as a source of a luteotrophin only during the first 12 days of gestation. After that time the placenta acquires competence to maintain pregnancy by producing a non-LH luteotrophic substance that can keep the corpora lutea producing progesterone.

Thus, LH has been identified as a major luteotrophic substance in the rat, and is apparently solely responsible for the steroidogenic competency of the rat corpus luteum. This fact is all the more important because in the discussion to follow we shall see that LH is also the primary luteotrophic substance in other mammals. As visualized by Moudgal, in the absence of prolactin LH causes the conversion of most of the progesterone into 20 α-dihydro-progesterone (20 α-OH-P), which is physiologically inactive. If both prolactin and LH are acting on the corpus luteum, as they are in the pregnant rat, prolactin partially blocks the rate of conversion of progesterone to 20 α-OH-P so that the net balance is in favor of progesterone, which is essential for progestational proliferation of the uterus and later for the maintenance of pregnancy.

Similar results have been obtained in the hamster. LH antiserum causes abortion or resorption in the hamster if it is injected between days 1 and 11 of gestation. Again, the implication is that both prolactin and LH are luteotrophic, although FSH also seems to participate in the maintenance of the corpus luteum. Prolactin alone is said to be luteotrophic in ferrets.

LUTEAL PHASE IN OTHER MAMMALS.   As in the case of the rat, many advances have been made in the physiology of the mechanism of formation and maintenance of the corpus luteum of nonmurine mammals so that revision of our thinking is necessary. We have gone into some detail (without exhausting the subject) on the hormonal control mechanism operating in the formation and the function of the rat corpus luteum. Now, let us take up the new and somewhat novel concept that perhaps no hormones are required for at least one phase of the formation of the corpus luteum. The evidence for that concept is as follows.

It is possible to introduce a fine needle into the preovulatory follicle of a rabbit and, by gentle suction, to remove the ovum without unduly disturbing the granulosa lining of the follicle. Ovectomy, as the procedure is called, has been performed on hundreds of individual rabbit follicles. When the oocyte from a single follicle is removed, the follicle promptly luteinizes and forms a corpus luteum. However, the corpus luteum thus formed is different from normally formed, LH-induced corpora lutea in that it is not solid (it has a hollow center). The histology of the luteinized wall is nevertheless indistinguishable from a normal corpus luteum (Figures 3-6 and 3-7). The significance of this finding is not only that the ovectomized follicles luteinize, but also that they secrete as much progesterone as the normally formed corpora lutea of pseudopregnancy. Because ovectomy affects only treated follicles, and not those in close proximity that were not ovectomized, it appears that the effect is purely local and does not involve a systemic feedback mechanism. The results of ovectomy also imply (at least, for the present) that no hormones other than those normally present in the blood of unmated rabbits are involved in luteinization after ovectomy. The most important implication of this work is that the oocyte produces a substance that it passes on to the follicular fluid, which is capable of preventing luteinization of the granulosa and theca cells of the follicle. That this phenomenon is not restricted to rabbits has been shown by the fact that ovectomized pig follicles also luteinize. However, the endocrine control mechanism involved in pigs is much more complex than that in rabbits, and more work is needed in pigs before an intelligent appraisal can be made.

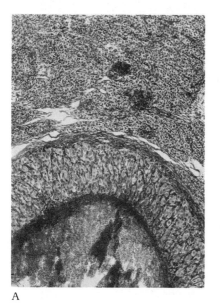
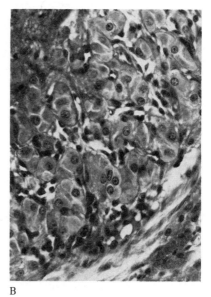

A   B

Figure 3-6 **A.** An ovectomized rabbit follicle; only the rim of the follicle luteinizes. These structures degenerate by five days after the oocyte has been removed. **B.** A higher magnification of the luteinized rim of the follicle shown in A. Note that the luteal cells are indistinguishable from those of the normal corpus luteum.

Figure 3-7 Intrafollicular injection of a subovulatory dose of LH causes luteinization of the follicle and entrapment of the oocyte.

As we have pointed out, the "autonomous luteinization phase" of the ovectomized follicles is short, lasting not more than five days, after which the hollow "corpus luteum" degenerates. However, it is possible, by the injection of LH directly into the ovectomized follicle, to convert a temporary structure into one that can live and function for at least the duration of pseudopregnancy. This suggests then that the formation of the corpus luteum may consist of two stages: a hormone-independent autonomous stage and a hormone-dependent postovulatory stage.

Let us now return to the more conventional aspects of the formation and function of the corpus luteum. It is surprising to find that corpora lutea form in pigs hypophysectomized immediately after ovulation. Not only do such corpora lutea persist as long as those in intact animals, but they also synthesize quantities of progesterone comparable to those found in intact control animals. The corpora lutea of both hypophysecto-mized and intact animals degenerate at approximately the same time. This suggests that in pigs the corpus luteum of the cycle is autonomous and does not require hypophyseal trophic substances for either mor-phologic development or steroidogenic competency. This fact is some-what reminiscent of the behavior of ovectomized follicles in rabbits and suggests that in pigs, as in rabbits, the mere act of ovulation is sufficient to cause formation of corpora lutea. In the pig, however, corpora lutea formed after ovulation survive for the normal life span of corpora lutea of the cycle, i.e., 20-21 days. This we shall call the "pig model" of corpus luteum formation.

In contrast to the pig, corpora lutea do not form in sheep hypophy-sectomized immediately following ovulation. Moreover, if a sheep is hy-pophysectomized as late as a few days after ovulation the partially formed corpus luteum immediately regresses. Obviously, in sheep the corpus luteum of the cycle is in constant need of support by a luteotrophic hypophyseal substance and thus is not autonomous. This we shall call the "sheep model" of corpus luteum formation.

The guinea pig model is intermediate between the pig and sheep models. In the guinea pig the corpus luteum requires hypophyseal luteotrophic support only two or three days after ovulation, after which it becomes autonomous for the remainder of the 16-day estrous cycle.

That corpora lutea in pregnant pigs and sheep do depend on hy-pophyseal luteotrophic substances for survival is demonstrated by the fact that hypophysectomy in either species causes immediate degenera-tion of the corpus luteum of pregnancy and abortion. Because in some species the corpus luteum of the cycle is either not formed or degenerates after hypophysectomy, it is possible to search for a hypophyseal luteo-trophic substance.

LUTEOTROPHIC SUBSTANCES.   There is no simple way to present the evidence. It is confusing in part because studies have used four types of animals: hypophysectomized-hysterectomized, hypophysectomized but uterus-intact, wholly intact, and hysterectomized but pituitary-intact animals. The reason for the use of hysterectomized animals is that in several species (cows, sheep, pigs) removal of the uterus prevents the degeneration of the corpus luteum; in the absence of the uterus the corpus luteum may persist for the duration of normal pregnancy. Con-sequently, it has seemed logical to conclude that the uterus normally secretes a substance that causes the destruction of the corpus luteum. In the absence of the uterus (and, hence, of a uterine luteolytic factor) the life span of the corpus luteum is extended. It is said, therefore, that the only proper way to study luteotrophic factors is in hysterectomized animals because luteotrophic factors would not overcome the effect of the uterine luteolytic factor. (For a more detailed discussion of luteolytic factors see pp. 79–80.) Some researchers are opposed to the use of hysterectomized animals in the search for luteotrophic substances on the grounds that the corpus luteum normally waxes and wanes in the presence of the uterus, not in its absence, and that for this reason the proper research animal is one with the uterus present.

In hypophysectomized sheep, as already stated, the corpus luteum either fails to develop or, if hypophysectomy is done after the corpus luteum is partially or fully formed, immediately degenerates. Thus, the hypophysectomized sheep is an ideal animal to be used in the search for luteotrophic substances. Previous work suggested that some hormones may not be maximally effective when injected at intervals of 12 or 24 hours. Accordingly, experiments were done in which hormones were continuously infused rather than injected. When this was done in hypophysectomized sheep it turned out that only crude LH was totally luteotrophic, whereas FSH, prolactin, and estrogen were not; highly purified LH was only partially luteotrophic since it was able to prevent degeneration of corpora lutea but not as consistently as crude LH (Table 3–3). The nature of the contaminant that makes LH totally luteotrophic remains unknown. The other important aspect of the luteotrophic effect of LH is the fact that it is effective only if it is infused. This suggests that the binding sites for LH in the corpus luteum are relatively unable to bind LH or do so in a very transient manner.

These studies were extended to animals in which both the pituitary gland and the uterus were present. Again, when LH was infused it was possible to double the normal life span of corpora lutea; if unlimited supplies of hormone had been available, the survival of corpora lutea might have been extended even further (Table 3–4).

In the studies with hypophysectomized or intact animals the minimal total dose of LH infused in 24 hours was 1.2 mg. Although this amount appears large, considering the fact that the infusions were systemic and that a sheep weighs about 50 kg, it is not unphysiologic; the amount of LH actually reaching the corpus luteum was probably in the nanogram range. In spite of the small amount of hormone reaching the ovary, enormous follicular development occurred, especially in pituitary-intact sheep. This fact suggested the possibility of competition for hormone between the corpus luteum and the follicles, and the question arose whether the LH dose could be further reduced in the absence of follicles. It is possible to eliminate all but a few of the primary follicles by X-irradiating the ovary, a treatment that apparently does not harm the corpus luteum, which retains its ability to synthesize progesterone. If

Table 3–3. **The effect of various hormones on maintenance of corpora lutea in hypophysectomized sheep. E, estrogen; P, prolactin; G, gonadotrophin**

TREATMENT	NUMBER OF ANIMALS	CORPUS LUTEUM WEIGHT[a] (grams)	PROGESTERONE		EFFECT ON PREGNANCY
			CONTENT (μg)	CONCENTRATION (μg/g)	
Hypophysectomized—Not Bred					
E	5	0.033 ± .006	Not measured		
P	3	0.050 ± .009	Not measured		
G[b]	4	0.688 ± .080	16.6 ± 3.0	23.6 ± 2.8	
G + P	2	0.574 ± .113	13.5 ± 4.2	23.5 ± 4.0	
G + P + E	2	0.810 ± .113	22.4 ± 4.2	27.1 ± 4.0	
Crude LH	4	0.727			
Hypophysectomized—Bred					
Control	1	0.033	Not detected		Not maintained
Control	1	0.185	Not measured		Not maintained
Control	1	0.076	Not detected		Not maintained
G	1	0.574	31.7	55.2	Maintained
G	1	0.780	32.0	41.0	Maintained
G	1	0.620	17.0	27.4	Maintained
Purified LH	5				Maintained[c]

[a]Mean ±SEM.
[b]Gonadotrophin contains FSH and LH, with small admixtures of other pituitary hormones.
[c]Maintained in 3 of 5 animals.
SOURCE: Kaltenbach *et al., Endocrinology,* 82:753 and 82:818, 1968.

this is done, both the weight of the corpus luteum and its ability to synthesize progesterone are maintained for a significantly longer period of time than when the follicles are present (Figure 3-8). Furthermore, it has been found that the minimal dose of infused LH required to double the life span of corpora lutea can be reduced to 300 μg, compared to the 1.2 mg required when follicles are present (Table 3-4).

This observation leads to the possible interpretation that there is normally competition between follicles and corpora lutea for a finite amount of LH. Assuming that toward the end of the follicular phase follicles are metabolically more active than corpora lutea (or have an increasing number of binding sites), the postulate seems reasonable that more LH will be incorporated in the follicles than in the corpora lutea, resulting in rapid degeneration of the latter at the end of the luteal phase. This hypothesis, admittedly unproven, is offered as one possible explanation for the precipitous decline of the corpus luteum at the end of the luteal phase. Other possible explanations will be presented shortly.

LH and such LH-like hormones as HCG appear to be more or less universal luteotrophins. They are certainly luteotrophic in cows, women,

Table 3–4. The effect of different doses of infused LH on maintenance of ovine corpora lutea to day 20

LH DOSE (mg/day)	NO. OF EWES	PERCENT WITH CORPORA LUTEA	CORPORA LUTEA WEIGHT (g)	PROGESTERONE CONCENTRATION[a] (μg/g)
Mid-Cycle Controls				
None	17	100%	0.668 ± 0.016	31 ± 3
Nonirradiated Ovaries				
2.5	6	67	0.868 ± 0.102	55 ± 6
1.2	4	75	0.837 ± 0.118	31 ± 8
0.6	5	0[b]	(0.143)[c]	
0.3	4	0[b]	(0.163)[c]	
0.0 (saline)	4	0	(0.147)[c]	
X-Irradiated Ovaries				
1.2	1	100	1.070 ± 0.207	40 ± 11
0.6	5	80[b]	0.610 ± 0.103	44 ± 6
0.3	3	67[b]	0.663 ± 0.145	29 ± 11

[a]Mean ± SEM of maintained corpora lutea.
[b]0% and 80% vs 0% and 67%, $p = .005$.
[c]Regressed corpora lutea.
SOURCE: Karsch et al., Endocrinology, 87:1228, 1970.

and pigs. There is an important distinction between the hypophyseal LH and placental HCG in that the former appears to have a low binding affinity for luteal tissue, thus requiring frequent injections or infusion, whereas HCG firmly binds to luteal tissue and remains active for intervals as long as seven days. Thus, if the aim is to maintain corpora lutea, HCG is the hormone of choice, although the half-life of LH can be substantially increased by injection with Freund's adjuvant or by mixing it with a vehicle that delays its absorption, such as beeswax.

All of these observations firmly point to the conclusion that LH is the normal luteotrophic hormone in a variety of species. However, one seemingly disturbing feature requires elaboration. The advent of the highly sensitive radioimmunoassay technique made it possible to sample blood from a variety of animals throughout the cycle and assay it for LH. When this was done one common feature of several species emerged: other than the ovulatory peak of LH, levels of LH during the cycle are very low and, even with the sensitive RIA, barely detectable. The argument is advanced that if LH is indeed the luteotrophic hormone, we should see a much more elevated level of LH, at least during the luteal phase and during pregnancy when the corpus luteum needs support. On the other hand, we have already noted that luteal tissue seems to have a limited binding capacity for LH, so that in order to produce a luteotrophic effect LH must be infused or injected in such a way as to prolong its half-life. We have also noted that seemingly physiological levels of infused LH can prolong the lifespan of corpora lutea almost indefinitely. In the absence of follicles, as little as 300 μg of LH/24 hours/ewe doubles the lifespan of luteal tissue. Thus, it appears that the minute levels of LH detected in the peripheral circulation of animals in the luteal phase and

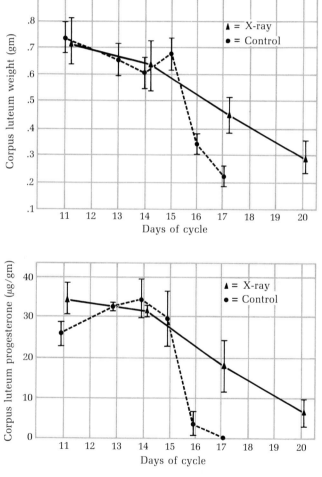

Figure 3-8 The progesterone concentration and weight of corpora lutea decrease precipitously at the end of the normal cycle of the ewe (dotted lines). In the absence of follicles after X-irradiation of the ovary, both the life span and the ability of corpora lutea to make progesterone are significantly prolonged (solid line). [From Karsch et al., 1970, Endocrinology, 87:1228.]

those pregnant may be sufficiently high to maintain the corpus luteum adequately.

ESTROGEN AS A LUTEOTROPHIN.   It has been known since 1937 that in pregnant and pseudopregnant rabbits the pituitary gland is essential for the survival of corpora lutea, thus suggesting that the adenohypophysis secretes a luteotrophic hormone. However, when FSH, LH, or extracts of rabbit hypophyses were injected into hypophysectomized pregnant or pseudopregnant rabbits, the corpora lutea degenerated. When such rabbits were treated with estrogen alone, the corpora lutea persisted and gestation continued in the pregnant animals. Much later, X-irradiation was applied to rabbit ovaries; as we have seen, this technique destroys all but a few primary follicles and leaves luteal tissue undamaged.* When both ovaries of pregnant rabbits were X-irradiated all the rabbits aborted very soon, whereas if only one ovary was treated pregnancy persisted. These observations implied that the presence of follicles was vital for the survival of corpora lutea. Accordingly, both ovaries were X-irradiated and injections of as little as 2–4 µg estradiol per animal were given. In this experiment pregnancy was maintained for as long as the daily estradiol injections were continued; it terminated almost immediately after the injections were stopped. Furthermore, progestin synthesis continued during estrogen treatment—progestins could be demonstrated in the ovarian effluent blood—whereas after cessation of the injections progestin synthesis abruptly stopped. Thus, the reason hypophysectomy of pregnant rabbits leads to interruption of pregnancy is that in the absence of the pituitary gland the follicles are not maintained, which in turn leads to estrogen deficiency and hence to luteal regression. This can be demonstrated in still another way. If LH is injected into pseudopregnant rabbits the corpora lutea degenerate, suggesting that LH is luteolytic. The explanation may lie in the fact that LH injections cause ovulations or luteinization of follicles, which are the source of the estrogen essential for maintenance of corpora lutea.

On the basis of these observations there is not doubt that the rabbit uses estrogen as a direct luteotrophic hormone, which may act systemically or perhaps even locally by diffusion from follicles to neighboring corpora lutea. However, it is still not clear whether estrogen is the substance that maintains the endocrine health and morphology of the luteal tissue, enabling it to synthesize progesterone, or whether it acts on luteal enzyme systems that turn on progesterone synthesis (steroidogenic effect). Recently, C. Lee and his colleagues (1971) showed that corpora

---

*This relatively easy technique is excellent for experiments in which destruction of follicles is desired; selection of appropriate X-ray doses permits destruction of follicles either permanently or temporarily.

lutea of rabbits possess receptor sites for estrogen, which disappear in pseudopregnant rabbits at about the time corpora lutea begin to regress.

Estrogen is luteotrophic in other species as well, notably in pigs, rats, sheep, and horses. However, in pigs and sheep it has been shown to be luteotrophic only in pituitary-intact animals, suggesting that its primary effect is its ability to cause LH release. (We have already discussed this property of estrogen in connection with the ovulatory peak of LH, which occurs just prior to ovulation.) Estrogen has been shown to be luteolytic in cows, hamsters, and probably women. However, this may be due to the fact that the doses of estrogen used were sufficiently high to block the hypophysis and thus prevent the release of its luteotrophic hormones—most probably LH.

LUTEOLYSIS. In practically all nonpregnant female mammals the corpus luteum degenerates precipitously and with its degeneration the synthesis and release of progesterone terminates abruptly at the end of the estrous cycle (Figure 3-8). The rapidity with which degeneration occurs has excited the imagination of research workers for a long time. We have already mentioned the classic studies by Leo Loeb, who clearly showed that in hysterectomized guinea pigs the corpora lutea were maintained for significantly prolonged periods. Subsequent work by others extended this fundamental observation to other species, notably cows, sheep, and pigs. Interestingly, hysterectomy of women and other primates does not interrupt normal cyclicity and does not cause prolonged maintenance of corpora lutea.

On the basis of the effect of hysterectomy in prolonging the life of corpora lutea in some species, the idea arose that the degeneration of corpora lutea at the end of the luteal phase may be due to a luteolytic substance secreted by the uterus (Anderson, 1973). Accordingly, many attempts have been made to extract such a substance from the uterus. So far no convincing evidence for the existence of such a substance, other than prostaglandin, has been brought out. Prostaglandin was originally obtained from the seminal plasma of rams and men, but since then prostaglandins have been demonstrated in a wide variety of tissues, including nerve endings. It has been found that systemic injection or intrauterine infusion of prostaglandin into pregnant females, rats, and primates causes destruction of corpora lutea, cessation of progesterone synthesis, and abortion. Thus, it appears that prostaglandin, particularly $PGF_{2\alpha}$, may be the best candidate for a uterine luteolytic substance. Work in progress strongly suggests that sheep $PGF_{2\alpha}$ reaches its highest concentration in the uterine endometrium toward the end of the cycle.

The mechanism of action of $PGF_{2\alpha}$ is presently not clear. If $PGF_{2\alpha}$ is incubated with slices of corpora lutea, the rate of progesterone synthesis is raised significantly. This makes it seem unlikely that prostaglandin in vivo reduces the rate of progesterone synthesis by interfering with biosynthetic pathways. It seems much more probable that it acts on the vascular system of the corpus luteum, so that the reduction in blood flow is the immediate factor responsible for the cessation of corpus luteum function and abortion. In addition to its effect on the corpus luteum, prostaglandin causes a variety of side effects, such as vomiting, diarrhea, muscular spasms, and respiratory embarrassment, in women in whom it is infused to induce abortion. (For details on prostaglandin, see p. 208.)

If prostaglandin is indeed the elusive uterine luteolysin, this would explain not only why corpora lutea persist after hysterectomy but also why in cases in which half of the uterus is removed only the corpora lutea on the side of the intact uterine horn degenerate whereas those on the side of the missing horn persist. It is interesting that the uterine luteolytic effect, if such exists, can be easily overcome in uterus-intact sheep by the infusion of physiological doses of LH, which prolong the life span of corpora lutea to at least double their normal life span. As stated earlier, this can be done in the presence of follicles with 1200 μg/24 hours/sheep or in their absence with as little as 300 μg/24 hours/ sheep. In either case the amounts of LH reaching the luteal tissue are certainly in the nanogram or perhaps even in the picogram range.

Pulsatile Releases of Hormones

If we now return to Figures 3–2 and 3–3 we note that both FSH and LH are found in the bloodstream of women and other mammals at a baseline level, except during preovulatory release of these hormones, when the level is considerably higher. These kind of data are obtained if the female is bled at infrequent intervals, say once a day. If blood samples are taken at more frequent intervals, every 30 minutes or less, a totally different picture emerges (Figure 3–9). In that case the LH and FSH undergo so-called pulsatile release—frequent spikes that return to baseline. In some animals (castrated monkey females) the intervals between spikes are very regular and occur hourly (circhoral); in others, although the spiking is much less regular, it nevertheless occurs (Figure 3–10). Even the ovulatory peaks of these hormones are not smooth ascents or descents of hormone release, but show spikes within the ascending and descending limbs of the peaks.

It now appears that all of the pituitary hormones may be released in such pulsatile fashion. If there is a physiological reason or advantage for pulsatile over steady release it is not obvious. Neither is the mechanism controlling this type of release clear. Knowing what we do about the role of releasing factors in the release of pituitary hormones, it may be that the releasing factors themselves are released in pulsatile fashion (due possibly to intermittent neuronal firing or biogenic amines), although what controls their ebb and flow is unknown. Finally, it is not even certain at this time whether the mechanism for the pulsatile release of pituitary hormones resides in the hypothalamus or the pituitary gland. It appears probable on the basis of fragmentary information that the mechanism resides in the hypothalamus. If LRH is infused into animals normally showing pulsatile releases of LH, the oscillations become less

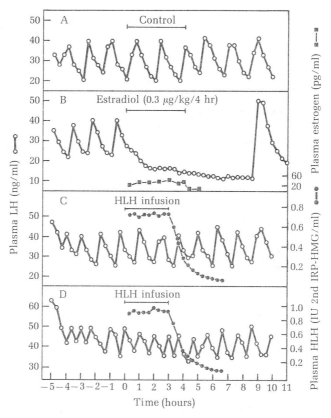

Time (hours)

Figure 3-9 **A.** Pulsatile (circhoral) release of LH in an ovariectomized monkey. **B.** Estradiol infusion completely blocks LH release. **C** and **D.** Infusion of human LH does not suppress pulsatile LH releases. These data throw doubt on the role of microcontrol systems (short feedback loops) discussed on p. 101. Pulsatile releases in intact animals do occur but are more difficult to demonstrate because of lower plasma LH and FSH levels. [From Knobil, 1974, *Recent Progress in Hormone Research*, 30:1-35.]

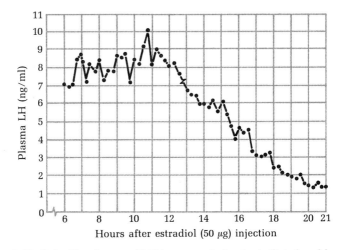

Figure 3-10 Pulsatile releases of LH in an ovariectomized gilt induced by estradiol. The animal was implanted with estradiol two days prior to 0 hour, when 50
μg of estradiol was injected. The plasma LH level at 0 hours was about 1.0 ng/ml,
and rose to the levels shown as the result of estradiol injection. [Unpublished data
from J. W. Hamilton *et al.*, University of Illinois.]

pronounced. This suggests that if a steady inflow of RH is provided, the
LH output also become more steady than it is under the influence of
endogenous RH. Obviously, much additional work is needed before
these interrelations are understood.

BREEDING SEASONS

Animals that reproduce when environmental conditions are most favorable for the pregnant or incubating mother and for the newborn have an
evolutionary advantage. Thus, most of our domesticated animals, before
domestication, had distinct breeding seasons, just as their wild proto-
types and relatives do today.

Continuous Breeders

Domestication tempers environmental exigencies to the point where
reproductive efficiency is less dependent on adequate food supplies and
proper temperatures than it had been in the wild. As food and shelter
were provided for domesticated species, their latent genetic potential
for prolonged breeding seasons asserted itself, even making it possible
for some species to become continuous breeders. The word "continuous"
should not be understood to mean unvarying; most domesticated species
exhibit peaks of reproductive activity. Man has distinct annual peaks of

prolificacy, which vary with the latitude. In the northern hemisphere conceptions occur significantly more often in May and June, the resulting births taking place in February and March. In the southern hemisphere these peaks of conception and parturition are reversed. Similar peaks and valleys of fecundity have been noted in other continuously breeding species, such as chickens (Figures 3-11 and 3-12). In chickens there also appears to be a definite correlation between the rate of reproduction and the amount of light. This correlation was noted by naturalists a long time ago, and light is still considered a prime factor in the reproductive activity of birds and mammals. (As we shall see shortly, however, the situation is not simple enough to be explained by changes in any single climatic factor, such as light, temperature, or precipitation.)

Before we consider the mechanisms that may control reproductive periodicity, a few general statements concerning breeding cycles are in order. We have already noted that all animals can be divided into two categories: the seasonally and the continuously breeding. The former include species in which the gonads of both sexes regress completely and become inactive. In other species, however, only the females become periodically sexually inactive; the males show continuous spermato- and spermiogenesis, independent of season. In species that have become continuous breeders more or less distinct peaks of prolificacy are discernible. In addition to the peaks already noted for man and chickens, similar peaks are known to exist in other species usually classed as continuous breeders.

On the North American continent horses breed during the spring and early summer; during the remainder of the year heats are sporadic and of unpredictable length. Conceptions, however, may occur throughout the year. Cows conceive at any time of the year, and their tendency to show seasonal peaks is disguised by the intervention of man, who, guided by such considerations as milk prices and the availability of pastures for calves, controls conceptions and creates somewhat artificial peaks of conception. If such restraints are not exercised, however, cattle show a decided peak of prolificacy in the fall, spring, and early summer, the chances of conception being distinctly less in the winter. Whereas wild rabbits show a distinct breeding season from December through July, domestic rabbits have become continuously breeding animals. Although they may conceive at any time of the year, under North American climatic conditions domestic rabbits show a definite low point in reproductive ability during July, August, and September.

There is no complete agreement whether reproductive efficiency varies seasonally in laboratory rats and mice, which are also continuously breeding species. (The sex ratios of laboratory rats and of all human populations studied are significantly higher during spring and summer.)

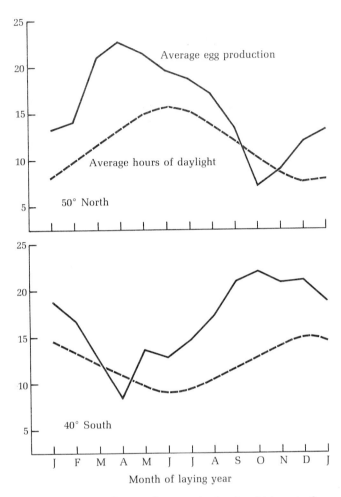

Figure 3-11 Comparison of rates of egg production by chickens in the northern and southern hemispheres. Annual production in the north was 196, and in the south 198. Note that the production and light curves for 50° N (border of Canada and U.S.A.) and 40° S (southern third of Argentina) are nearly mirror images of each other. Note also that at both latitudes the rate of egg production begins to decrease or to increase somewhat before the corresponding change in the light curve. [Data from Wetham, 1933, *J. Agric. Sci.,* 23:383.]

There is, however, no doubt that the wild rat shows seasonal periodicity; its peak of prolificacy occurs from December through June. During the remainder of the year both the number of pregnant females and litter size are significantly smaller; in this species, therefore, both the efficiency of conception and the rates of ovulation and fertilization may be affected.

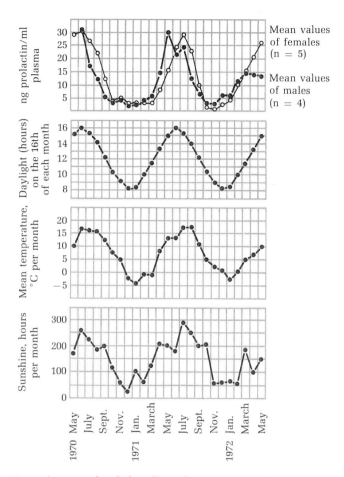

Figure 3–12 An example of the effect of environmental factors on hormonal levels: changes in plasma prolactin levels in nonlactating cattle and in bull calves from birth to 24 months of age. An important parameter in this cause-and-effect relationship is probably light (number of daylight hours), although temperature cannot be excluded. Such interrelations have been noted for other pituitary hormones, but are not as well documented. [From Schams and Reinhardt, *Hormone Res.*, 5:217, 1974, by permission of S. Karger, A.G., Basel.]

The examples cited leave no doubt that in most if not all continuous breeders there are peaks and troughs of prolificacy. The peaks roughly coincide with the periods of maximum length of daylight.

In this connection it is interesting to note that for females living in the wild the only normal reproductive states are either pregnancy, pseudopregnancy, or lactation; estrous cycles are rare. Females trapped in the wild during the breeding season usually fall naturally into one of these three categories. Wild animals are therefore not exposed to the constant and frequent alternation of hormones, estrogen, and progesterone, as are laboratory animals or animals belonging to one of the domestic species whose breeding is controlled by man. As a rule, girls in economically developed nations do not get pregnant until long past puberty. Where control over family size is exercised, intervals between pregnancies may also be long, and toward the end of the reproductive life of women the nonpregnant state (continuous cycles) may last for years. Assuming that the normal state of the female is pregnancy and/or lactation, the question arises whether the frequent cycles (hence, hormonal changes) to which civilized women are exposed during the major parts of their lives may not be contributing factors to the frequency of such debilities as cancer of the reproductive duct system and the breasts. Data on this possibility are not available, but experiments on laboratory species would not be difficult, even if time consuming.

Seasonal Breeders

Animals that breed seasonally may show one or two annual breeding seasons separated by periods of complete sexual inactivity. During this anestrous period the gonads of females—and in some species those of males—involute completely. The testes of males are retracted from the scrotum into the body cavity; gametogenesis stops in both sexes, and the gonads resemble the gonads of sexually immature juvenile members of the species. All structures dependent on sex hormones also involute during the anestrous period.

Migratory birds are the best-known example of seasonally breeding animals, and their urge to migrate seems to be intimately connected with the functional state of the gonads. The whole problem of bird migration and its causes is too little understood and too complex to warrant discussion here. Migration is certainly not caused by one factor alone (such as gonadal recrudescence), but is probably due to an interplay of a whole series of factors, which are probably endocrine but nevertheless certainly not simple.

The complexity of the problem of seasonal breeding in mammals can be illustrated with domestic sheep. A hundred sheep were checked daily for heat without being allowed to conceive over a period of three years. The average date of onset of the breeding season in each of the three years, together with the standard deviation, was found to be:

1951: July 8, ±40 days
1952: August 27, ±26 days
1953: September 1, ±9 days.

The average dates of onset in 1952 and 1953 do not differ significantly from each other or from the average dates for other groups of sheep observed since that time at the University of Illinois. But the year 1951 was outstanding with regard to earliness of onset of the breeding season—not only for this experimental flock but for other flocks in this and neighboring states. Climatological data were obtained for these three years, but no correlation was found between the onset of the breeding season and temperature, average daily cloud cover, or precipitation.

The failure to find significant correlations in these data, and in similar but more extensive data covering 24 years and four different breeds, was disappointing because of the following experimental data. (1) It has been shown that one can control (prevent or hasten) the onset and termination of the breeding season in sheep by manipulating the duration of the light to which they are exposed. (2) Exposure of sheep to an environmental temperature of about 45°C and to normal daylight caused them to come into heat and to conceive 30–40 days earlier than the controls, which were kept at a normal environmental temperature (Dutt and Busch, 1955). The failure to find similar cause-and-effect relations under natural conditions is puzzling and further emphasizes the difficulty of the problem involved.

Sheep also furnish an outstanding example of the fact that the breeding season is a very labile characteristic and can be modified by genetic selection. At least two breeds of sheep, one Asiatic and one South American, are known to be continuous breeders, and the Merinos and the Dorset Horn have more prolonged breeding seasons than other common breeds.

The lability of the breeding season is further illustrated with birds. Mallards in captivity easily lay 60–120 eggs yearly (if they are not given a chance to incubate them), and sparrows and flickers may lay 50–80 eggs consecutively if the eggs are removed from the nests daily and the females are not allowed to brood. Similarly, the wild ancestors of domestic chickens normally lay 12–20 eggs and then incubate them, but they can be induced to lay many more eggs if they are deprived of the chance of incubating those they have laid. Because of the lability of the genetic mechanism controlling reproductive cycles in birds and mammals, it has been possible to convert seasonally breeding animals into continuously breeding ones. But it has not been possible to eliminate all the manifestations of the initial pattern of seasonal breeding, which continues to manifest itself in greater prolificacy at certain times of the reproductive year.

Effects of Light on Reproduction

There can be no doubt that light plays an important role in determining the reproductive periodicity of animals. In some animals light governs only macroperiodicity by determining the onset of the breeding season, as it does experimentally in sheep, but apparently does not play a major role in governing estrous cycles within the breeding season. In chickens, and presumably in other birds, light governs microperiodicity as well; not only the breeding season itself but also the laying of eggs within clutches is controlled by light. Just as there are "long-day" and "short-day" plants, so there are "long-day" and "short-day" animals. Breeding activity in long-day breeders is initiated by increases in the length of day, and in short-day breeders by decreases.

LONG-DAY BREEDERS.   Before we consider in detail the role of light in reproductive events, a few general comments are in order. One can easily demonstrate the effect of light by exposing seasonally breeding animals—such as sparrows, ferrets, or goats—to additional light during their non-breeding season. Their gonads enlarge to the size usually seen only during the breeding season, and complete gametogenesis occurs. In a series of experiments begun by Bisonette and continued by Benoît and others it has been established that light usually acts on the retinas of the eyes and that its effect is relayed from them via the optic nerves to the hypothalamus, which responds by releasing a substance that stimulates the pituitary gland. This chain is neurohumoral. In ducks it has also been shown that neither eyeballs nor intact optic nerves are necessary for the transmission of the light stimulus to the hypothalamic region, and that the transmission to photoreceptors in the hypothalamus occurs directly through the tissue of the orbit. But if the pituitary stalk (connecting the hypothalamus and the pituitary gland) is cut, the ducks can no longer respond to light. This leads to the postulate that in normal animals light acting on the hypothalamus causes it to secrete a substance that reaches the pituitary gland via the portal system and stimulates that gland to produce or release the gonadotrophic complex.

It does not follow, however, that the reproductive behavior and organs of all seasonally breeding animals is totally dependent on stimulation by light. Ferrets kept in complete darkness and blinded ferrets show estrus at the same season as control animals kept under normal light conditions. Cows that are blind because their optic nerves are completely atrophied, as a result of vitamin A deficiency, show perfectly normal reproductive cycles, with normal rates of ovulation and fertilization. The reproductive activity of Eskimos is not reduced during the long winter night.

Male chickens raised in total darkness show a growth of testes and combs that is, on the average, significantly below that of control birds kept under normal daylight conditions; but they do achieve spermatogenesis, though significantly later in life than the controls. When 99 percent of the control males have testes with active spermatogenesis, only 25 percent of the males kept in the dark may have reached that stage.

From these observations it appears that although light is not essential for the complete development of reproductive organs, it does have the ability to hasten and synchronize development.

The most important role of light in governing reproductive functions may lie in its ability to synchronize reproductive events in all the members of a local population, causing them to be in similar reproductive states at any time of the year. Light is one environmental factor that can be expected to affect both sexes of a species in a uniform manner. Without the synchronizing effect of light, reproductive rhythms would still be possible, but reproductive efficiency would probably be greatly impaired; periods of reproductive readiness in different individuals would be out of phase and their coincidence would be purely a result of chance.

SHORT-DAY BREEDERS. The species briefly discussed above, including most birds and the smaller seasonally breeding mammals, all seem to have one factor in common: they respond to increasing light by increased reproductive activity. They can be properly called the "long-day" species. By contrast, the short-day species respond to increasing light by decreased reproductive activity (anestrus); in fact, their breeding season is brought on by decreasing light. A prime example of such animals is the domestic sheep. We have already commented on the fact that in sheep the control of seasonal breeding seems to be very labile. We find a complete gradation from continuous breeding in some breeds, through prolonged seasonal breeding in other breeds, to very restricted breeding seasons, which, in at least one South American breed, is said to consist of 3–5 estrous cycles each year.

A series of experiments initiated by J. F. Sykes and extended by N. T. M. Yeates showed that one can hasten the beginning of the breeding season by reducing the amount of light normally available in the spring, and that one can end the breeding season before its normal end by increasing the amount of light. It was found that the breeding season starts 13–16 weeks after the change is made to decreasing length of day, and that the season ends 14–19 weeks after the length of day is increased. These findings may seem hard to reconcile with the idea that light "stimulates" pituitary gland activity (as it does in chickens), for the reproductive activity of sheep follows decreasing amounts of light and hence decreasing stimulation of the pituitary gland.

On the basis of the finding that the total gonadotrophic potency of a sheep's pituitary gland is significantly higher during the nonbreeding season than during the breeding season, it is postulated that the pituitary glands of all birds and mammals respond to light by greater activity. In sheep the rate of secretion of the gonadotrophic complex that is compatible with reproductive functioning is below the maximum capacity of the gland for such secretion. Therefore, with increasing light the pituitary gland of sheep secretes more than the compatible amount of gonadotrophic hormones, possibly leading to an imbalance between the FSH and the LH and making normal reproductive performance impossible. The endocrine limits within which the different species can function normally are apparently quite flexible and subject to modification by genetic selection. This may account for the fact that in ovines we find the whole gamut of continuous breeding, prolonged breeding seasons, and very short breeding seasons.

To explain why sheep begin to show cyclic activity of ovarian function with the onset of the breeding season, we should recall that with decreasing length of daylight the amount of gonadotrophic hormone synthesized by the hypophysis drops to a point compatible with follicular growth and estrogen production. Here we must invoke the ovary-pituitary feedback mechanism, which is responsible for cyclic behavior of all ovarian function. That the feedback mechanism does not function perfectly at the beginning of the breeding season (or at its end) as shown in Table 4-1 (p. 124). The reason why the oscillations are not normal at these periods—deviating significantly from 16 days—may be because both at the beginning and the end of the season not enough follicular estrogen is produced to cause a more precise feedback relationship between the ovaries and the hypophysis. That the pituitary gland of sheep is competent to release at least LH during the nonbreeding season is seen from the fact that injection of estrogen can cause LH release (Figure 3-5). If LRH is injected into anestrous ewes, some of them can be caused to ovulate, showing that the ovary too is able to respond to trophic hormones.

One further example should be cited to point out that there is significant variation between species with regard to the response elicited by various light regimes. Some birds reproduce essentially normally under a great variety of light regimes, from almost total darkness to continuous light. Nevertheless, there is a high positive correlation between the amount of light received and the egg-laying activity of the females. In sheep increasing amounts of light seem to cause an imbalance in the gonadotrophic hormones, resulting in a sexually quiescent period (the nonbreeding season). If rats are exposed to continuous lighting, normal cycles disappear within a very short time after exposure and rats show what is called "continuous estrus." Their ovaries contain no corpora lutea but only large follicles that do not ovulate, the uteri are enlarged, and the vaginas show constant cornification—all testifying to continuous secretion of estrogen. The margin of tolerance in birds is such that the light-induced increase in gonadotrophin results in increased egg layings; in sheep it leads to an imbalance of hormones to which the ovary is incapable of responding; and in rats it leads to an overstimulation of the ovaries and apparently a deficiency in LH, since the enlarged follicles are incapable of ovulating.

We may note here that the identical condition of constant estrus can be produced in at least two other ways. One is to place electrolytic lesions in the anterior region of the hypothalamus; the other is to administer to prepubertal female rats (about five days old) a single dose of a steroid hormone (for instance, 1.0 mg of testosterone). When these animals reach an age at which their controls have begun to show normal ovarian and vaginal cyclic activity, the steroid-treated females show constant estrus, with large ovaries containing many large vesicular follicles; these follicles secrete estrogen, evidenced by continuous vaginal cornification and the absence of corpora lutea. In rats with electrolytic lesions—as well as those in which constant estrus was induced by prepubertal treatment with a steroid—cyclic activity can be induced by injecting them with progesterone, which presumably inhibits LH release and allows it to accumulate to the level where it can cause ovulation. In both types of treated rats ovulation can also be induced quite easily by the injection of exogenous LH. It thus appears that in rats in constant estrus, regardless of how this condition was induced, the breakdown of cyclic reproductive activity occurs because insufficient ovulation-inducing hormone is stored in the pituitary gland.

Finally, it seems important to point out again that no single factor can explain the phenomenon of seasonal breeding or of reproductive rhythm. Light, temperature, adequacy of the food supply, neural stimuli (such as the presence of males or other animals), time of feeding, and probably a host of other factors—singly or, more probably, in combination—are responsible for the reproductive rhythm of birds and mammals.

THE GONADOTROPHIC COMPLEX IN NONESTROUS STATES

Pregnancy

Our interest in the gonadotrophic potency of the pituitary gland of pregnant females was originally due to an attempt to explain suspension of the estrous or menstrual cycle. It was also important to see if there was any correlation between the gonadotrophic potency of pituitary glands and the reproductive efficiency of polytocous animals. In neither cycling nor pregnant pigs is there any correlation between the gonadotrophic potency of the pituitary and the number of eggs ovulated or corpora lutea formed. This is in sharp contrast to the high correlation between hormone content and the number of ripening follicles and the number of follicles destined to ovulate (Table 3-2). This dissociation between gonadotrophic potency and efficiency of ovulation may have several explanations. The amounts of LH released may be the same, regardless of the number of follicles to be ovulated, or small amounts of LH may be as effective as large amounts in inducing ovulation. If release of LH is a neurohumoral phenomenon even in spontaneously ovulating mammals, as it is known to be in some, one would not expect to find a correlation between efficiency of ovulation and the quantity of LH released. Thus, in the woman, a single-egg ovulator, the preovulatory LH peak may approach 500 ng/ml of peripheral plasma and is usually above 100 ng/ml. In the rat, which ovulates multiple follicles, these values are lower than in women. Among domestic animals, the polytocous pig has an LH peak of 5–8 ng/ml, whereas the monotocous sheep and cow have plasma values of around 20–30 ng/ml. It is obvious that there is no correlation between the amount of LH released prior to ovulation, the number of eggs to be ovulated, and the size of the animal.

The gonadotrophic potency of the pituitary glands of both swine and cows is high at the time of conception, but it diminishes gradually during the course of pregnancy and is lowest just before parturition. This seems to be the general rule for all species investigated with the exception of the rat, in which the potency increases. In the cow and the sow the steady decrease in gonadotrophic potency is coupled with a progressive decline in follicular development in the ovaries. In both species the decrease in pituitary potency is inversely related to the increased production of estrogen by the placenta during pregnancy. Placental estrogen may be responsible for the inhibition of the pituitary gland during pregnancy. It also suggests that the effect of estrogen is to prevent the *formation* of gonadotrophic hormone rather than its release from the gland.

Since during the first trimester of pregnancy the total gonadotrophic potency of pituitary glands does not sink below that found in cycling animals, an explanation for the absence of cycles during pregnancy must be looked for elsewhere. The answer may be an imbalance of the two members of the complex, an imbalance that may be brought about by the rising tide of progesterone, of estrogen, or both.

Castration

The gonadotrophic potency of the pituitary glands of both sexes is always higher in castrates than in intact animals. Here again the power of estrogen (and to a lesser extent of androgen) to inhibit the pituitary function may explain the increase in potency. After the inhibiting influence of the gonadal hormones is removed by castration, the pituitary gland is no longer restrained in its formation of gonadotrophins. This interpretation is supported by experimental evidence. The rise in the gonadotrophic potency of castrates of both sexes can be prevented or returned to normal by the injection of either estrogen or androgen. In such experiments it is found that estrogen is much more efficient than androgen in inhibition of the pituitary, and androgen is more effective than the other hormones classified as steroids.

Nonbreeding Seasons

The mechanism of control of breeding seasons will be taken up in detail elsewhere (pp. 86–91). It is appropriate to point out here, however, that the failure of sheep to show estrus during part of the year is not due to a reduction in pituitary function, as has been commonly assumed. On the contrary, the pituitary glands of sheep contain significantly more total gonadotrophic hormone during the nonbreeding season than during the estrous cycle. This is reflected in ovarian activity during the nonbreeding period; the average diameter of all follicles and the diameters of the largest follicles are the same then as during the estrous cycle. There are, however, significantly fewer follicles during the nonbreeding season. Nothing is known about the proportion of FSH and LH in the pituitaries during the two reproductive stages, but it is tempting to assume that the anestrous period is caused by an imbalance of the two hormones. This imbalance could be brought about by a disproportionate increase in FSH and a decreased or unaltered rate of LH secretion. This contention is supported by the finding that ovulation can be caused in some anestrous sheep by the injection of LH alone.

Prepuberty

Because the gonads of most immature mammals respond to exogenous gonadotrophins, even producing fertilizable and viable eggs, and because spontaneous precocious sexual maturity sometimes occurs, it seems that the onset of sexual maturity is primarily determined by the pituitary function and does not depend on the ability of the gonads to respond to hormonal stimulation. In a recent study in which the pituitary glands of pigs ranging in age from one to 300 days were assayed for gonadotrophic potency, it was found that the glands are significantly more potent before puberty than after. There is a steady decrease in potency from the age of one day to sexual maturity. After the onset of sexual maturity the gonadotrophic potency remains almost unchanged throughout the reproductive life of the animal. Nothing is known about changes in the FSH-LH ratio during this period, but it appears that the prepubertal

period (the nonbreeding season of the young!) may be due to an imbalance between the two components of the gonadotrophic complex.
The assumption that the pituitary glands of the young secrete primarily
FSH is supported by the fact that the ovaries of immature females show
considerable follicular development, whereas the estrogen-dependent
duct system shows insignificant growth. The prepubertal period may
thus be compared to the anestrous period of seasonally breeding mammals, such as sheep. In both instances the high gonadotrophic potency
of the pituitary glands is correlated with a nonbreeding period and must
be lowered (change in rate of LH secretion?) before normal reproductive
cycles can begin.

CLASSIFICATION OF GONADOTROPHINS

Chorionic Gonadotrophin

In 1927, when P. E. Smith opened the field of endocrinology by his initial
experiments with hypophysectomy and replacement therapy, Ascheim
and Zondek discovered a gonadotrophic hormone in the urine and the
blood of pregnant women. They assumed that this hormone was of
pituitary origin and called it "prolan" (later subdivided into prolan A
and B). Subsequent work showed that this hormone does not come from
the pituitary gland but is made by cells covering the chorionic villi (the
cytotrophoblasts). It was then called "anterior-pituitary-like" hormone
(APL). Still later this too was found to be a misnomer, and both terms
were discarded in favor of "chorionic gonadotrophin" or "hormone of
human pregnancy." Human chorionic gonadotrophin (HCG) shows
largely LH-like effects. Older studies had shown that it appears only during the early stages of pregnancy in women, and that it had no demonstrable effects in hypophysectomized rats and mice. This led to the
supposition that it showed gonadotrophic activity by its effect on the
pituitary gland of assay animals, causing that gland to release its own
gonadotrophic complex.

However, the work of Lyons and his colleagues (1955) has shown
that chorionic gonadotrophin is (1) present in appreciable amounts
throughout pregnancy, the peak of its production occurring during the
early weeks; (2) that it does have gonadotrophic action in hypophysectomized females if it is given in sufficiently high doses; (3) that it changes
qualitatively from early to late pregnancy (Table 3-5). The fact that this
hormone is found in the urine of pregnant women is the basis of most
pregnancy tests. In pregnant women it first appears in the urine at about
30 days, reaches a peak at about 60 days, and decreases in amount at
about 80 days (based on bioassay; see p. 189f. for further details). This
hormone has been found in the urine of other primates (rhesus monkeys
and chimpanzees) but not in other mammals.

Table 3-5. Qualitative changes in human chorionic gonadotrophin throughout pregnancy, as assayed in rats

DAYS AFTER OVULATION	M.U.U.[a] 24 HOURS	EFFECT ON REPRODUCTIVE SYSTEMS OF RATS			
		FOLLICLES	CORPORA LUTEA	OVARIAN WEIGHT (mg)	UTERINE WEIGHT (mg)
5–16	35–100	None	None	8–24	27–43
19–23	150–200	S, M, L[b]	Lut F[c]	18–29	38–53
26–33	5,000–15,000	L, cysts	CL	44–88	81–164
35–50	75,000	L, cysts	CL	58–129	117–154
121–238	20,000–40,000	S, M, L	None or lut F	22–60	50–109
Control	. . .	S, few	None	10	22

[a]Mouse uterine units.
[b]S, M, L = small, medium, large.
[c]Lut F = luteinized follicles.
SOURCE: Lyons et al., Endocrinology, 53:674, 1955.

Urinary Gonadotrophins

Gonadotrophic hormones are found in the urine of normal nonpregnant women, but at a much lower concentration than in pregnant women. A rise in the concentration of gonadotrophic hormone occurs midway between two menstrual periods. This "ovulatory peak" is thought to coincide with or to precede ovulation. The hormone is probably of pituitary origin and shows predominantly FSH-like effects when given to test animals.

Surgical or physiological castration (menopause) causes an increase in urinary gonadotrophins. This hormone, too, shows mostly FSH-like activity and is of pituitary origin.

A small quantity of an FSH-like hormone is found in the urine of normal men; the amount greatly increases after castration.

In women high urinary gonadotrophic titers are also characteristic of certain malignancies, such as chorioepithelioma and hydatiform moles. It is also noteworthy that substantial amounts of gonadotrophic hormone (largely FSH), which can be detected by the standard pregnancy tests, appear in the urine of men with testicular neoplasms.

Pregnant Mare's Serum (PMSG)

In 1930 Cole and Hart made the important discovery that the blood of mares between the 40th and 140th days of pregnancy contains large quantities of a gonadotrophic hormone (called by them equine gonadotrophin). Unlike the hormone of human pregnancy, which is formed by placental tissue, the equine gonadotrophin of pregnancy is probably formed in the endometrial cups of the pregnant uterus. And, whereas human chorionic gonadotrophin is found in high concentrations in the urine, PMSG occurs almost exclusively in the blood. PMSG remains in the blood stream not only in mares but also in animals into which it is injected. For this reason a single injection is as effective as the same dose divided into multiple injections—in distinct contrast to other gonadotrophins, which are metabolized rapidly and are more effective in multiple doses.

Whereas human chorionic gonadotrophin shows predominantly LH effects, PMSG shows both FSH and LH effects. When PMSG is given to hypophysectomized female rats in small doses, it has a predominantly FSH effect; when the doses are increased, the LH effect asserts itself and produces ovulation or luteinization, depending on the conditions under which the PMSG is injected. However, the PMSG complex is not comparable to the pituitary gonadotrophic complex. All attempts to fractionate PMSG chemically, and to divide the complex into FSH and LH, have thus far failed.

PMSG is a useful tool in endocrine research. It is easily available commercially, and one can prepare it under ordinary laboratory conditions by bleeding mares at the right stage of gestation. It can be easily standardized, a quality that makes it especially valuable for certain therapeutic or experimental purposes. It produces primarily follicular growth when given subcutaneously, and ovulation when subcutaneous injection is followed by intravenous injection. Because it has high FSH activity, it frequently produces cysts instead of follicles, especially when the doses are large and the period of injections is prolonged. By use of the minimum effective doses over the shortest possible time, this difficulty can frequently be overcome.

Pituitary Gonadotrophins

The pituitaries of all vertebrates studied are known to secrete a hormone complex consisting of FSH and LH. However, there are significant differences among the species in the ratio at which the two components of the complex are produced. We have already pointed out how this ratio differs within the estrous cycle of one species. When the FSH and LH potencies of the pituitary glands of different species are compared during the same stage of the cycle, considerable differences are found. The pituitary glands of cattle are very rich in LH and low in FSH; the reverse is true of the pituitary glands of humans and horses. The significance of these differences is not known; cattle, sheep, and pigs manage to reproduce quite well despite a high LH content, and rabbits, rats, and horses despite a high FSH content. The duration of heat, or sexual receptivity, is generally longer in animals in which the FSH-LH ratio is high. The ratio increases successively in cattle, sheep, pigs, horses, and women; the duration of sexual receptivity is 18 hours for cattle, one day for sheep, 2–3 days for pigs, 5–10 days for horses, and the whole 28 days of the menstrual cycle for women. There seems to be an inverse relationship between the amount of LH found in (and secreted by?) the pituitary gland and the duration of sexual receptivity.

INTERRELATION OF THE HYPOTHALAMUS AND PITUITARY

During the infancy of endocrinology scientists were quite content to think simply in terms of hormonal feedback mechanisms and ascribed to them the major role in controlling the rate of flow of this or that hormone, which in turn governed the rate of function of the glands controlled by these hormones. A classical example of such a hormonal feedback mechanism is the interrelationship between hypophyseal FSH and follicular estrogen. Increasing amounts of FSH cause growth of follicles, which secrete estrogen; as the blood level of the latter rises, due to follicular enlargement, estrogen inhibits FSH production and release and follicular growth slows down. When follicles reach ovulatory size, the same hormonal feedback mechanism releases hypophyseal LH, resulting in ovulation. Similar feedback mechanisms were postulated and successfully demonstrated for the rate of thyroid function—in which the blood level of thyroxine governs the rate of release of TSH—and for the rate of adrenal function—in which the blood level of adrenal steroids signals for increased or decreased ACTH release.

For many years these mechanisms for maintaining hormonal homeostasis were thought to be perfectly adequate to explain the phenomena of the cyclic behavior of the ovaries and the normally steady state of the thyroids and the adrenals. However, it has developed that the hormonal feedback system represents only half of the control system that regulates the endocrine events of the body. The other half is the nervous system— the nervous and endocrine systems are closely interlocked.

For the time being (qualifying statements will be made later), let us say that no hypophyseal hormone is released without a direct or indirect signal from the internal peripheral environment. Such signals may be intended for the adenohypophysis (anterior lobe) or the neurohypophysis (posterior lobe).

If they are intended for the posterior lobe the problem is relatively easily solved because the latter is innervated, so that the demand for the release of either oxytocin or vasopressin can be and is obeyed instantly.

It has been established beyond reasonable doubt that the posterior (neural) lobe of the pituitary serves only as a reservoir for the hormones that are actually synthesized in the paraventricular and supraoptic nuclei of the hypothalamic region (see Figure 3–13), from where they migrate along the supra-optic pituitary tract that connects the hypothalamus and the neural lobe. If this tract is cut and blocked, an accumulation of neurosecretory granules occurs on the hypothalamic side of the block, and depletion occurs on the neurohypophyseal side as the neural lobe is being emptied of its hormone stores. This arrangement probably explains why, as we shall see later, the removal of the neural lobe does not always lead to symptoms of deficiency of oxytocin or vasopressin.

If the signals from the periphery are intended for the anterior lobe of the hypophysis, relaying the signal is much more complicated because that gland is not innervated. Instead, it is connected to the median eminence and the hypothalamic nuclei by a very elaborate blood system called the portal system of the hypothalamo-hypophyseal unit. In general, it is correct to say that blood flows mainly in one direction, from the hypothalamus down to the adenohypophysis. The afferent nerve endings from the periphery are known to terminate at the neurosecretory cells of the hypothalamus. (At this point Figure 3–14 should be studied carefully and fully understood.) The demand for the release of a hormone of adenohypophyseal origin is in most cases transmitted via the afferent nerve endings, and the hypothalamic neurosecretory cells respond by elaborating releasing factors (RF). These factors are transported through the portal system directly to the anterior lobe, which responds by the release of the appropriate, previously synthesized hormone.

It appears probable that all the hypothalamic RFs have now been identified and isolated. All of them are polypeptides of small molecular weight, probably less than 1000. (A more detailed discussion is presented in Chapter 6.)

In earlier studies on the separation of the various RFs it was thought that two different RFs could be isolated from hypothalamic extracts for LH and FSH. However, recent work has shown that the two releasing factors (LRH and FRH) are identical and should be renamed GNRH (gonadotrophin-releasing hormone). Apparently, both FSH and LH are released in response to the action of a common RF; this double release is best illustrated by the twin peaks of FSH and LH that occur just prior to ovulation (see White, 1970).

It is now becoming more certain that in addition to the function of hormone release the RFs are also able to cause the synthesis of their respective hormones. For this reason it has been proposed to call them *releasing hormones* (RH) rather than releasing factors.

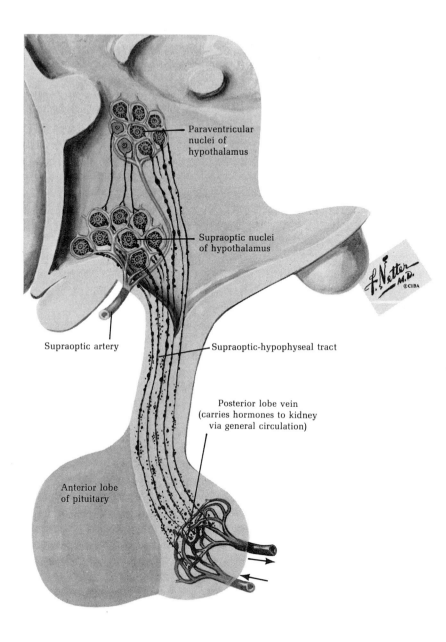

Paraventricular
nuclei of
hypothalamus

Supraoptic nuclei
of hypothalamus

Supraoptic artery

Supraoptic-hypophyseal tract

Posterior lobe vein
(carries hormones to kidney
via general circulation)

Anterior lobe
of pituitary

Figure 3–13 Relation of the hypothalamus to the posterior lobe of the pituitary gland. In contrast to the anterior lobe, the posterior lobe is innervated by the supra-optico-hypophyseal tract. The secretion products are conducted along the nerve fibers from the hypothalamus to the capillaries. [Copyright 1956 CIBA Pharmaceutical Co., division of CIBA-GEIGY Corp. Reproduced with permission from *Clinical Symposia*, illustrated by Frank H. Netter, MD. All rights reserved.]

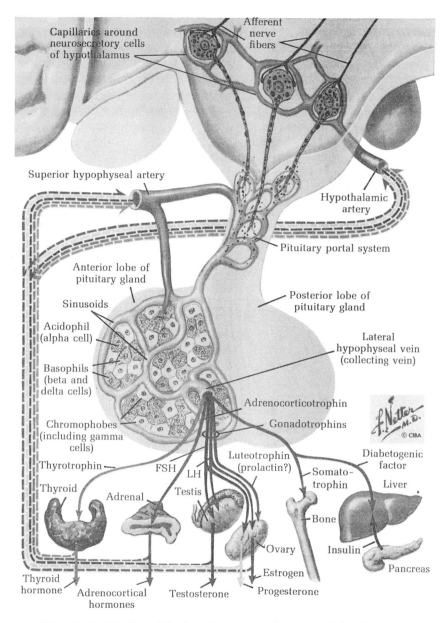

Figure 3-14 Relation of the hypothalamus to the anterior lobe of the pituitary gland. Note that the humoral substance from neurosecretory cells in the hypothalamus passes to the anterior lobe of the pituitary gland through the portal system. Compare with Figure 3-6. [Copyright 1956 CIBA Pharmaceutical Co., division of CIBA-GEIGY Corp. Reproduced with permission from *Clinical Symposia*, illustrated by Frank H. Netter, MD. All rights reserved.]

At least one of the hypothalamic factors, the one concerned with the control of prolactin, does not fall into the general classification of releasing factors. In mammals the hypothalamus appears to make two factors: the major one *inhibits* prolactin release (PIF) and the second (which is still in contention) *causes* prolactin release (PRF) (see Meites, 1970). In contrast to mammals, the predominant factor in birds seems to be a true prolactin releaser. It is interesting to note that the hypothalamic factor TRF (either natural or synthetic), in addition to controlling the release of TSH, also causes release of prolactin. No explanation for this dual and disparate role of TRF is available.

Recently it has been found that in addition to a growth hormone releasing factor (GRH) there is also an inhibiting factor, which has been synthesized and named Somatostatin. The purpose of a dual control system for at least two of the pituitary hormones, prolactin and growth hormone, is not known. However, this fact does raise the question whether there may not be such dual control systems for all the pituitary hormones. It appears likely that thus far only the releasing hormones have been found for most pituitary hormones, and that the inhibitors remain to be discovered.

One other peculiarity of the hypothalamo-hypophysial system deserves special mention. Ever since the early hypothalamic extracts were prepared for assay it has been known that the hypothalamic fragments contain substantial quantities of all the hormones normally associated with the pituitary gland. The source of these pituitary hormones in the hypothalamus is not clear. One hypothesis is that these hormones reach the hypothalamus as a result of backflow from the pituitary gland when the animal is killed prior to removal of the hypothalamus. Another, perhaps more plausible explanation is the following. It is well known that the bloodflow from the hypothalamus to the adenohypophysis via the portal system is overwhelmingly unidirectional. However, it is also known that there are minor blood vessels, called the short feedback loops, which originate in the adenohypophysis and reach into the hypothalamus. In them the bloodflow is in the direction of the hypothalamus. According to one theory the short feedback loops may transport pituitary hormones back to the hypothalamus and may serve as an internal microsystem that controls the rate at which RFs are synthesized. Thus, if too much LH is accumulated in the adenohypophysis, LH may reach the hypothalamus and thus signal (via LRH?) that the rate of LH synthesis should be reduced. That this theory has some merit can be shown experimentally by implanting small quantities of a pituitary hormone, say ACTH, in the median eminence; the pituitary content of ACTH is significantly reduced, whereas the other hormones remain unaffected.

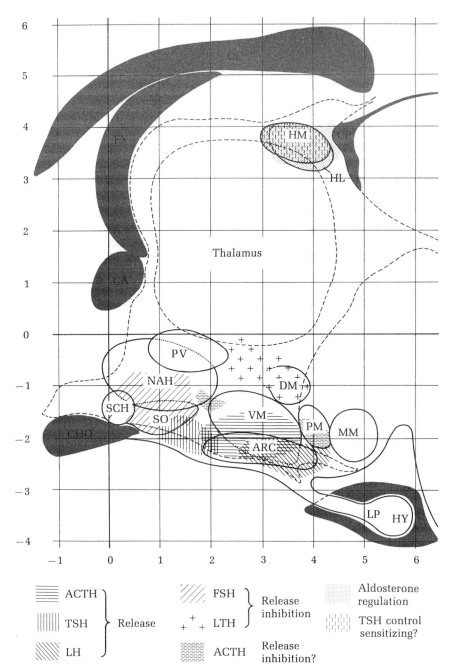

Figure 3–15 Stimulation or electrolytic lesions in the areas shown leads to the release or the inhibition of release of the hormones shown. [After Szentagothai et al., *Hypothalamic Control of the Anterior Pituitary*, Akademiai Kiado, Budapest.]

ARC—arcuate nucleus   SCH—suprachiasmatic nucleus
PM—premammilary nucleus   NAH—tanterohypothalamic nucleus
DM—dorsomedial nucleus   HM—habenular nucleus
VM—ventromedial nucleus   CHO—optic chiasma
SO—supraoptic nucleus

In addition, there is an external macrosystem that depends on the levels of hormones in the peripheral blood. The sensing devices, whatever they may be, respond to the levels of the steroids or thyroxin in the peripheral blood system, and the feedback seems to be either at the hypothalamic or hypopheseal level. Some of the experimental data suggest that a major site of the feedback mechanism is in the hypothalamus, which is involved in the regulation of the synthesis or elaboration of RFs. However, some of the feedback information is also registered directly by the adenohypophysis, which responds by the required adjustment of the rate of elaboration of its hormones. How the adjustment is mediated either at the hypothalamic or pituitary level is not clear, and much work is required before the system can be understood.

Evidence for the existence of RFs has been obtained in many ways, including extraction of hypothalami and the separation and purification of these factors (as we noted earlier, at least three RFs have already been synthesized). One of the newest and most sophisticated approaches to the identification of RFs involves the cannulation of the portal system of anesthetized animals (rats and monkeys), which permits direct demonstration of RFs in the portal circulation. This blood can be extracted and added to bits of incubated pituitary tissue; if RFs are present, pituitary hormones will be released into the incubation medium and can be demonstrated by appropriate assay methods.

Other approaches to the study of the hypothalamo-hypophyseal axis involve electrical stimulation or the placement of electrolytic lesions in specific hypothalamic regions. These procedures cause the release or inhibition of release of specific hormonotrophic substances, such as TRH, LRH, or CRH. It is then possible to construct maps like the one in Figure 3–15, showing which regions are thought to control what pituitary hormones. There is considerable overlap between the regions, but there is also little agreement concerning the exactness with which the regions can be localized—considering the smallness of the hypothalamus this is not surprising. For at least one hormone, LH, two separate hypothalamic control areas have been postulated. The ventromedialarcuate nucleus is thought to control the tonic release of LH, whereas the massive cyclic or ovulatory release of LH is triggered in the preoptic area.

Originally, the site of formation and storage of RFs was thought to be the hypothalamus. However, it has recently been shown that considerable quantities of TRH (and perhaps other RFs) can be extracted from other portions of the brain of the rat. In fact, the hypothalamus contained only 31 percent of the total TRH found, while the forebrain and the brainstem contained 26 percent and 17 percent respectively (Winokur and Utiger, 1974). There are unpublished suggestions that the same is

true for all other RHs. These findings raise a host of questions regard-
ing the roles that RFs may play in the brain other than their control of
the pituitary gland, as well as questions concerning the origin of these
substances in the brain.

Interrupting the Connections
Between the Hypothalamus and Pituitary

STALK SECTIONS.   Because both the neural and the vascular connections between the hypothalamus and the pituitary gland run through the
pituitary stalk, it is fairly easy to study the interrelation of the two
bodies by cutting the stalk, thus interrupting humoral and neural inter-
communication. However, it was learned early that mere cutting of the
stalk is not enough to break all connection, for the stalk regenerates,
and both the neural and the vascular connection may be reestablished.
One can prevent this by inserting a disk of metal, paper, or plastic between the cut ends of the stalk; the pituitary functions abnormally or
even stops completely. Rats and rabbits thus treated fail to have estrous
cycles; in both mammals and birds the testes and accessory glands
shrink; and in chickens the ovaries and combs become completely
atrophic.

In mammals, although there is good evidence that section of the stalk
impairs the pituitary function with regard to all pituitary hormones
(specifically, TSH and ACTH, and possibly the lactogenic and growth
hormones), it is the secretion of the gonadotrophic hormones, and
therefore the function of the gonads, that is affected most spectacularly.
In the chicken only the gonadotrophic function of the pituitary appears
to be affected. This fact suggests that pituitary malfunction in the chicken
is not due to an ischemia caused by the interruption of the portal system,
for it does not seem probable that ischemia would affect the formation
and release of gonadotrophic hormones only, and not other pituitary
hormones. Moreover, the portal system is not the only blood supply to
the pituitary in mammals and birds, and stalk section does not reduce
the systemic circulation to the gland (Figure 3–13). Nevertheless, profound
histological degenerative changes do occur in the gland after stalk section.

Confirmation of the conclusions drawn from the results of stalk section
can be obtained by other techniques, such as electrostimulation of the
hypothalamic region, which causes ovulation in rats and rabbits, and
milk ejection in lactating sheep, goats, and rabbits. Direct stimulation of
the pituitary gland itself is either ineffective or effective only in a
minority of the animals stimulated. The use of drugs that block nervous

impulses and prevent stimuli from reaching the hypothalamus furnishes additional evidence of neurohumoral control of the pituitary. The adrenergic blocking drug Dibenamine prevents ovulation in rats, rabbits, and hens by blocking the neurohumoral LH-releasing mechanism. Atropine blocks or delays ovulation in rats, rabbits, and cows, but has only a slight effect in chickens. Similarly, Nembutal anesthesia blocks ovulation in rats; in chickens, however, it is completely without effect and may, in fact, hasten ovulation.

A number of nonspecific substances, such as picrotoxin, copper acetate, cadmium salts, and Metrazole, cause ovulation when injected into rats or rabbits. How these substances work (while some closely related ones do not) is not quite clear, but it is possible that some of them work directly on the anterior lobe of the pituitary, causing it to release LH, and that others work on the nervous system, which, in turn, acts via the hypothalamus, the portal system, and the anterior pituitary.

PITUITARY AUTOGRAFTS. The interdependence of the anterior pituitary gland and the hypothalamus has been further demonstrated by experiments in which the pituitary gland is taken out of its natural location, away from its normal humoral and neural connections with the hypothalamus, and transplanted to the subcapsular pocket of the kidney or to the eye chamber, where it usually becomes well established. However, findings from these experiments are not nearly as clear-cut as one would want them, especially when effects in different species are compared.

In female rats if the pituitary is transplanted from the sella turcica to a blood-rich organ, such as the kidney capsule, it becomes vascularized; after a time it regains some semblance of normal histological organization, but the organization is not typical of the pituitary in its normal location. Such relocated pituitary glands do not synthesize and release the amounts of hormones typical of the adenohypophysis; they shift to the synthesis of predominantly one hormone, prolactin. In female rats with new corpora lutea the copious release of LTH leads to pseudopregnancy (and formation of deciduomata if the uterus is challenged), or to pregnancy if the autotransplant was performed at the proper time. The usual feedback mechanisms do not operate because pseudopregnancy is found to continue almost indefinitely and pregnancy will last beyond the normal 21-day period, until the young die in utero and are resorbed. No detailed studies of the rates of synthesis have been made, but the consensus is that synthesis of both TSH and ACTH is markedly decreased, although the production of growth hormone seems to continue.

It is also possible to transplant the pituitary into the third ventricle of the brain in such a way that the gland either does or does not touch the median eminence of the hypothalamus, the hypophyseotrophic area. If the gland does touch, that part of it immediately abutting the median eminence regains normal histological appearance and produces the expected hormones, some of them cyclically. Glands not touching the hypothalamus behave like those transplanted to the kidney capsule; they remain histologically abnormal and do not synthesize and release the hormones normally expected from them.

Finally, it has been shown that the adenohypophysis of the male rat normally does not function in the distinctly cyclic manner of the adeno-hypophyses of females in causing the release of FSH and LH. However, when the adenohypophysis of a male is transplanted to the sella turcica or to the third ventricle of a female it is able to change pace and release the gonadotrophic complex in such a manner as to support normal ovulatory cycles of the female host.

These facts seem to boil down to this: so far as the pituitary gland is concerned, there is some virtue in being in the proximity of the hypo-thalamus, from which it derives signals (releasing factors? other factors?) that tell it whether it is supposed to be a "male" or "female" pituitary, and what hormones it is supposed to synthesize, in what quantities, and when. It is also probable that the hypothalamus normally produces a substance that inhibits the synthesis or release of LTH (see also Chapter 6).

Unfortunately, this clean picture became muddied when R. Courrier made autotransplants of the pituitary to the kidney in males instead of female rats. Quite unexpectedly he found that in many subjects the relocated pituitaries continued to secrete enough gonadotrophic hor-mone to maintain testes weight, completely normal spermatogenesis, and normal weight and function of all the male accessory glands. Whether thyroid function is depressed is not clear, but it is known that the adrenals become atrophic or function below the level of those of normal males. There is no evidence as to whether in males the relocated pituitary secretes abnormal quantities of prolactin. The growth rates of intact males and those with pituitary autografts are identical.

At present there is no rational basis for deciding why the pituitary in one sex is utterly dependent upon the hypothalamus for directions as to what hormones to secrete, whereas in the other sex it depends for guid-ance on the hypothalamus for only one hormone (or perhaps two). Is this difference due to the "sex" of the hypothalamic centers?

If we now consider the situation in another species, birds, we realize that we are far from being able to generalize on the problem of the inter-relation between the hypothalamus and the pituitary gland. In chickens separation of the pituitary from the influence of the hypothalamus, by such means as cutting the stalk to the anterior lobe, causes rapid and complete involution of the gonads, but has no apparent effect on the secretion rate of TSH or ACTH. If the adenohypophysis of growing male chickens is autotransplanted to the kidney, the testes and the comb regress to the size typical of hypophysectomized birds. But in such birds neither the thyroids nor the adrenals nor the growth rates are affected. After hypophysectomy the iodine uptake by the thyroid becomes minimal, but the adrenal continues to produce corticosterone, provided that the stalk and the median eminence have not been damaged during the operation.

From these experiments it appears that in the chicken the pituitary gland is dependent on hypothalamic proximity only for the release (and perhaps the synthesis) of FSH and LH. TSH and somatotrophin are apparently secreted at normal or near-normal levels by the relocated pituitary. The relocated pituitary also responds to injections of graded doses of thyroxine by graded reduction of TSH production; just like the pituitary in situ. Although there is an extrahypophyseal source of ACTH, it is not certain whether the relocated pituitary secretes ACTH. No comparable studies have been made in mammals, and the degree of dependence of the pituitary on proximity to the hypothalamus (for the synthesis and release of hormones other than the gonadotrophins) remains unknown.

Regulation of Pituitary Cyclicity

The pituitary gland of females releases its hormones in a cyclic manner, whereas that of the male does not. This difference accounts for the cyclic function of the female gonad and the relatively steady functioning of the testicles of the male. Why is there a difference? The answer is rather complete for rats, and of great interest.

Experiments have shown that if the pituitary gland of a male rat is transplanted into the sella turcica of a hypophysectomized female, the male pituitary gland will behave cyclically and release the proper hormones consonant with the genetic cyclicity of the female. If the pituitary gland of a female is placed into the sella of a male it obeys all the signals appropriate in the male and loses the cyclicity that it had in the female. This suggests that cyclicity is not a property of the pituitary gland but of the hypothalamus.

The next question is why the hypothalamus of the female becomes cyclic while that of the male does not. Light is shed on this problem by the following experiments. If female rats 5 days of age are given a single large (1 to 1.5 mg) injection of androgen, 99.8 percent of the subjects do not show normal estrus cycle when they mature sexually. If the injections are administered before or after day 5 the proportion of sterile rats is either very low (before the age of 5 days), or has no effect at all and the females show normal cyclicity at puberty. The ovaries of these prepuberally androgenized females show follicular development of the cystic type and they rarely ovulate normally. However, injection of LH into such rats causes ovulation and corpus luteum formation. Furthermore, if one blocks the pituitary gland by the injection of large doses of progesterone and then withdraws the steroid, normal ovulations will occur.

These findings suggest that there is adequate tonic release of the proper pituitary hormones to cause maturation of the ovaries. They also suggest that the adenohypophysis is unable to store adequate amounts of LH that could be released cyclically to cause ovulation. In other words, what was originally intended to be a "cyclic" female-type hypothalamus has been converted into a "noncyclic" male-type hypothalamus by the single injection of androgen. The fact that delay of treatment to day 6 or later has no effect on the "sex" of the hypothalamus—females treated after day 5 remain normal females—suggests that the time of programming the hypothalamus (to become cyclic or noncyclic) occurs at a very precise time. The nature of the programming remains unknown, but the fact that a single injection of androgen is sufficient to deliver the message implies that whatever "settings" in the hypothalamus are made by the androgen occur very rapidly. The conclusion appears valid that in rats the hypothalamus is normally "set" to be cyclic, and that in order to become acyclic it must receive a message to that effect by day 5 of life. In male rats this message is delivered by testicular androgen, as is shown by the fact that testicular androgen is produced in sufficient quantities by the young males' own testes to accomplish the task.*

The implications of this work are far reaching and indeed important. For this reason similar studies have been carried out on several other mammalian species and on birds. Postnatal treatment of lambs, calves, pigs, guinea pigs, and rabbits with androgen had no effect, and the treated females showed normal cyclicity at maturity. Chicks treated at a wide variety of postnatal periods also showed normal cyclicity at maturity, and only hamsters showed responses similar to those in rats. This failure does not necessarily mean that the phenomenon is restricted to rats, but could mean that the differentiation of the hypothalamus into "male" and "female" occurs in different species at different times, in some of them perhaps in utero. Since very few experiments involving androgenization of fetuses in utero have been undertaken, it would be

---

*There are many interesting details of this phenomenon that cannot be discussed here; the interested reader should consult Barraclough, 1973.

interesting to see whether hypothalamic derangement of the type described for rats can be duplicated in other animals. The system of hypothalamic differentiation as it operates in rats is just too pretty to be restricted to rats.

Induced Ovulation, Pseudopregnancy, and Delayed Implantation

Two reproductive events that are clear-cut examples of neurohumoral phenomena are induced ovulation and pseudopregnancy, both of which are discussed in some detail elsewhere (pp. 166f. and 277f.). In induced ovulation the nerve impulse resulting from copulation is known to culminate in release of a luteotrophic substance. But the triggering mechanism, especially in rabbits, is extremely sensitive; it is not confined to the genitalia, and events leading to release of LH may be initiated by external contact of females penned together. Spontaneous pseudopregnancy, on the other hand, does not occur in the absence of genital stimulation. A phenomenon comparable to this is found in the hen, and possibly in birds in general. The presence of a foreign body (or an ovum) in the oviduct prevents the release of LH and hence ovulation; both of these events occur in their proper order as soon as the foreign body is removed or the egg is laid (p. 149). Here, to be sure, the nerve impulse appears to inhibit hormone release rather than stimulate it, as it does in the previous two cases.

Implantation

An age-old and important question (to be further discussed in Chapter 11) is: How do the ovaries "know" that the uterus is pregnant and that the corpora lutea should be maintained for the duration of pregnancy rather than being allowed to degenerate as they normally do when a female does not become pregnant?

One explanation is that the uterine contents may in some way signal the pituitary to secrete a luteotrophic substance that maintains the corpora. This theory is based on the finding that implantation of beads into the uterine lumen of sheep sometimes causes, through nondegeneration of the corpora lutea, a significant prolongation of the estrous cycle. In the presence of beads the corpora were maintained for an average of 21 days instead of the normal 9–10 days. The short-term effect of the beads is not surprising if we remember that beads are inert and of a fixed size, whereas the embryo and its fluid-filled membranes are growing constantly and establishing an intimate union with the maternal uterine tissues. The degree and intensity of the neural stimuli originating from a living embryo and its uterine attachments are certainly much greater than the stimuli coming from beads.

In the guinea pig uterine contents (beads) also significantly modify the life span of corpora lutea, causing a prolongation of the cycle. A balloon inserted into the uterine lumen of the cow shortly after heat caused a significant shortening of the subsequent cycle. Since this experiment was tried only shortly after heat, it remains unknown whether a foreign body inserted later would have led to prolonged maintenance of corpora lutea, as happened in sheep. In contrast to results in sheep, cattle, and guinea pigs, foreign bodies placed into the uterine lumina of pigs at different times of the cycle have no effect whatever on the life span of their corpora lutea. The reasons for these species differences remain unknown (see Anderson et al., in Nalbandov, 1963).

Of significance is the finding that partial resection of the pituitary stalk or total section of the nerves going to the uterus prevents implantation and pregnancy in sheep (Nalbandov and St. Clair, unpublished). If either of these operations is performed, estrous cycles remain normal in length (about 16 days) and ovulation occurs at normal times. However, if the females subjected to such operations are mated with fertile rams, even as many as five times, most of them do not become pregnant. The ovulated eggs are fertilized and cleave normally, but implantation apparently cannot proceed normally. One operation interrupts the neural pathway from uterus to hypothalamus, the other the pathway from hypothalamus to pituitary gland, and either prevents the signal for release of luteotrophic hormone from reaching the pituitary gland. That this explanation is probably correct is suggested by one case in which a degenerating blastocyst 18 days old was recovered from the uterus and a 16-cell egg was flushed from the oviduct of the same ewe.

We have already discussed the fact that infusion of LH into either hypophysectomized or intact pregnant and nonpregnant sheep significantly prolongs the life span of corpora lutea. It is tempting to infer that LH is the normal luteotrophin of sheep and that LH release is the factor responsible for the maintenance of corpora lutea in the event that the egg is fertilized and pregnancy is initiated. If this postulate is correct, the problem is to locate a mechanism by which the pituitary gland is signaled that the uterus is pregnant and that the corpus luteum should be maintained rather than allowed to degenerate. A very nice mechanism of this type was discovered by Moor (1968). He found that if he removed the embryo from a pregnant sheep prior to day 12 of the cycle (or even earlier), the ewes showed normal heat and ovulation about 16 days after the previous heat at which mating had occurred. If the embryos were removed on days 13, 14, or 15 of pregnancy, a significant lengthening of the cycle resulted, suggesting that the presence of the embryos in the uterus had initiated a signal causing the prolongation of luteal function.

Finally, Moor has shown that by transplanting sheep embryos into unmated ewes by day 13 (or earlier), the corpus luteum is maintained and so is pregnancy. If the embryo transfer is delayed to days 14, 15, or 16, a large proportion of the sheep show normal heat and ovulation at the expected time.

That LH is very probably involved in this mechanism can be seen from experiments by Karsch (1971), who showed that if infusion of LH into normal sheep was begun on day 12 of the cycle, the life-span of the corpus luteum could be prolonged for the duration of infusion. However, if onset of infusion was delayed to days 13, 14, or 15, regression of the corpus could be prevented in only three out of 12 ewes. If we collate this information with that obtained on embryo removal or transplant, we come up with the conclusion that the embryo may be responsible for the turning on of hypophysial LH, which in turn preserves the corpora lutea.

The nature of the signal emanating from the embryo is not known. Moor found that if minces of previously frozen sheep embryos were infused into the uterus of unmated sheep at the optimal time (day 13), the corpus was maintained in some subjects; but if the minces were denatured by boiling, the corpus was not maintained. Similarly, if minces of pig embryos or of other nonspecific materials of other organs were infused into the uteri of sheep, corpora lutea were not maintained. Thus, it seems that the embryo may be producing a humoral substance that may act via the general circulation as a signal to the adenohypophysis to release LH. Of course, there are other ways in which these data can be interpreted.

One difficulty faced by the above hypothesis is the fact that a central mechanism, the hypothalamo-hypophyseal system, is presumed to release LH in response to a signal received from the embryo. If an embryo in one uterine horn is isolated by tying and resecting the opposite non-pregnant horn, the corpus luteum on the side of the conceptus is maintained, whereas the corpus luteum in the resected nonpregnant horn degenerates. This observation implies that the conceptus has a local "antiluteolytic" or local luteotrophic effect and provides evidence against the concept of a central luteotrophic hormone release system. This is just one example of the extreme complexity of the whole problem of corpus luteum maintenance beyond its life span in the cycle.

Milk Ejection

Another complex neurohumoral mechanism is associated with milk ejection and, to some extent, with milk formation. Stimulation of nipple or breast by suckling or milking initiates the release of certainly one and possibly two hormones from the pituitary gland. In lactating females of many species a corpus luteum of lactation is formed, which persists, as a rule, until the young are weaned. It is postulated that the act of suckling causes the release from the anterior pituitary lobe of a luteo-trophic substance that maintains the corpora lutea of lactation. The suckling reflex causes the flow of several pituitary hormones (among them prolactin). That the formation of milk in a breast is not due to the local irritation of suckling or to the emptiness of the breast resulting from removal of milk can be shown, for example, in rats. If most of the glands are taped and the young are allowed to suckle at only one or two of them, milk formation in the unused glands appears to be as copi-ous as in the used ones.

Much better understood is the role of the posterior lobe in milk ejec-tion, commonly called milk "let-down." Among the earliest hints that a neural pathway is involved in this mechanism were the observations that anesthesia of lactating females reduces the milk yield to a small fraction of the normal amount, and that embarrassment in women and emotional stress in all lactating females prevent the let-down of milk temporarily or even permanently. Recent work has succeeded in ex-plaining the relation between stimulation of the nipple and the let-down of milk. In classic studies, first Gaines and then Ely and Petersen showed that oxytocin is the hormone that is responsible for the ejection of milk. It has been found that the nursing stimulus causes the release of oxytocin from the posterior lobe, and that this hormone acts on the myoepithelial "basket cells" around the alveoli, causing them to constrict and to squeeze out the milk contained in them.

If oxytocin is injected into lactating females under deep anesthesia, milk ejection is normal; in the absence of exogenous oxytocin it is not. Similarly, injection of oxytocin into cows with cannulated teats causes unrestrained and immediate gushing of milk. Milk ejection can also be initiated in lactating sheep and rabbits by the electrical stimulation of the paraventricular nucleus of the hypothalamus. All these observations strengthen the theory that oxytocin is the normal milk ejection hormone, and that the pathway from nipple through hypothalamus to posterior pituitary lobe is neural. The posterior lobe responds by immediate re-lease of oxytocin into the pituitary vein.

Transport of Sperm in the Female

Finally, a few words are in order about the neurohumoral mechanism involved in the transport of sperm in the female reproductive tract. It has been known for a long time that the propulsion of semen from the site of ejaculation to the oviduct (where fertilization takes place) is not due to the automotive ability of sperm cells, but is completely accomplished by contractions of the female duct system. Since in all species in which such propulsion has been timed it has been found to require not more than a few minutes (less than one minute in the rat and about two and a half minutes in the cow), there is no relation between the rate at which semen is propelled and the distance separating the cervix and the vagina from the oviducts. It is also known that lactating cows and women eject milk from the mammary glands during coitus. In view of the preceding discussion of the mechanism of milk ejection, this fact suggests that the act of mating also causes release of oxytocin. The question whether oxytocin is involved in uterine contractions during mating was answered affirmatively by VanDemark and Hays (1952), who found that mating causes a rise in intramammary pressure in the cow and that strong uterine contractions were induced in cows through stimulation of the genital system by natural mating, by artificial insemination, by manipulation of the vulva, or even merely by the sight of the bull. Furthermore, increased activity was noted in isolated uteri perfused with oxytocin, and even in such preparations semen deposited in the cervix was transported to the oviduct in five minutes.

All these observations taken together are interpreted to mean that coitus induces nervous impulses that reach the posterior lobe of the pituitary via the hypothalamus, activating the release of oxytocin, which then causes the uterine and oviducal contractions that are responsible for the rapid movement of semen from the site of ejaculation to the oviduct.

The Neurohypophysis and the Hypothalamus

We are accumulating more and more evidence that hormones from the neurohypophysis are really formed in the hypothalamus, and that the posterior lobe of the pituitary serves merely as a reservoir. We have seen that the demand for these hormones may be sudden and may require an immediate response. Birds are especially well suited to the study of the effects of stalk section on the posterior lobe because a septum separates the two lobes and it is possible to cut either or both of the two stalks. If the posterior lobe is removed from laying hens the results are immediate and dramatic. There is no apparent effect on the rate of ovulation or the time of oviposition, but the water intake increases rapidly until birds weighing 1,000–2,000 grams may drink 900–1,000 grams of water daily. Both the polydipsia and the diuresis seem to be permanent, but both can be controlled very effectively by pitressin. Oviposition, which is known to be due to oxytocin-induced contractions of the shell gland and the vagina, is not affected. It is possible that the amount of hormone required to induce oviposition is small and that the hypothalamus may produce and release enough oxytocin to take care of this need. Neuro-hypophysectomy in mammals does not usually prevent normal delivery of young, but it makes lactation impossible. Young mice attempting to suckle from neurohypophysectomized mothers starve to death because the mechanism for milk let-down has failed.

The role of oxytocin in controlling the output of pituitary hormones related to reproduction is not clear. There are several claims in the litera-ture that precocious puberty has been induced in male and female rats and rabbits injected with oxytocin. However, there are also data (and many unpublished experiments) showing that no effect is produced by this treatment. The claims cannot be dismissed without considering the fact that in the early days of study and isolation of the hypothalamic RFs oxytocin and vasopressin were considered promising candidates for the role of RF. Daily injections of minute quantities of oxytocin into the third ventricle of rats induced a premature opening of the vagina and a very significant increase in the weights of the reproductive organs. The uterus and ovaries of treated females weighed 386 and 74.2 mg, those of control females weighed 229.8 and 8.9 mg. When the same comparison was made in hypophysectomized female rats there was no difference between treated and control females. It was concluded that oxytocin acts on the anterior pituitary lobe, causing it to release gonadotrophic hormones (Corbin and Shottelius, 1960).

In experiments with dairy cattle Armstrong and Hansel (1959) found that subcutaneous injections of oxytocin immediately after ovulation cause these animals to form small and histologically abnormal corpora. What is more significant, all oxytocin-treated animals come in heat and ovulate about seven days after the last heat and ovulation. Experiments on both the rat and the cow can be interpreted to mean that oxytocin stimulates the release of gonadotrophic substances from the pituitary gland. However, the clue to the possible pathway of action of oxytocin lies in the observation that oxytocin loses its ability to shorten the es-trous cycle of cattle if it is injected into hysterectomized females; in this case the corpora lutea formed are perfectly normal. Since in cattle the uterus appears to be essential for the action of oxytocin, it seems that oxytocin acts on the uterus rather than directly on the anterior pituitary lobe, and that (neural?) signals originating in the uterus are responsible for the modification in pituitary function. This is reminiscent of the effect of foreign bodies in the uterus on the length of estrous cycle in sheep, cattle, and guinea pigs, discussed before.

The other posterior pituitary hormone, vasopressin, is of less direct importance to reproduction. It too is formed in the hypothalamus and only stored in the posterior lobe. Its main function is the regulation of water balance in animals; it is called the antidiuretic hormone, or ADH. Although both ADH and oxytocin are stored in the posterior lobe, the mechanisms controlling their release appear to be different. By appropriate experimental manipulation of animals, it is possible to deplete the posterior lobe of one of the hormones without greatly affecting the concentration of the other.

THE THYROID GLAND

It is not within the scope of this book to discuss the entire endocrine economy. However, the thyroid gland and its hormones play such an important role in reproduction, as well as in regulation of the metabolic rate of the body, that they cannot be overlooked. Thyroid hormones probably influence the various reproductive functions at the cellular level. Although there is some evidence that thyroid hormones influence the pituitary-adrenal relation, we shall be concerned mainly with the thyroid-pituitary axis.

In studies of the relation of thyroid hormones to reproduction the classic procedure of surgical extirpation of the thyroid gland followed by replacement therapy has been used. This procedure has the disadvantage that in most mammals and birds extra thyroid tissue (so-called thyroidal rests) is usually present in the thymus gland, making complete thyroidectomy virtually impossible in any animal more than one week old. For this reason it is frequently better and easier to study thyroid physiology by the injection of radioactive iodine (^{131}I), which is selectively picked up by all thyroid tissue (including the rests) and so destroys the tissue by radioactivity. It is sometimes more expedient to use goiterogenic agents (such as thiouracil and thiourea), which prevent the synthesis of thyroid hormones and, if given in sufficiently large doses, can stop all thyroid activity.

That thyroid hormones affect the pituitary-gonad axis can be seen from the facts that administration of these hormones increases the gonadotrophic potency of the pituitary glands of rats and that thyroidectomy decreases the gonadotrophic potency of the pituitary glands of rats, rabbits, and goats. There is considerable evidence that the alteration in gonadotrophic potency in the change from euthyroidism to hypo- or hyperthyroidism is due to a shift in the FSH–LH ratio and in the rate at which these hormones are secreted.

Of considerable interest are the data obtained by the University of Michigan workers (Table 3–6) who rendered male rats and mice hypo- or hyperthyroid and then treated them with a standard dose of a gonadotrophic hormone. The two species responded in exactly opposite ways. Hypothyroidism increased the ability of rats, and decreased the ability

Table 3-6.　Effect of thiouracil (TU) and thyroprotein (ThP) feeding on the response of immature rats and mice to a constant dose of gonadogen

TREATMENT	NO. RATS	TESTES WEIGHT (mg/100 g)*	SEM. VES. WT. (mg/100 g)* (± SE)	NO. MICE	TESTES WEIGHT (mg/100 g)*	SEM. VES. WT. (mg/100 g)* (± SE)
			Thiouracil Feeding			
	5	549.1	23.6 ± 0.7	8	693.3	58.5 ± 14.7
Gonadogen	5	856.0	49.0 ± 8.2	6	643.3	321.7 ± 4.5
TU (0.1%) + gonadogen						
for 20 days	5	1,177.2	115.8 ± 24.2	6	139.6	256.5 ± 22.3
			Thyroprotein Feeding			
Control	11	1,003.0	29.2 ± 1.9	9	547.6	142.7 ± 10.6
Gonadogen	10	798.8	80.2	9	572.6	226.9 ± 19.5
ThP (0.32%) + gonadogen						
for 20 days	10	871.4	36.0	5	647.7	288.2 ± 24.6
ThP (0.64%)	5	770.8	28.3 ± 2.9	5	630.9	195.9 ± 11.7

*Per 100 g of final body weight.

SOURCE: Data abridged from Meites and Chandrashaker, *Endocrinology*, 44:368, 1949.

of mice, to respond to gonadotrophin. Conversely, hyperthyroidism decreased the ability of rats, and increased the ability of mice, to respond to gonadotrophin. This observation shows that the ability of end organs to respond to trophic hormones depends on the thyroid state of the animals. More important, it also shows that the "euthyroid" mouse secretes suboptimal amounts of thyroid hormone (exogenous thyroid hormone increases the responsiveness of its testes and accessories to gonadotrophin), and that the "euthyroid" rat is really hyperthyroid, for "hypothyroidism" improves its ability to respond to gonadotrophins. Whether these differences pertain only to gonadotrophic hormones remains unknown, but they should serve as a warning against applying conclusions across species unless experimental proof warrants it.

The effects of the various thyroid states on the reproductive performance of males and females of various species are summarized in Tables 3-6 and 3-7.

Effect of Thyroid States on Reproduction in Females

That female reproductive performance can proceed normally only in euthyroid individuals has been shown in many experiments. In one such experiment it was found that the estrous cycles of rats were length-

Table 3-7. Effect of hypothyroidism on the litter size of rats

TREATMENT	NUMBER OF RATS	NUMBER OF YOUNG BORN		
		DEAD	ALIVE	TOTAL
Thiouracil	10			
Before		6	83	89
		Avg. 8.3 live young (no resorptions)		
After		6	33	39
		Avg. 3.3 live young (2 females resorbed litters)		
Thyroidectomy	9			
Before		1	79	80
		Avg. 8.9 live young (no resorptions)		
After		10	21	31
		Avg. 2.3 live young (2 females resorbed litters)		

SOURCE: Krohn and White, *J. Endocrinol.*, 6:375, 1950.

ened by 1.27 days after thyroid inhibition, and by 2.52 days after thyroidectomy. The deleterious effect of hypothyroidism on embryonal survival and on total litter size are summarized in Table 3-7. Even if the four hypothyroid females that resorbed their whole litter are omitted from the final calculation, it is seen that the hypothyroid females produced only half as many young as the controls. By contrast, the reproductive performance of most guinea pigs remains unaffected over a wide range of hypo- and hyperthyroidism. In some strains, however, in which stillbirths and abortions are conspicuously high, administration of thyroid hormone alleviates the condition. These observations further emphasize the fact that it is not possible to generalize from one species to another or even to draw conclusions that will hold true within the same species.

An estimate of the different rates of thyroid activity during the estrous cycles of rats and mice shows that the thyroid of female rats is most active during heat, whereas that of female mice is most active during proestrus (Table 3-8).

In normal chickens thyroxine injection can completely inhibit estrogen-induced hyperlipemia, hyperprotenemia, and increase in blood biotin, but it does not interfere with the response of the oviduct to estrogen.

The stimulating effect of thyroid hormone on spermatogenesis and the ovarian function in such a variety of species as chickens, mice, rabbits, swine, sheep, and cattle, plus the fact that egg production by hens usually declines during the summer, when thyroid secretion is low, instigated work on the effect of feeding thyroid hormone (in the form of

Table 3-8.   **Mean uptake of ^{131}I as percentage of injected dose during estrous cycle of rats and mice**

STATE OF CYCLE	RATS $\overline{X} \pm SE^*$	MICE $\overline{X} \pm SE^*$
Proestrus	9.17 ± 0.64	13.74 ± 1.13
Estrus	18.23 ± 0.65	9.14 ± 0.81
Metestrus	10.79 ± 0.67	9.29 ± 0.56
Diestrus	8.99 ± 0.78	10.62 ± 0.66

*Significantly different at 1% level of probability.
SOURCE: Soliman and Reineke, *Am. J. Physiol.,* 178:89, 1950; *J. Endocrinol.,* 10:305, 1954.

thyroprotein) to laying hens. An evaluation of the available data suggests that the treatment does not benefit all birds; for some, however, the summer slump was prevented or alleviated. The feeding of thyroprotein or of thiouracil to laying hens increases the incubation period of the eggs and increases the size of the thyroids of chicks hatched from those eggs.

Effect of Thyroid States on Reproduction in Males

The response of the secondary sex organs of castrated male mice and the comb of capons is increased by thyroxine. In the castrated male rat, however, the response of the seminal vesicles to androgen is reduced in the hyperthyroid condition. In the intact mouse thyroid hormone injections do not affect the response of the seminal vesicles to exogenous androgen.

It is well known that the environmental temperature affects the thyroid activity of various species. At high temperature and high humidity the secretion of thyroid hormone is generally reduced; at low temperature it is increased. In several species (mice, rabbits, sheep, and swine) exposure to high temperatures induces degenerative changes in the testes and reduces fertility. Testicular changes in the ram may not be a direct effect of the high temperature on the testes, as in cryptorchism, but an effect of decreased thyroid secretion. Changes in the ram induced by the feeding of thiouracil are similar to those obtained in "summer sterility"; on the other hand, the administration of thyroid hormone sometimes prevents changes in the testes during the summer. Similar testicular changes were found in aged rabbits and in mice and rabbits kept under intermittent high temperatures.

In addition to its direct or indirect action on spermatogenesis, the thyroid has an effect on the mating desire, at least in some species. Guinea pigs and rats are apparently not affected in their sex drive by thyroidectomy. Thyroidectomized bulls lose sexual interest in cows in heat, but their libido returns to normal not only with thyroid hormone therapy but also after treatment with dinitrophenol, which is known to increase the basal metabolic rate.

The beneficial results of thyroid hormone treatment of male swine, sheep, and cattle with poor fertility and libido seem to indicate that there is a definite threshold below which thyroid secretion cannot drop if reproduction is to remain normal.

The effect of thyroidectomy on the reproductive functions varies from a rather mild effect in the guinea pig to severe interference in young rats and chickens. In the latter, if thyroidectomy is performed during the first two days after birth or hatching, the gonads are severely affected and resemble those of hypophysectomized animals. After spermatogenesis and oogenesis have been completed the thyroid appears in most species to have the function of modulating gonadal performance. Thyroidectomy of female rats causes degenerative changes in the ovary.

The addition of thyroxine to semen increases the consumption of oxygen if the sperm concentration is above $800,000/cm^3$. The fertility of the semen of some bulls, as measured by the nonreturn rate of the cows bred with such semen, is also increased by the addition of thyroxine.

SELECTED READING

Anderson, L. L. 1973. Effect of hysterectomy and other factors on luteal function. In "The Female Reproductive System, Part 2" (Vol. 2, Sec. 7, *Handbook of Physiology*). Williams and Wilkins.

Armstrong, D. T., and W. Hansel. 1959. Alteration of the bovine estrous cycle with oxytocin. *J. Dairy Sci.* **42:**533.

Asdell, S. A. 1957. Reproductive Hormones. In *Progress in the Physiology of Farm Animals*, Vol. 3. Butterworth.

Barraclough, C. A. 1973. Sex steroid regulation of reproductive neuroendocrine processes. In "The Female Reproductive System, Part 1" (Vol. 2, Sec. 7, *Handbook of Physiology*). Williams and Wilkins.

Benoît, J., and I. Assenmacher. 1953. Rapport entre la stimulation sexuelle pré-hypophysaire et la neurosecretion chez Poiseau. *Arch. Anat. Microscopique Morphol. Exp.*, **42:**334.

Cole, H. H., and H. Goss. 1943. The source of equine gonadotrophin. In *Essays in Biology*. University of California Press.

Corbin, A., and B. A. Schottelius. 1960. Hypothalamic neurohumoral agents and sexual maturation of immature female rats. *Am. J. Physiol.*, **201:**1176.

Cowie, A. T., and S. J. Folley. 1955. Physiology of gonadotrophins and the lactogenic hormone. In *Hormones: Physiology, Chemistry, and Applications*, G. Pincus and K. Thimann, eds., Vol. 3. Academic Press.

———. 1957. Neurohypophysial hormones and the mammary gland. In *The Neurohypophysis*. Academic Press.

De Groot, J. 1952. The significance of the hypophysial-portal system. *Van Gorcum's Medische Bibliotheek,* Part 118.

Emmens, C. W., ed. 1950. *Hormone Assay.* Academic Press.

Fitzpatrick, R. J. 1957. On oxytocin and uterine function. In *The Neurohypophysis,* H. Heller, ed. Academic Press.

Goldman, B. D., and V. B. Mahesh. 1969. A possible role of acute FSH-release at ovulation in the hamster as demonstrated by utilization of antibodies to LH and FSH. *Endocrinology,* **84:**236.

Hammond, J., Jr. 1954. Light regulation of hormone secretion. In *Vitamins and Hormones,* Vol. 12. Academic Press.

Harris, G. W. 1955. *Neural Control of the Pituitary Gland.* Arnold.

Hays, R. L., and N. L. VanDemark. 1953. Spontaneous motility of the bovine uterus. *Am. J. Physiol.,* **172:**553.

Hill, R. T. 1937. Ovaries secrete male hormone. *Endocrinology,* **21:**495.

Jones, E. E., and A. V. Nalbandov. 1972. Effects of intrafollicular injection of gonadotrophins on ovulation or luteinization. *Biol. of Reprod.,* **7:**87.

Lee. C., P. L. Keyes, and H. I. Jacobson. 1971. Estrogen receptor in the rabbit corpus luteum. *Science,* **173:**1032.

Maqsood, M. 1952. Thyroid functions in relation to reproduction of mammals and birds. *Biol. Rev.,* **27:**281.

Meites, J., ed. 1970. *Hypophysiotropic Hormones of the Hypothalamus: Assay and Chemistry.* Williams and Wilkins.

Moor, R. M. 1968. Effect of embryo on corpus luteum function. In "VIII Biennial Symposium of Animal Reproduction," *J. of Animal Sci.,* Vol. 27, Suppl. 1, p. 97.

Moudgal, N. R., A. J. Rao, R. Maneckjee, K. Muralidhar, V. Mukku, and C. S. S. Rani. 1974. Gonadotropins and their antibodies. In *Recent Progress in Hormone Research,* Vol. 30. Academic Press.

Nalbandov, A. V. 1961. Comparative physiology and endocrinology of domestic animals. In *Recent Progress in Hormone Research,* Vol. 17. Academic Press.

———, ed. 1963. *Advances in Neuroendocrinology.* University of Illinois Press.

———. 1972. Interaction between oocytes and follicular cells. In *Oogenesis,* edited by J. D. Biggers and A. W. Schuetz, p. 513. University Park Press.

———. 1973. Control of luteal function in mammals. In "The Female Reproductive System, Part 1" (Vol. 2, Sec. 7, *Handbook of Physiology*). Williams and Wilkins.

Rowlands, I. W. 1949. Serum gonadotrophin and ovarian activity in the pregnant mare. *J. Endocrinol.,* **6:**184.

Villee, C. A., ed. 1961. *Control of Ovulation.* Pergamon Press.

White, W. F. 1970. On the identity of LH- and FSH-releasing hormones. In *Mammalian Reproduction,* edited by H. Gibian and E. J. Plotz. Springer-Verlag.

Winokur, A., and R. D. Utiger. 1974. Thyrotropin-releasing hormone: regional distribution in rat brain. *Science,* **185:**265.

Yeates, N. T. M. 1954. Daylight changes. In *Progress in the Physiology of Farm Animals,* Vol. 1. Butterworth.

4

Reproduction in Female Mammals and Birds

The events of the estrous cycle of nonprimates and those of the menstrual cycle of primates are compared. In primates and nonprimates all the physiological and morphological changes in the ovaries, the vagina, and the uterus taking place during these cycles are identical, except that nonprimates show periodic peaks of sexual receptivity (heat, or estrus) whereas some primates do not. In primates a sloughing off of the uterine endometrium is accompanied by bleeding; in nonprimates no comparable bleeding is seen. Menstruation in primates is due to an absence of hormones (usually progesterone). Pseudomenstruation in nonprimates (bitches and several laboratory species) is due to diapedesis and is in no way comparable to the menstrual bleeding of primates, for it is usually brought on by the injection of estrogen and occurs normally when the follicle is at the height of its development.

Birds, in which eggs are ovulated successively but laid in regular clutches, are contrasted with polytocous mammals, such as rats, in which several eggs are ovulated simultaneously. Obviously, the endocrine control mechanisms for the two types must be different—how they differ is not known, but several tentative explanations are offered for consideration. The reader is challenged to find his own interpretation of the facts and to design experiments that may answer the questions.

Events associated with the estrous cycle can be hastened, delayed, or otherwise drastically modified by exteroceptive factors, such as odor, sound, and change in environment. Little is known about the "carriers of excitation," known as pheromones. Since they can cause profound shifts in the neuroendocrine control mechanisms, which are generally considered to be very stable, further study is important.

THE ESTROUS CYCLE

The seasonal patterns of vertebrate reproduction and breeding behavior are remarkably varied. In man and most domestic animals both sexes breed continuously throughout the year; some domestic species show seasonal peaks of fecundity, but only sheep show breeding seasons in the strict sense of the term (the males are continuous breeders but the females have a distinct breeding season). In the majority of wild mammals and birds the sexes have synchronous, alternating breeding and nonbreeding seasons. Although most animals can be classified as seasonal or continuous breeders, some, such as the bitch, fit neither category.

Seasonal-breeder females go through a nonbreeding (anestrous) period during which they are sexually less active. In continuous-breeder females the sexual cycles are repeated more or less continuously throughout the year. All females except the higher primates permit copulation only during a definite period within each sexual cycle. These periods of psychological and physiological readiness are called periods of heat, or estrus (from the Latin *oestrus,* gadfly, frenzy). The period from the beginning of one heat to the beginning of the next heat is called the estrous cycle. When in heat, a female is in a psychological state that is distinctly different from her state during the rest of the cycle. The male ordinarily shows no sexual interest in the female outside heat, and if he does his advances are repelled.

The psychological, physiological, and endocrine events of reproduction are all correlated. In addition to the externally visible manifestation of sexual receptivity, certain changes in the vaginal histology make it possible to follow, without recourse to surgery, the ovarian events that are primarily responsible for the physiological and psychological changes. The uterine endometrium also undergoes cyclic changes. These are, indeed, correlated with ovarian events, but only after surgery can they be used as guides to stages in the estrous cycle.

Heat

Heat coincides with the greatest development of ovarian follicles. The psychological manifestations of heat are brought about by a female sex hormone, estrogen, which is produced by the ovarian follicles. Complete heat can be brought about by estrogen even in ovariectomized females. It is important to keep this fact in mind; even though heat is caused by an ovarian hormone, it is in a sense independent of ovarian activity. In intact females exogenous estrogen causes heat at almost any time during the estrous cycle; heat can thus be completely divorced from the most important ovarian event, ovulation. This factor in the therapeutic use of estrogen is frequently overlooked in veterinary practice.

In the female guinea pig a trace of progesterone is necessary before estrogen causes the female to show full mating response. In the rat estrogen alone can bring about heat, but less estrogen is required for the full response if the female is pretreated with progesterone. The same seems to hold in sheep and possibly in females of other species. One is tempted to assume that priming with progesterone is essential for the full copulatory response in all females, especially in view of the fact that the follicles secrete progesterone before ovulation and before becoming luteinized.

In some domestic animals, especially cows and mares, *quiet heat* occasionally occurs. In quiet heat all the histological and physiological phenomena of normal heat are observed, including ovulation, but the psychological mating response is lacking. The need for estrogen may be greater in some individuals than in others, and quiet heats may be caused by the failure to secrete estrogen in large enough quantities to bring about the mating response. About 10 percent of mares show quiet heat, especially during March and April. Quiet heat also occurs in dairy cows, and becomes greatly important when it occurs in an unduly large proportion of the population. The prevalence of quiet heat in a Swedish breed of dairy cattle has brought these animals close to extinction; the bulls are unable to tell when the cows are in heat.

Frequently in mares, and occasionally in cows, *split heat* is observed. In split heat the initial period of sexual receptivity is interrupted by a period of nonreceptivity (one or two days in the mare, a few hours in the cow), which is then followed by another period of receptivity.

Table 4-1.   Changes throughout the year in the length of the estrous cycle in mature sheep

MONTH	NO. OF SHEEP	AVERAGE CYCLE LENGTH		CYCLES		
		DAY	SD*	NORMAL	LONG	SHORT
May	12	33.5	8.43	33%	67%	0%
June	61	58.5	8.75	16	75	8
July	49	32.5	4.17	14	76	10
August	99	19.0	1.12	55	26	19
September	330	16.5	0.43	72	11	17
October	375	16.8	0.29	78	9	13
November	365	16.8	0.29	84	6	10
December	394	16.8	0.27	82	6	11
January	373	16.5	0.50	54	15	31
February	156	18.3	0.79	38	35	27
March	21	22.3	2.30	10	43	48

*Standard deviation.

SOURCE: Williams et al., J. Animal Science, 15:984, 1956.

The duration and intensity of heat are variable (Table 4-4). Parous cows have more intense heat and a slightly longer cycle than virginal cows. There is no difference between beef and dairy cattle in length of cycle or duration of heat, and the cycle generally lengthens with increasing age. In sheep, in the middle of the breeding season, both the length of the cycle and the duration of heat are remarkably uniform, but early and late in the season the length of the cycle is extremely variable (Table 4-1). In mares both the cycle and heat vary in length, the cycle being somewhat shorter late in the season (August and September) than in the spring; the cycle of the mare seems to be governed largely by the characteristics of the breed. In swine heat usually lasts two days in gilts and three days in parous females, but deviations from these norms are common; increasing age generally lengthens the cycle.

One aberration of heat—common in cattle, less so in mares—is nymphomania. The causes of this condition, which is characterized by sterility and a more or less continuous manifestation of the psychological desire to mate, will be discussed later. It does not occur in sheep, goats, or swine, but is said to be known in bitches.

PHENOMENA RELATED TO HEAT. Many animals in heat show greatly increased activity. Pigs and cows in heat walk four or five times as much as they do during the rest of the cycle. This increased activity is caused by estrogen. Rats in activity cages run spontaneously much more at the height of heat than during diestrus or after castration. Spontaneous

Table 4-2. Vaginal pH of rats in different circumstances

CONDITION	pH
In diestrus	6.1
At beginning of proestrus	5.4
In heat	6.1
Ovariectomized	7.0
Ovariectomized and injected with	
8 I.U. of estrogen	4.1

SOURCE: Asdell, 1946.

activity in rats can be increased during the inactive phase by the injection of estrogen. During the menstrual cycle women show two peaks of spontaneous walking: one (for unknown reasons) during the menses and the other at the time of ovulation.

Pigs in heat also show very significant changes in bioelectric potentials measured externally over the ovarian region of the body; pigs in heat registered 23.2 millivolts, nonestrous females only 6.6. Similar changes in electric potential have been recorded at the time of ovulation in women. In most women ovulation is preceded by a slight dip in basal body temperature; this is followed, either at or shortly after ovulation, by a significant rise in temperature. The fact that a rise in temperature can be produced by the injection of progesterone may imply that this phenomenon is due to the preovulatory increases in secretion of progesterone. Similar changes in temperature have been looked for in cows, sheep, pigs, and rhesus monkeys, but the results have been inconclusive, possibly because of the difficulty of obtaining basal body temperatures in animals other than humans.

During heat (or the greatest follicular development) the cervix secretes the greatest amount of mucus, which is least viscous during heat or, in women, at the time of ovulation. Cervical mucus has a pH of 6.6–7.5 (the average for cows is about 6.9), and this remains fairly stable throughout the cycle. Sperm survive in the cervix (up to 72 hours in women) much better than they do in the vagina, where they become nonmotile within a few hours. The vaginal pH is generally alkaline, but it varies greatly among individuals and also within the cycle. In cows the vaginal pH varies from 7.5 to 8.5. In all animals investigated (cows, mares, women, rats) the vagina is more alkaline in diestrus and becomes more acid during heat (or during the greatest follicular development). That the change in pH is due to estrogen has been shown by the injection of this hormone into ovariectomized women and cows. The vaginal pH of rats is acid but changes during the cycle as well as under different experimental conditions, as shown in Table 4-2.

POSTPARTUM HEAT.   Females of several species come into heat shortly after parturition. Rats have a postpartum heat accompanied by ovulation within 48 hours after parturition. As soon as the young rats begin to suckle, no further heats occur until the young are weaned. Mating at the postpartum heat results in pregnancy, but the interval between mating and parturition is usually significantly longer than normal, for reasons that will be discussed in another connection (p. 276ff). An anovulatory condition is said to exist in women as long as they breastfeed their babies, but there appears to be much individual variation in this respect. Sows come into heat within a few days (usually from three to seven) after parturition, but the endocrinology of this event is not clearly understood. The ovaries of the sow show practically no follicular development at this time, no ovulations occur, and therefore mating during the postpartum heat does not result in pregnancy. However, sows do come into heat and ovulate if the litter dies or is removed at parturition or a few days later. This fact may be responsible for the general but erroneous belief of farmers that mating during postpartum heat results in pregnancy. Mares show a "foal heat," which begins from five to 10 days after parturition and lasts from one to 10 days. Ovulation may or may not occur at that time; more frequently than not, mating during foal heat does result in pregnancy.

No postpartum heat is observed in cows. The first heat occurs between 30 and 60 days after parturition; subsequent heats are not inhibited by lactation, as they are in most other animals. In most domestic sheep parturition occurs at a time when sheep do not normally show estrous cycles. For this reason few observations on the occurrence of postpartum heat in sheep are available. When ewes were caused to lamb early in the fall, however, most of them showed postpartum heat from one to 10 days after parturition, providing the lambs were not allowed to suckle. No postpartum heat is noted in suckled ewes. The relation of postpartum heat to uterine morphology and to lactation and its endocrinology is discussed in Chapter 10.

Vaginal Changes During the Cycle

Although the main physiological events of the estrous cycle occur in the ovaries, these events are reflected in the changes that take place in the vagina under the influence of the ovarian hormones, estrogen and progesterone. This finding was a significant contribution to the development of the physiology of reproduction, for it made it possible to diagnose ovarian events by simple techniques not involving repeated surgery. Stockard and Papanicolau, and also Long and Evans, observed

Table 4–3.   Ovarian activity and vaginal histology of rats during the estrous cycle

STAGE	DURATION OF STAGE	OVARIAN EVENTS	TYPES OF CELLS IN VAGINAL SMEAR
Diestrus	Half of whole cycle	Corpora lutea	Nucleated epithelial and leucocytes
Proestrus	12 hours	Follicles growing fast	Nucleated epithelial
Early estrus	12 hours	(Mating)	Cornified
Late estrus	18 hours	Ovulation	Cornified
Metestrus	6 hours	Corpora lutea formed	Leucocytes among cornified
Beginning of diestrus or anestrus		Functional corpora during early part	Cornified disappearing

that the histology of the vaginal epithelium does not remain constant during the estrous cycle. The vaginal epithelium is cyclically torn down and rebuilt, fluctuating between the stratified-squamous and low-cuboidal types. These cyclic changes can be followed by means of the vaginal-smear technique, in which the debris accumulated in the vaginal lumen is swabbed out and the cells obtained are examined under the microscope. The types of cells predominating in the smear give a clue to whether the vaginal epithelium is or is not being stimulated by estrogen.

Changes in vaginal histology during the estrous cycle are found in all mammalian females. The vaginal-smear technique is most useful, however, with species having short estrous cycles (mice and rats), for in them vaginal histology reflects ovarian events most accurately. In species with longer cycles, such as women and all domestic animals, vaginal changes lag from one to several days behind ovarian changes, and vaginal smears are therefore less reliable indicators of ovarian events. Furthermore, females with long cycles show considerable individual variation, and this also makes the application of the technique less precise and less useful. For rats, in which the cycle lasts about four days, very careful comparisons have been made between ovarian morphology and vaginal histology, and the cycle has been broken down into its parts (Table 4–3).

The rapid growth and cornification of the vaginal epithelium during early and late estrus have been found to be caused by estrogen. When in the normal cycle the level of estrogen drops after ovulation, or when in castrate females injection of estrogen is stopped, the cornified vaginal epithelium begins to break down, the scales disappear, and leucocytes become predominant. The vaginal epithelium changes histologically from the thick stratified-squamous type produced by estrogen to a thin low-cuboidal epithelium typical of the anestrous phase of the cycle (Figure 4–1).

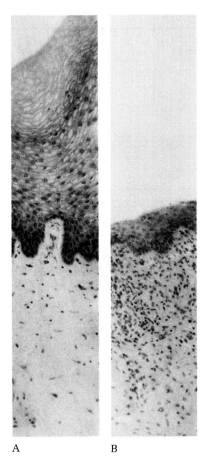

                                    A   B

Figure 4-1 Histology of the vaginal epithelium of a ewe in heat. **A.** The epithelium is thick and stratified squamous on day 1 of the cycle. **B.** The epithelium is
low-cuboidal on day 10.

Uterine Changes During the Cycle

If one follows the histological and morphological changes of the uterus
throughout the cycle, one finds that neither the size nor the histology of
that viscus is ever static. The most striking changes occur in its endometrium and its glands. During the follicular phase of the cycle the
uterine glands are rather simple and straight, with few branchings
(Figure 4-2*a*). This appearance of the glands is typical of estrogen
stimulation; in fact, it can be duplicated by the injection of estrogen into
castrated females, in which the epithelium is much lower, the endometrium thinner, and the glands fewer than they are when estrogen is

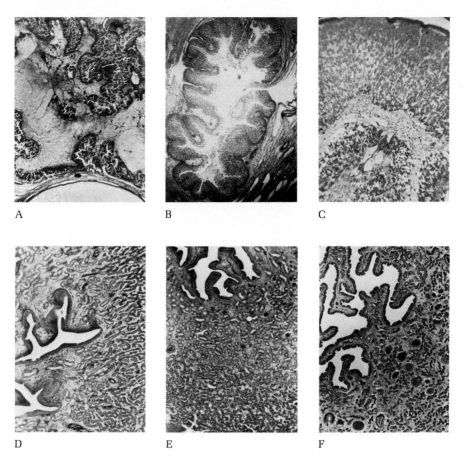

A

B

C

D

E

F

Figure 4–2 Comparison of the uterine endometria and the corpora lutea of swine during the follicular luteal phases of the cycle. **A.** A follicle one or two days after ovulation. Note the folded granulosa layer and the lumen filled with lymph and blood. **B.** An ovulated follicle on the way to becoming a corpus luteum, day 5 of cycle. The granulosa layer has proliferated but is still folded. **C.** Almost mature corpus luteum on day 9. The granulosa layer is much thicker, almost filling the former follicular cavity. **D, E, F.** Uterine endometria corresponding to the ages of the corpora lutea shown above. Note that the glands in D are still straight but that the glands in E are more convoluted. Compare the epithelial linings of the uterine lumina of D and F, and note that in F (height of luteal function) the epithelium has become crinkled. Note also that in F there are many more blood vessels than in E, and more than in D than in E.

acting on the uterus. Histological sections through estrogen-stimulated uterine endometria show a multitude of holes, which are the lumina of the simple, nearly unbranched glands. Such endometrial cross sections resemble a cut through a piece of Swiss cheese, and the estrogen-stimulated endometrium is frequently called a "Swiss-cheese endometrium."

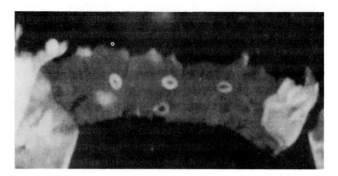

Figure 4–3 Uterine cast flushed from the uterine horn of a normal ewe shortly after mid-cycle. The holes and projections in the cast (top layer of uterine epithelium) are openings to the uterine glands.

During the luteal phase, when progesterone is acting on the uterus, the endometrium increases in thickness conspicuously. The glands grow rapidly in diameter and length, becoming extremely branched and convoluted (Figure 4–2b). In bitches, rabbits, women, and some other species the endometrium appears to be perforated by irregularly shaped, fringed openings, which connect with openings from other glands. Progesterone-stimulated endometria in histological sections resemble a lace curtain, giving rise to the descriptive term "lace-curtain effect."

One of the main phenomena of menstruation in primates is the sloughing off of the uterine endometrium and its complete replacement by a new endometrial lining (described in the discussion of the menstrual cycle of primates, p. 139). It is not generally known that in nonprimates, too, there is a cyclic sloughing off and regeneration of the uterine endometrium, a process that seems to be hormone-controlled, just as in primates, although the details of it are not yet clear. Endometrial destruction and regeneration in nonprimates involve no bleeding, possibly because only the epithelial layer of the endometrium is involved. In sheep this occurs during the early follicular phase and is completed in about three or four days (Figure 4–3). In cows and pigs a similar process takes place late in the luteal phase and is completed by the onset of the follicular phase. Rats show an almost continuous sloughing off of the endometrium, and the presence of uterine debris does not seem to be correlated with any hormonal state.

Effects of Estrogen and Progesterone on the Castrate Uterus

The changes occurring in uterine endometria during the normal cycle can be duplicated in castrated or infantile females, both of which have similarly unstimulated uteri. Estrogen causes increased vascularization and greater mitotic activity of the uterus, which result in a great increase in the weight of that organ. In rats and mice estrogen therapy leads to an accumulation of water in the uterine lumen; in other mammals there is no significant accumulation of water in the lumen, but the uterine interstitium becomes very edematous. Myometrial smooth muscles undergo hyperplasia and hypertrophy. The increase in the weight of the uterus is proportional to the amount of estrogen injected and is due to the combination of effects mentioned.

Small physiological amounts of progesterone have very little if any effect on the uteri of castrate or infantile females. Although very large amounts can produce the effects typical of that hormone, progesterone alone (without estrogen) is not very effective. For this reason the effects of progesterone are best studied after "estrogen priming." The priming is accomplished by pretreatment of the female with a dose of estrogen that by itself is too small to cause any morphological or histological changes ascribable to estrogen. After estrogen priming much smaller doses of progesterone are required to produce typical progestational effects, including an increase in the thickness of the endometrium, which is primarily caused by coiling and convolution of the endometrial glands. (Other typical effects of progesterone on the uterus are discussed later, p. 176.)

The estrogen and progesterone effects discussed here can be produced in castrated females of all species. The same uterine changes are produced by these hormones when they are secreted by the ovaries during the estrous cycle. But the uterine changes occurring during the cycle are not always as clear-cut as they are in castrates treated with estrogen alone or with progesterone after estrogen priming, probably because during the normal cycle there is considerable overlapping in the secretion and action of these two hormones, so that the uterus is never acted upon by either alone. Since the uterine endometrium of rats and mice does not undergo the changes typical of females with long cycles, it is thought that the corpora lutea of rats and mice do not become functional during the short cycles and do not secrete progesterone. Estrogen and progesterone injected into castrated rats and mice cause the typical changes in the uterine endometria, just as they do in other mammals.

In species with longer cycles the change from the follicular to the luteal phase is distinct and typical. However, there is great variation within species in the time at which the change occurs. In a homogeneous group of sheep (in which there is very little variation in length of cycle after the breeding season becomes established), it is possible to separate the ewes in luteal phase from those in follicular phase by the histology of their endometria. But there is great variation in the onset of the luteal phase. The endometria of some ewes still show the full progestational effect on day 10 of the cycle, but others, at the same time and with the same length of cycle, show the full estrogenic effect. Some ewes may have progesterone-stimulated uteri for the whole 16 days of the cycle even though the corpus has regressed morphologically and histologically. After considerable experience it is possible to evaluate the reproductive state of women, ewes, and pigs from the histology of the corpora lutea and the uterine endometrium, and this method is very useful in the analysis of the reproductive physiology of domestic animals. The evaluation of uterine histology in cows is even more difficult than in sheep and pigs because the differences between the luteal and follicular phases in cows are not as clear-cut as they are in the other two species. Even with the limitations cited, however, uterine histology provides a good guide to ovarian activity, especially in cases of abnormality, such as cystic ovaries or nonfunctional corpora lutea.

Ovulation

The culminating event of the estrous cycle is the rupture of the follicle and the shedding of the ovum. Because the endocrine control of ovulation and the physiology of the phenomenon are discussed elsewhere, we shall confine ourselves here to the relation of ovulation to the rest of the estrous cycle.

All the events of the cycle are related to ovulation in anticipation of the possibility that the ovum will be fertilized and that pregnancy will ensue. It is in preparation for this latter possibility that the uterine changes described in the preceding sections take place. During heat the female permits or encourages frequent matings and thus increases the chance that viable sperm will be present in the reproductive tract for fertilization when the ova arrive. In the great majority of spontaneously ovulating females ovulation occurs shortly before or shortly after the end of heat. In induced ovulators the ovum is shed within a few hours after copulation. Table 4–4 gives the length of cycle, duration of heat, and time of ovulation in relation to the onset of heat (or, in induced ovulators, to the time of copulation) in several animals. Only ranges are available for some species, generally because in these species the intervals vary with breed, strain, age, and possibly other factors, such as season. The time of ovulation given for most species is for the majority of individuals studied. For instance, of the cows studied, 75 percent ovulated 12–14 hours after the end of heat; the others ovulated as early as $2\frac{1}{2}$ hours before the onset of heat or as late as 22 hours after the end of heat. Similarly, most women ovulate about day 14 of the cycle, but there are cases on record in which ovulation occurred during the menstrual flow or very early or very late during the cycle. For this reason the figures given in the table should be regarded simply as a guide to the

Table 4–4. **Length of cycle, duration of heat, and time of ovulation in some animals**

	CYCLE (days)	HEAT OR SEXUAL RECEPTIVITY	TIME OF OVULATION
Mare	19–23	4–7 days	1 day before to 1 day after heat
Cow	21	13–17 hours	12–15 hours after end of heat
Sow	21	2–3 days	30–40 hours after start of heat[a]
Ewe	16	30–36 hours	18–26 hours after start of heat
Goat	19	39 hours	9–19 hours after start of heat
Guinea pig	16	6–11 hours	10 hours after start of heat
Hamster	4	20 hours	8–12 hours after start of heat
Mouse	4	10 hours	2–3 hours after start of heat
Rat	4–5	13 or 15 hours	8 or 10 hours after start of heat
Woman	28	Continuous	Days 12–15 of cycle
Dog[b]	——	7–9 days	1–3 days after start of heat
Fox[c]	——	2–4 days	1–2 days after start of heat
Rabbit	——	——	$10\frac{1}{2}$ hours after copulation (induced)
Mink	8–9	2 days	40–50 hours after mating (induced)
Cat	——	4 days	24 hours after mating (induced)
Cat	15–21	9–10 days	If no mating occurs
Ferret[d]	——	——	30 hours after mating (induced)

[a]Some breeds may ovulate as early as 18 hours after heat starts.
[b]Two heats yearly, fall and spring, no cycles.
[c]Season December–March, no cycles.
[d]In absence of male, continuous heat from March to August.

reproductive behavior of a majority of females, and should not be considered as absolute figures that apply to all females under all conditions.

In a few animals variability in the time of ovulation is insignificant. The vast majority of all rabbits, regardless of breed, ovulate 10–11 hours after copulation or after the injection of an ovulation-inducing hormone. In rats and mice the length of cycle and time of ovulation are quite constant within each of the various strains, in most of which the cycle lasts four days and in a few of which it lasts five days.

The time required for the completion of the ovulatory process in polytocous animals, after the first follicle has ruptured, is not accurately known. In pigs ovulation may be completed within six hours; in rats and rabbits it takes much less time. In pigs the intervals between ruptures of follicles are very uneven; unovulated follicles can be found as late as four days after the end of heat. It is probable that these lagging follicles persist as cysts in the next cycle.

The mare has the longest heat of all spontaneously ovulating non-primate females. Among the wild Equidae repeated matings throughout the entire heat, lasting four to seven days, are probable. Under domestication, however, the long heat presents a problem; it is difficult to time the mating in such a way as to have viable sperm in the oviduct at all the times ovulation could occur. To increase the chances of fertilization, one can take advantage of the fact that LH-containing hormones (such as chorionic gonadotrophin) cause ovulation of the mature follicle well before the normal time. The time of ovulation can thus be controlled by the breeder, who can avoid repeated breeding of the mare by causing ovulation through hormone injection and by timing the breeding to ensure the presence of viable sperm when the ovum arrives in the oviduct. According to some authors, stallion sperm remains motile in the reproductive tract of the mare for as long as five days, but it is not known how long it retains its ability to fertilize the ovum.

The dog and the fox (and possibly other Canidae) are unique in that the first polar body is not extruded at the time of ovulation (as it is in all other mammals). Fertilization of the ovum in these females does not occur for a considerable time after ovulation, but it is not known whether the delay in fertilization is due to the delay in the extrusion of the polar body.

Bizarre Phenomena Related to the Cycle

We have described the female reproductive cycle as rather unvarying, delimited by such cyclical events as psychological estrus, menstruation, ovulation, corpus luteum formation, etc. Since these phenomena are controlled by interrelated neuroendocrine mechanisms, one might think that once a cycle has begun it must run its course in an orderly fashion before the next cycle can begin. Nevertheless, counterexamples to neuroendocrine stability, which is held generally responsible for the predictability of the length of the estrous cycle, time of ovulation, onset of menstruation, duration of heat, etc., are observed from time to time. These examples of derangement of reproductive phenomena are quite drastic but common enough and too well documented to be ignored.

We are presently at the very beginning of studies of reproduction in relation to behavior, but the fragmentary information available makes it clear that much future effort can be expended profitably to increase our understanding of the mechanisms that permit psychogenic factors to modify the function of the neuroendocrine mechanism. Such studies have much theoretical and practical importance. For these reasons a brief summary of the best-documented phenomena will be made here in the hope that this review will stimulate further study of animal behavior as it relates to reproduction.

It should be remembered that each of the effects discussed here is due to a significant modification of one or several of the basic neuroendocrine control systems discussed earlier. The reader is further urged to think about the drastic endocrine shifts that must have occurred to produce results deviating so greatly from the expected.

It all began with the finding that such exteroceptive signals as the odor of the male can significantly modify the length of the estrous cycle, hasten or delay ovulation, prevent implantation of fertilized eggs, etc. These phenomena were first studied in detail by Whitten and by Parkes and Bruce, and are appropriately named after their discoverers. To them we shall add a few other effects that seem to fall into the category of deranged cycles.

THE WHITTEN EFFECT.   Whitten housed 10 to 30 female mice in a single box. Subsequently, each mouse was placed singly in a clean box with one vigorous male. Because in mice the cycle lasts four to five days, one would expect about one-quarter or one-fifth of all the mice to mate on each of the succeeding four or five nights after pairing. Instead, there was a highly significant shift in the number of mice mating on each of the five nights. In one typical experiment, involving 317 females, the observed mating frequencies (as judged by the presence of vaginal plugs) were distributed as follows:

> First night: 43 females mated
> Second night: 44 females mated
> Third night: 146 females mated
> Fourth night: 42 females mated
> Fifth night: 14 females mated
> Remainder: 28 did not mate at all

Assuming that a cycle of four to five days is typical for this group, on a purely random basis 58 mice should have mated each night (ignoring the 28 mice that did not mate at all). It is of great interest that in the mice that should have mated on days 1 and 2 heat and ovulation must have been delayed, whereas in those mice that should have mated on days 4 and 5 these events must have been hastened. Whitten found evidence that the odor of the male (or perhaps ingestion of the male feces by the female) was the immediate cause of this modification of the cycle.

THE BRUCE EFFECT.   The experiments involved two strains of mice, which for the sake of simplicity will be identified as A and B. Bruce allowed females of Strain A to mate with males of Strain A. Half of the females exposed were then put with other females of the same strain, and most of them, as expected, became pregnant or pseudopregnant. The other half were exposed to males of an alien strain B within 24 hours after they had mated with males of their own strain. These females *remated* with males of Strain B and, what was most amazing, most of them gave birth to young that were genetically identifiable as having been sired by B males. It is postulated that the corpora lutea that formed after the mating with A males degenerated after exposure to B males, that a new crop of follicles matured and ovulated (fertilized by sperm from B males), and, finally, that the eggs from the previous ovulation (fertilized by sperm from A males) degenerated and yielded to the newcomers from the second mating. It is important to pause here and to reflect on all the endocrine upheavals that must have occurred to make these events possible.

Bruce was able to initiate some of these events by placing A females, following their mating with A males, into boxes that previously housed B males. This was sufficient to interrupt the original (A $\times$ A) pregnancy, and most A females either resumed normal cycles or became pseudopregnant. The contraceptive effect of B males was traced to their odor, which is able to initiate the whole train of events described.

The action of an "odor" on the central nervous system can profoundly affect sexual behavior in several species. To contrast these external agents to internal ones (hormones) the term *pheromones* or "carriers of excitation" was introduced. That pheromones play an important role in the sexual integration of invertebrates has been known for some time. For instance, the queen bee secretes a substance that, when ingested by the workers, keeps their sexual apparatus underdeveloped. It appears now that pheromones may play an important role in vertebrates as well.

Signoret and Mouleon removed the olfactory bulbs from sexually mature pigs and found that the subsequent cycles became either very irregular or ceased completely. The ovaries of such females contained no corpora lutea but many vesicular follicles, which were somewhat smaller than the follicles of ovulatory size. In no sense were the ovaries of these females atrophic, and they bore a striking morphological resemblance to the ovaries of constant-estrus rats. In anosmic pigs the pituitary is unable to provide sufficient quantities of LH for ovulation.

Much additional work will be needed before we can understand the neuroendocrine pathways involved in this phenomenon. The importance of olfaction is underscored by a congenital genetic defect malformation of the olfactory bulbs. This condition is frequently accompanied by genital agenesis.

Part of the explanation of the Bruce effect lies in the nonmaintenance of the corpora lutea formed after the initial mating of A females with A males. The contraceptive effect of B males can be halted if immediately after mating with A males the A females are injected with prolactin (the LTH of mice and rats), which prevents the degeneration of the first set of corpora lutea. From this we may infer that the odor of B males is able to block secretion of the LTH in mice and rats.

That olfaction may also play an important role in the mating desire of male primates may be seen from the following series of elegant experiments by R. P. Michael (1973). Male monkeys fitted with nasal plugs would not mate with sexually receptive females until the plugs were removed. This suggests that in order to recognize a sexually receptive female and to respond by concupiscence the male must be able to detect an odor. In experiments with castrate female monkeys mating frequencies were, as expected, near zero. When these females were treated with estrogen, mating frequencies rose very sharply. Next, Michael and his colleagues collected vaginal secretions from estrogen-treated castrate females and smeared them on the rumps of untreated castrate females and again noted a very significant increase in mating frequencies in all females. Gas chromatography of the vaginal secretion of estrogen-treated females showed that it contained the following aliphatic acids: acetic, propionic, isobutyric, butyric, isovaleric, and isocaproic. Finally, they prepared synthetic mixtures of these acids and found that a mixture of the first five completely duplicated the effects of vaginal secretions. When applied to the vulvas of untreated castrate monkeys, the mixture aroused males to mate with them as frequently as they did with estrogen-treated castrates. This is the first demonstration of a pheromonal effect in higher mammals and the first successful demonstration of the chemical composition of a mammalian pheromone.

MISCELLANEOUS EFFECTS. The Bruce and the Whitten effects could be classified according to their triggering mechanism, i.e., olfaction. Other effects to be discussed here have not been studied in sufficient detail to permit us to classify them according to triggering mechanisms. To what extent modifications of reproductive behavior are caused by stress, proximity of males (olfaction, actual physical contact), or other environmental factors remains unresolved.

Reminiscent of the Whitten and Bruce effects is the observation that in sheep the onset of the breeding season may be advanced by several weeks if the ewes are exposed to rams several months prior to the expected onset of the breeding season, as compared to ewes not exposed to males.

In the discussion on sterility (Chapter 12) it is shown that sexually mature pigs that are "sterile" on their home farms, despite normal estrous cycles and frequent mating with fertile males, will conceive the first mating after they have been moved from the home farm to a new location. The move by truck is apparently the only change undergone by these animals, since they did not remain in the new location sufficiently long to benefit from any possible amelioration in management or diet. Similarly, normal, sexually mature pigs moved from one location to another show a tendency to synchronize their sexual activity (both heat and ovulation) in relation to the moving day. A significantly higher proportion show early heat (five to eight days after being moved) than would have if they had been left in the original habitat.

There also appears to be a "travel" effect on the onset of puberty in swine. In a well-documented study by du Mesnil du Buisson and Signoret, 1043 prepubertal gilts were shipped to an experiment station and checked for heat twice daily. Another group of 858 females of the same breed and age and of similar weights was slaughtered. Seventy percent of the slaughtered animals had infantile ovaries and reproductive tracts, while 30 percent had "some follicular development" or had ovulated one or more times. On the basis of this sample it is presumed that the 1043 experimental animals were in a similar stage of reproductive development. Of the experimental group, 277 females (26.5 percent) showed heat four to six days after arrival at the station; an additional 188 females showed heat either one to three or seven to nine days after arrival. Thus, 465 females (45 percent of the total) showed heat one to nine days after arrival at the station. It should be kept in mind that these animals were predominantly prepubertal and that, according to the evidence of the slaughtered group, only about 190 of them should have shown sexual activity had they remained on their native farms.

It appears that both hastening of the onset of puberty and synchronization of cycles occurred in this sample of pigs. It remains unknown whether the stress of transportation, the new environment, or the daily exposure to males produced this remarkable shift in the reproductive behavior of these females.

The failure of menstruation in women is a well-known phenomenon. The failure may last for months because of fear of pregnancy; in many such cases menstruation resumes within a few days after the woman has been informed of a negative pregnancy test. An extreme result of psychosomatic modification of the menstrual cycle is false pregnancy (pseudocyesis), in which the desire for pregnancy may lead not only to complete cessation of menstrual period but also to an enlargement of the breasts and swelling of the abdomen. In extreme cases the woman may even experience labor pains at the expected time, as if a fetus were present.

In several species females in heat show the lordosis reflex. When touched on the back they become rigidly immobile. When the pudendal region is touched they adopt a copulatory posture: the back arches concavely, the hind legs are rigidly extended, and the pudendal region is everted. Females not in heat try to escape the touch of the experimenter and do not show the lordosis reflex. This reflex is a reliable means of detecting heat in laboratory mammals, such as guinea pigs, cats, and rats. (In addition to the above symptoms, rats characteristically show a twitching of the ears.)

Domestic mammals show a variant of the lordosis reflex, which has been analyzed in some detail in pigs by Signoret. He has confirmed the observations of many practical pig breeders that a sow in heat will become rigidly immobile if, in the absence of a boar, a man sits on her back or even simply exerts manual pressure on the back. Apparently, this response is genetically controlled. In cross-bred swine used for experimental work at the University of Illinois 90 to 100 percent of all females responded to the test by rigid immobility. In the Large White breed used by Signoret only about 50 percent of the females in heat responded to the signal. The other 50 percent of the females in heat escaped the experimenter's attempts to sit upon them or to exert manual pressure. Signoret noted that the proportion of females responding could be greatly increased if a male was present, and he proceeded to fractionate the role of the male in eliciting the reflex. He found that 90 percent of the females showed the reflex if they could smell the males, the odor being produced by infiltration of the area with male urine, and if they could listen to a recording of his typical rutting call. If, in addition to the odor and the call, they could also see the male, a further 7 percent showed the reflex; if they could also have physical contact with him, the remaining 3 percent showed the reaction. The recorded call alone raised the response from 46 to 71 percent, but if the rhythm of the call was altered only an additional 9 percent of the females responded. Odor alone raised the percentage of positive responses from 50 to 81 percent.

How the phenomena of sound and odor can intensify psychological heat, which is normally attributed to the action of estrogen on the central nervous system, remains completely unknown and invites further study. Of great interest is the fact that there is a very significant relationship between the ease with which the rigidity reflex can be induced and the fertility of swine (this is discussed in greater detail in Chapter 12, pages 318–319).

MENSTRUATION

The estrous cycle of nonprimates is characterized by short heats, during which the female permits mating; ovulation takes place either shortly before or shortly after the end of heat. Primates do not have defined peaks of sexual desire, and as a rule permit copulation throughout the entire cycle. Ovulation in primates occurs midway between two menstrual periods. In both the estrous and the menstrual cycles the endometrial linings of the uterus are sloughed off; bleeding accompanies this breakdown only in primates and not in other mammals. Before discussing the physiological basis of menstruation we shall look at the results of experiments that formed the basis for the understanding of the mechanism of menstruation.

One of the earliest observations showed that the removal of ovaries from primate females precipitated menstrual flow within a day or two. This observation led to the assumption that menstruation was caused by the absence of ovarian hormones, which proved to be correct when it was found that the administration of estrogen or progesterone could prevent or stop postcastration menstruation. Menstruation can be prevented by these hormones for as long as they are injected, but menstrual flow resumes within 48 hours after the withdrawal of progesterone and within four days after the withdrawal of estrogen. This flow is known as withdrawal bleeding.

A few basic experiments illustrating the relation of hormone states to menstruation are shown in Figure 4-4. Menstruation is induced by ovariectomy and can be stopped (graph A) or prevented (graph B) for as long as estrogen or progesterone is injected. It can also be prevented following castration if progesterone is substituted for estrogen at some point during the injections (graph C), but it is not prevented by continuing the estrogen after termination of progesterone, even if the dose of estrogen is increased (graph D). From these facts it is inferred that normal menstruation is the result of the withdrawal of progesterone and that it occurs from a progestational endometrium. Under experimental conditions progesterone, estrogen, androgen, and certain adrenal hormones are capable of preventing or stopping menstruation. In the normal cycle of primate females (graph E) menstruation takes place while the corpus luteum is regressing (withdrawal of progesterone) and before rapid follicular growth has begun. It is not quite clear why the menses stops during the normal cycle, but we may assume that the slowly rising level of estrogen coming from growing follicles is responsible.

Morphological Changes in the Uterus

In animals with long estrous cycles (all mammals except rats and mice) profound uterine changes take place under the influence of sex hormones. In both primates and nonprimates the presence of ovarian hormones causes a thickening of the various uterine structures. In the absence of these hormones the lush growth attained during peak hormonal action can no longer be maintained, and a gradual breaking down of the uterine tissue takes place; a thinning of the mucosa, leucocytic infiltration, and a decrease in the rate of blood flow occur during the phase of hormonal withdrawal.

Primate females (except the New World monkeys) differ from nonprimates in that they have spiral arteries in the endometrium. When in the absence of progesterone the mucosa becomes thinner, the spiral arteries become exposed and bleeding from them begins. It is equally

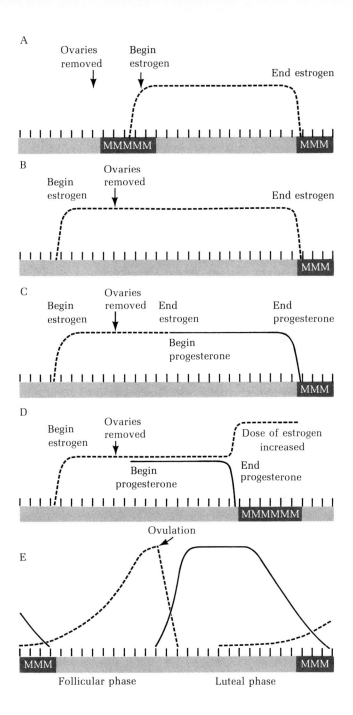

Figure 4–4 Relation of hormonal states to menstruation in primates. The broken line indicates the estrogen level, the solid line the progesterone level. **A–D.** Effects of castration and hormone administration on menstruation (MM); **E.** Postulated hormone levels of primate females during the normal cycle.

plausible, however, to assume that the absence of progesterone causes vasoconstriction, which leads to local ischemia, to the subsequent break-down of the endometrium, and hence to menstruation. In any event, menstrual flow is preceded by a thinning of the endometrium, vaso-constriction, and an increased coiling of the spiral arteries. These events lead to local necrosis and to desquamation. Starting with a few isolated patches, this process of gradual necrosis and renewal eventually in-volves the entire uterine surface. Hemorrhages in an area in which local necrosis is in progress last from 30 to 60 minutes.

New World monkeys and all nonmenstruating mammals have no spiral arteries. (Although the former do menstruate, the menstrual flow is microscopic.) Domestic animals do not menstruate, but the uterine endo-metrium is periodically torn down and built up. In the ewe, for instance, big parts of the uterine epithelium, resembling casts of the uterine horns, are sloughed off (Figure 4–3). This occurs midway between two ovula-tions and approximately at the time the corpus luteum wanes. Similar casts of the uterine endometrium are formed and sloughed off by cows and swine during the late luteal and early follicular phases. During the remainder of the cycle uterine debris is composed of individual cells and is not nearly so profuse as it is midway between two heats. The de-tails of the formation of these uterine casts have not been worked out, but it appears likely that the tearing down of the endometrium is caused by the same hormonal deficiencies that are responsible for menstruation in primates.

Pseudomenstruation in Nonprimates

The vaginal discharge of blood frequently observed in bitches and cows is incorrectly referred to as menstruation. It is in no way comparable physiologically and endocrinologically to the bleeding seen in primate females. In the bitch the bleeding starts in proestrus and is coincident with the most rapid growth of follicles. It lasts for seven to 10 days, and copulation is usually not permitted until after hemorrhage stops. Such bleeding is caused by diapedesis in the uterine endometrium and is not associated with endometrial breakdown. It can be caused in the bitch by the injection of estrogen and stops when the injection is discontinued. It therefore seems to be associated both experimentally and normally with high estrogen titers rather than with the low hormone levels that bring on menstruation in primate females.

In the majority of cattle (50 percent of the cows and 80 percent of the heifers) blood is discharged within about 48 hours after the end of heat, or from 15 to 20 hours after ovulation. This bleeding is also due to diapedesis, but in cattle it is associated with some destruction of the endometrial epithelium, especially in the intercaruncular areas. The hormonal causes of this uterine bleeding are not clear; ovariectomy of cows in heat causes hemorrhages, but withdrawal of injected estrogen does not. In cattle neither the presence nor the absence of uterine bleeding seems to be associated, as has been assumed occasionally, with ease of conception or with partial or total sterility.

Microscopic uterine bleeding has been noted in a variety of animals, such as elephant shrew, guinea pig, sheep, and sow, but in none of them is the bleeding comparable to the menstruation of primates. In primates bleeding is due to hormone withdrawal and occurs from a progesterone-stimulated endometrium. In nonprimates bleeding is due to diapedesis and occurs from an estrogen-stimulated endometrium; at least in the bitch it seems to be definitely associated with the presence of estrogen.

THE REPRODUCTIVE PHYSIOLOGY OF THE HEN

Genetic selection has made the domestic hen one the most remarkable and potentially most prolific reproductive organisms among the higher vertebrates. The reproductive pattern of hens differs from that of mammals in several important respects. Hens do not show a cyclic reproductive pattern, at least not in the sense in which this term is applied to mammals. Hens lay eggs in *clutches,* which consist of one or more eggs laid on consecutive days, followed by a day of rest. A prolific hen lays five or more eggs in a clutch. Clutches of 50–100 eggs are not uncommon, and many hens have laid 365 eggs in one year.

Before discussing the details of reproduction in the hen we shall do well to understand events leading up to the laying of an egg. Under ordinary daylight conditions ovulation occurs in the majority of hens in the morning hours and rarely after 3:00 P.M. The ovulated egg spends about $3\frac{1}{2}$ hours in the magnum portion of the oviduct acquiring the albumen coat. Then $1\frac{1}{4}$ hours are spent in the isthmus, where the soft shell membranes are formed, and 21 hours are needed for the formation of the calciferous shell in the uterus, or shell gland. Egg formation requires 25–26 hours, and ovulation of the next egg in the clutch occurs 30–60 minutes after oviposition of the previous egg. Thus, since ovulation does not occur in regular 24-hour cycles, the time of ovulation is later each successive day of the clutch. Eventually it will be as late as 2:00 or 3:00 P.M. When that point is reached ovulation is held in abeyance and laying is interrupted for one or sometimes several days before a new clutch is started. The first ovulation of the new clutch occurs very early in the morning.

Once the length of a clutch has been determined by observation—and a hen's clutch length varies little—it is easy to time oviposition and to predict from that when the next ovulation is most likely to occur. If a hen has a clutch length of four eggs and has laid the first egg of the clutch at 8:00 A.M., she is likely to ovulate the next egg about 9:00 A.M. on the same day. Because about 26 hours are needed to finish the ovulated egg, she should lay this egg about 11:00 A.M. the next day. Similarly, the time of laying can be used as a clue to the probable time the laid egg was ovulated the day before. After a hen's clutch length has been determined over a period of two or three weeks, and oviposition of the first one or two eggs of a clutch has been accurately timed, both ovulation and oviposition of succeeding eggs of the clutch become highly predictable. For this reason the hen is an unequaled experimental animal for the study of ovulation and related phenomena.

The mechanisms controlling ovulation (LH release), oviposition, and the egg's rate of travel through the oviduct have been worked out in part, but some of the details remain unknown. The productivity of a hen is generally determined by the length of the clutch, and that depends on the interval between oviposition and the following ovulation. The larger that interval, the fewer eggs will be produced in one clutch. Hens laying 20, 30, or more eggs in one clutch accomplish this by two means: by shortening the interval between oviposition and ovulation to a few minutes (hens with very long clutches ovulate even before the finished egg is laid), or by shortening the time the egg spends in the shell gland to as little as 18 hours.

Interrelation of the Pituitary and Ovary

The classic concept of the regulation of gonadal function by pituitary hormones has been amply confirmed in chickens by Fraps and others. In general, the mechanism of this control is approximately the same as in mammals. The gonadotrophic complex governs follicular growth and ovulation directly and egg formation indirectly. A few avian deviations will be emphasized here.

The ovaries of immature mammalian females can be stimulated to endocrine and gametogenic activity long before the females would normally reach sexual maturity. By contrast, the ovary of the immature chicken is amazingly unresponsive to stimulation by any gonadotrophin of mammalian origin until 20 or 30 days before the chicken would normally reach sexual maturity. Injections of crude chicken pituitary hormones are much more effective; they cause the formation of eggs of almost ovulatory size in sexually immature pullets. This indicates that normal follicular maturation in the chicken is stimulated by a pituitary hormone not found in mammalian gonadotrophic preparations. Additional evidence on this point is given elsewhere.

Injections of chicken pituitary hormones are also more effective than mammalian gonadotrophins in causing secretion of androgen and estro-

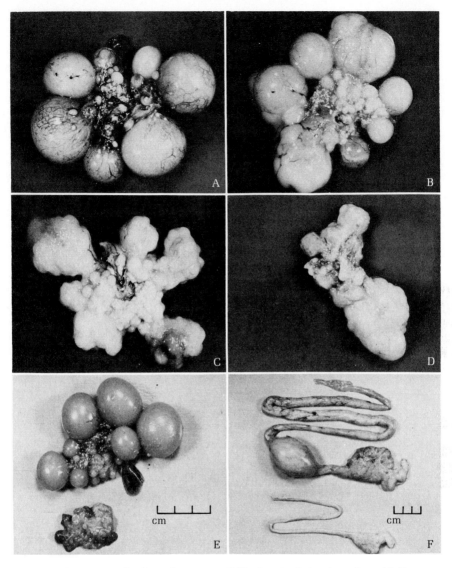

Figure 4-5 After hypophysectomy, follicular atresia is extremely rapid. Compare the normal hen's ovary (**A**) with those 18 hours (**B**), 24 hours (**C**), and 48 hours (**D**) after hypophysectomy. Comparison of the ovaries (**E**) and oviducts (**F**) of a normal and a hypophysectomized hen six days after operation (normal organs at top). [H. Opel and A. V. Nalbandov, 1961, *Prov. Soc. Exp. Biol. Med.*, 107:233.]

gen by the ovaries of immature chickens, as the subsequent growth of oviducts and combs demonstrates.

Hypophysectomy of adult laying hens leads to rapid regression of ovary, oviduct, and comb (Figure 4-5). Normal laying, or at least ovulation, may be maintained in hypophysectomized laying hens for not more than seven days after the operation if mammalian gonadotrophins are injected, and not more than 15 days if chicken pituitary hormones are injected. If ovarian atresia is permitted to occur after hypophysectomy, mammalian gonadotrophins do not stimulate the follicles to return to normal ovulatory size (even though they grow substantially), regardless of the dose. The proportion of FSH and LH in mammalian gonadotrophins may be such that these hormones can bring about only partial follicular growth in chickens.

Of some interest are studies of the effect of starvation on the egg-laying ability of hens. Withdrawal of feed from laying hens caused an immediate cessation of laying, and the rates of follicular regression of oviducts and comb very much resembled the effects of hypophysectomy (Figure 4–5). On the day following withdrawal of feed most hens laid the finished egg, the yolk for which had been ovulated the day preceding onset of starvation. After this terminal egg no further ovulations occurred unless the hens were injected daily either with PMSG or a crude chicken pituitary powder. With such treatment most hens were able to continue laying eggs for as long as 11 days in the complete absence of feed. Considering the chemical composition of the components of yolk and white and the total mass of materials produced, this is indeed a remarkable feat, involving the mobilization of 14.5 percent of the hens' total body reserves. Since only a gonadotrophic hormone was injected, and since for the production of a complete egg a functional oviduct is prerequisite, this means that the single daily GTH injections were able to maintain the flow of ovarian steroid hormones, which maintained the oviduct in a functional state. The ovarian steroids were also able to mobilize the necessary precursors for the yolk and the proteins for the albumen from the body reserves (fat stores and muscle tissue), to direct them to the liver, remobilize them from there, and finally, under the influence of the GTH, to direct the lipoproteins to the follicles and the albumen protein to the oviduct. The complexity of this whole metabolic performance is impressive.

Hormonal Control of Clutches

Experiments with hypophysectomized chickens clearly show that in chickens as well as mammals the gonads totally depend on pituitary gonadotrophins for support and maintenance. The number of eggs laid in a clutch depends on the genetic setting of the pituitary gland for the amount of gonadotrophin it produces. This is supported by the fact that hens with one- or two-egg clutches can be induced to lay three- or four-

egg clutches by the injection of either mammalian or avian gonadotrophic preparations.

One of the most intriguing questions in the field of avian reproduction
concerns the control in the laying hen of the so-called follicular hierarchy—the gradation of weights and sizes of the follicles. Only one follicle, the largest, is mature and can ovulate on any one day; as soon as
this follicle ruptures, the second largest matures to become the largest,
and so on. Control of the flow of gonadotrophic hormones in the hen is
thus totally different from that in mammalian females. In mammals
the flow is set to provide just enough gonadotrophin to mature one follicle (or several, in polytocous females). (The hypothalamo-hypophyseal
systems that control cyclic spacing of ovulations in mammals are discussed on p. 97ff.) In the chicken, on the other hand, the gonadotrophic
hormone flow is set not only to insure that only one follicle is of ovulatory size at any one time, but also to maintain the follicular hierarchy.
How this is accomplished is presently unclear, although it is suspected
that the arrangement of the ovarian vascular system may have some-
thing to do with the distribution of the gonadotrophins and thus with
the gradation in size of the follicles. This gradation can be easily dis-
rupted by the injection of gonadotrophic hormones, for instance, PMSG;
as many as 10 or 20 follicles can be grown to ovulatory size and caused
to ovulate simultaneously by the intravenous injection of an LH-containing hormone.

The details of the hormonal interplay between the ovary and the
hypothalamo-hypophyseal system of birds are not at all clear, except
that we do know that the avian ovary totally depends on gonadotrophic
hormones of pituitary origin (Figure 4-5). The laying hen is very unusual
in that bioassays of its pituitary and plasma for FSH and LH show three
peaks of these two hormones during the period immediately following
oviposition and preceding the ovulation of the next egg (see Figure 4-6
for the data on LH). (Radioimmunoassay for LH shows only one peak,
coinciding approximately with the latest peak demonstrated by bioassay.) How these hormonal peaks relate to follicular growth and ovulation
is presently not understood, nor is it known what causes them.

It is known that the hypothalamus participates in hypophyseal control of the release of LH and FSH. Surgical lesions of the hypothalamus,
particularly in the preoptic nuclei of the paraventricular region, can
block ovulation, (Ralph and Fraps, 1960). Furthermore, ovulation can be
induced by small doses of progesterone injected either systemically or
directly into the hypothalamus. That progesterone acts directly on the
hypothalamus and not on the pituitary gland is shown by the fact that
progesterone does not induce ovulation if the paraventricular region is

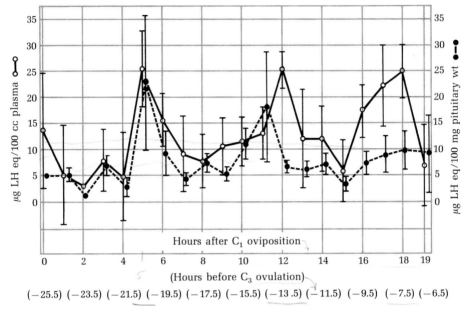

Figure 4-6 Changes in pituitary and plasma LH during the laying cycle of hens. Note that there are three plasma LH peaks, the first occurring about 20 hours and the last about seven hours prior to ovulation. These data were obtained by bioassay (OAAD) and disagree with radioimmunoassay data, which show only one peak, at about seven hours. Compare with Figures 4-7 and 4-8. [After D. M. Nelson et al., 1965, Endocrinology, 77:889.]

lesioned. Further support for hypothalamic involvement is provided by the fact that ovulation in the laying hen can be blocked by the nerve-blocking drug Dibenzyline.

Thus, it appears that progesterone is involved in the feedback mechanism between the ovary and the hypothalamal-hypophyseal axis. It is known that this hormone is produced by the ovarian follicle; progesterone found in chicken plasma was shown to come from the ovary. We can only assume that the synthesis of progesterone is not constant, but that it fluctuates in some kind of sequence that results in the kind of cyclic release of the gonadotrophins shown in Figure 4-6. Curiously, estrogen does not seem to participate in the feedback mechanism. Even milligram doses of estrogen injected into hens that lay daily do not interrupt the clutch sequence, suggesting that in the female neither the hypothalamus nor the pituitary gland can be blocked by estrogen. By contrast, the male is very sensitive to estrogen; microgram doses easily suppress the pituitary gland, causing complete regression of the comb and testes.

The question of what constitutes the ovulation-inducing hormone (OIH) in hens is also inadequately researched. Any LH-containing hormone (of mammalian or avian origin, with the possible exception of HCG) can cause ovulation in either intact or hypophysectomized hens. Ovulation can be induced in the hen by direct injection of LH-containing hormones into the follicle wall. However, LH injected alone will not ovulate the follicle unless it is injected systemically. This implies that the LH becomes somehow modified when injected systemically, acquiring a property that transforms it into an OIH. A mixture of LH and FSH injected into the follicle wall will induce ovulation, provided there is more FSH than LH in the mixture (Ferrando and Nalbandov, 1969). We shall return to the problem of OIH when we discuss the mechanism of ovulation in mammals (pp. 171–174).

Yet another control mechanism may be involved in triggering the release of OIH from the pituitary gland of the hen. In the majority of laying hens in which a foreign body has been placed in the magnum portion of the oviduct the ovaries are maintained in their normal laying condition but no ovulations occur. Once the foreign body is removed ovulations resume. The question arises whether the yolk traveling down the oviduct is treated like a foreign body and normally prevents the release of OIH until after the egg is sufficiently near completion and the next ovulation may take place. The interval from injection of exogenous LH to ovulation is 8 to 10 hours and a single injection is sufficient to cause ovulation, which still does not explain why the hen may normally have three LH releases. It is, of course, possible that two of the releases are concerned with follicular maturation rather than ovulation.

RIA methods for measuring chicken LH have recently been developed (Furr et al., 1973). The results disagree with those obtained from bioassay. RIA detects a single plasma peak of LH occurring roughly four to eight hours prior to ovulation; this peak coincides with a similar peak detected by bioassay (ovarian ascorbic and depletion method). The disagreement between the two assay methods can not be reconciled at present.

A recent study shows that the chicken follicle synthesizes estrogen, progesterone, and testosterone (Shahabi et al., 1975). The results of this study are presented in Figures 4–7 and 4–8. Figure 4–7 compares the change in levels of the three steroids in the plasma during the ovulatory cycle; Figure 4–8 shows the changes in the rate of synthesis of these steroids in the three largest follicles. While there is much interesting detail in the data, only three aspects will be mentioned here (the interested reader is referred to Shahabi et al., 1975).

Figure 4–7 shows the concentration of the three steroids in the plasma; all three show significant peaks at about six hours prior to the next

150

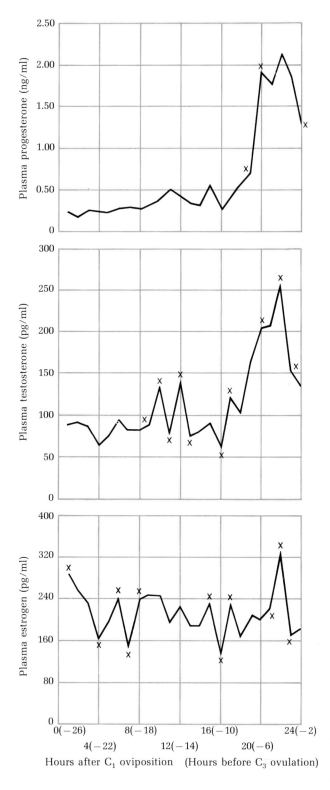

Figure 4-7 Concentration of progesterone, testosterone, and estrogen in the plasma of laying hens during the laying cycle (significant changes in concentration occur between Xs). Note the different profiles for the three steroids. All three show significant peaks about six hours before the next expected ovulation. These peaks coincide roughly with the LH peak detected both by OAAD (Figure 4-6) and radioimmunoassay. [Data from N. Shahabi et al., 1974, *Endocrinology*, 96:962.]

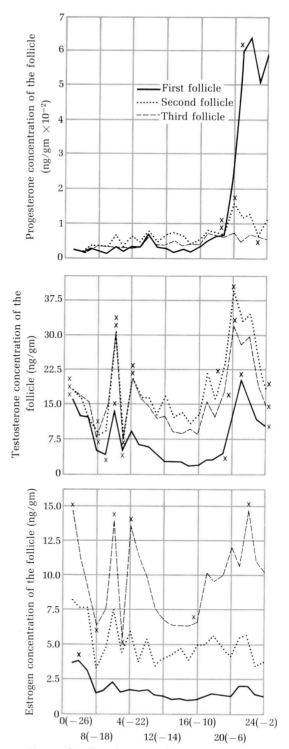

Figure 4-8 Rates of steroid synthesis by the three largest follicles during the laying cycle of the hen. Note that about six hours prior to ovulation only the largest follicle shows increased progesterone synthesis, all three follicles show an increase in testosterone, and only the smallest shows an increase in estrogen. Note also that both testosterone and estrogen show increased rate of synthesis about 22 hours prior to ovulation, at about the time when OAAD detects an LH peak (compare Figure 4-6). [Data from N. Shahabi, et al., 1975, *Endocrinology,* 96:962.]

expected (C_3) ovulation. Assuming that the OIH-releasing mechanism in chickens is similar to that of mammals, it is reasonable to suppose that one of the steroids is responsible for LH release—but which one? Experimental evidence shows that premature ovulation may be induced in laying hens by injecting progesterone or testosterone either directly into the hypothalamus or systemically. The question whether all three steroids participate in the sensitization of the hypothalamo-hypophyseal system remains unresolved. The second problem that remains to be investigated is the fact that, by both RIA and bioassay, the LH peak immediately preceding the next ovulation coincides with the peaks of plasma steroids. Unless one assumes that LH causes ovulation in the hen very rapidly—and all experimental evidence is against this assumption— the question can be asked whether the steroid peaks are the cause or the result of the preovulatory gonadotrophin peak.

Next we will examine the concentrations of steroids in follicle walls (Figure 4-8). The largest follicle, the one destined to ovulate shortly, has the lowest concentration of estrogen and testosterone. This may mean either that only the largest follicle is able to release its steroids into the peripheral circulation or that it shows a lower rate of synthesis than do the other follicles. Further, all three follicles show significant peaks in the accumulation of testosterone shortly before the next ovulation; only the third largest follicle shows such a peak for estrogen, and only the largest follicle shows a peak for progesterone, about six hours before it is due to ovulate. Finally, there are significant coincident peaks of estrogen and testosterone between four and 10 hours after oviposition, while there are none for progesterone.

If one makes the reasonable assumption that steroid synthesis is controlled by hypophyseal trophic hormones, then it follows that the peaks shown in the follicle walls must be due to impingement of gonadotrophins. Of course, this raises the question of how it is possible for the trophic hormone to discriminate between the follicles and at the same time be able to affect different steroids differently.

The reason for this seemingly detailed but actually incomplete analysis of the specialized control system of the chicken is to call attention to a relatively unexplored field: questions concerning the responses of follicles of different sizes to trophic substances. For instance, do small mammalian follicles lie relatively dormant, steroidogenically speaking, or do they contribute the major part of one steroid but not of another, as is apparently the case in the chicken? Most studies have been concerned with determining steroids in the peripheral blood in relation to the stages of the mammalian cycle, or with injecting trophic substances and measuring the resulting changes in steroid levels in the peripheral circulation or the ovarian effluent blood. What the trophic hormones do at the follicular level remains unknown, but is obviously of great interest and importance.

Another point that is also unresolved concerns the difference between rates of steroid synthesis and release. Inspection of the data in Figures 4-7 and 4-8 shows that there is no coincidence between the peaks of synthesis and the peaks of the same hormones in the plasma except immediately prior to ovulation. The data on chickens seem to suggest that at certain stages of the ovulatory cycle steroids are synthesized but not released into the peripheral circulation. Whether they are essential in the internal metabolism of the follicle remains unknown. The only time the rates of synthesis of all three steroids and their release into the peripheral circulation coincide to a great and obvious degree is immediately preceding ovulation. But, here again, cause and effect remain to be studied.

The Avian Oviduct

After ovulation the ovum is picked up by the fimbria of the oviduct and conducted into the magnum. Here it acquires the albumen layers. Albumen secretion in the magnum is controlled by two hormones. One is estrogen, which has the primary function of causing the anatomical and glandular development of the whole oviduct. In immature female chicks estrogen causes a spectacular growth of the magnum and its glandular elements, but estrogen alone cannot cause formation of albumen antecedents in the glands, nor can it cause the secretion of albumen proper into the lumen of the magnum. A second hormone is required for both the formation and the secretion of albumen.

Both androgen and progesterone, acting on an estrogen-developed magnum, can cause the formation of albumen granules and the release of these granules into the lumen. The attempt to decide which of these hormones is actually responsible for these effects in the laying hen presents a dilemma. Because androgen is known to be an endogenous avian hormone, it is tempting to ascribe the albumen-secreting effect to it rather than to progesterone, whose experimental effectiveness in this reaction and in other physiological phenomena of the hen cannot be doubted.

After the growth of the magnum has been accomplished by estrogen and the formation of albumen granules has been caused by either androgen or progesterone, there still remains the actual secretion of albumen from the glands into the lumen. This is normally elicited by the presence of any foreign body in the magnum, be it an ovum, ping-pong ball, or even a cockroach (which on one occasion had found its way in some unexplained manner into the lumen of the magnum, was later laid, neatly encased in albumen and shell, and almost found itself part of breakfast). The similarity in physiological response of the magnum and the mammalian uterus is striking, even though in one the result of trauma or the presence of an implanting embryo (the "foreign" body) is the secretion of albumen, in the other the secretion of "uterine milk" and the formation of placentomata.

Having acquired albumen during the two and a half or three hours in the magnum, the egg moves on to the isthmus, where the shell membranes are secreted. This part of the oviduct is histologically distinguishable from the magnum but is controlled by the same hormones, which act in the same manner and in the same sequence as they do on the magnum. The egg spends one and a half hours in the isthmus and, having acquired the soft shell membranes and some water, moves on to the shell gland, or the uterus.

It is more appropriate to call this part of the oviduct the shell gland rather than the uterus, for it is not homologous and in no way comparable to the mammalian uterus. The egg spends about 22 hours here while the calcareous shell is secreted around it. The endocrine control of the shell gland differs from the control of the upper regions of the oviduct in that estrogen alone causes the gland's growth and development as well as the mobilization and secretion of the calcium salts needed for the formation of the shell. Here also lies one of the most interesting aspects of avian endocrinology, one that needs additional work before it is completely understood. It is well known that estrogen causes the mobilization of calcium from bones. This leads to hypercalcemia, which is typical of laying hens and which can be induced in nonlaying birds and males by injection of estrogen. Equally well established is the fact that estrogen encourages osteoblastic activity in the bone and hence deposition of calcium. It remains unknown how estrogen simultaneously accomplishes these opposite effects—deposition and withdrawal of calcium—but it is probable that it does so with the cooperation of another hormone (parathyroid?).

It is generally assumed that oviducal motility is caused by posterior pituitary hormones. Additional work on this subject is needed in view of the fact that the egg moves through the various segments of the oviduct at different rates. According to earlier studies, oviposition itself is caused by oxytocin, but subsequent attempts to confirm this finding with more purified preparations of this hormone have not been uniformly successful. The effectiveness of oxytocin is inversely related to the length of time the egg spends in the oviduct and the shell gland.

Moreover, neurohypophysectomy of laying hens does not interfere with normal oviposition, nor does it alter the length of time spent by the egg in the various portions of the oviduct. Of course, this does not prove that the posterior lobe and its hormones play no role in oviducal motility, for it is known that that lobe may be merely the storage place for oxytocin and pitressin and that both substances are secreted in the hypothalamic region of the brain. It is possible that once the hen is adjusted to the absence of the posterior lobe the amount of posterior-lobe hormones coming from the hypothalamus is enough to control all functions normally associated with them except polydipsia and polyuria. Neurohypophysectomy probably does not alter the function of the anterior lobe.

For all these reasons attention has been focused on the possible role of prostaglandins in controlling oviducal motility. Both $PGF_{2\alpha}$ or PGE can cause premature expulsion even of soft-shelled eggs from the shell gland, and appear to be much more efficacious in this respect than is oxytocin. Presumably the source of these prostaglandins is the oviduct, although this is presently not established.

SELECTED READING

Altmann, M. 1941. Interrelations of the sex cycle and the behavior of the sow. *J. Comp. Psych.,* **31**:481.

Asdell, S. A. 1946. *Patterns of Mammalian Reproduction.* Comstock.

Breneman, W. R. 1955. Reproduction in birds: the female. In *Comparative Physiology of Reproduction and the Effects of Sex Hormones in Vertebrates.* Cambridge University Press.

Burger, J. F. 1952. Sex physiology of pigs. *Onderstepoort J.,* Suppl. No. 2, pp. 2–217.

Eckstein, P., and S. Zuckerman. 1956. The oestrous cycle in the mammals. In *Marshall's Physiology of Reproduction,* Vol. 1, Part 1, 3rd ed. Longmans.

Farris, E. J. 1954. Activity of dairy cows during estrus. *J. Am. Vet. Med. Assoc.,* **125**:117.

Ferrando, G., and A. V. Nalbandov. 1969. Direct effect on the ovary of the adrenergic blocking drug Dibenzyline. *Endocrinology* **85**:38.

Fraps, R. M. 1955. Egg production and fertility in poultry. In *Progress in the Physiology of Farm Animals,* J. Hammond, ed., Vol. 2. Butterworth.

Hansel, W., and S. A. Asdell. 1952. The causes of bovine metestrous bleeding. *J. Animal Sci.,* **11**:346.

Michael, R. P. 1973. The effects of hormones on sexual behavior in female cat and rhesus monkey. In "Female Reproductive System, Part 1" (Vol. 2, Sect. 7, *Handbook of Physiology*). Williams and Wilkins.

Ralph, C. L., and R. M. Fraps. 1960. Induction of ovulation in the hen by injection of progesterone into the brain. *Endocrinology,* **66**:269.

Reynolds, S. R. M. 1949. *Physiology of the Uterus,* 2nd ed. Hoeber.

Richardson, K. C. 1935. The secretory phenomena in the oviduct of the fowl. *Philos. Trans. Roy. Soc. London (B),* **225:**149.

Shahabi, N. A., H. W. Norton, and A. V. Nalbandov. 1975. Steroid levels in follicles and plasma of hens during the ovulatory cycle. *Endocrinology,* **96:**962.

Shahabi, N. A., J. M. Bahr, and A. V. Nalbandov. 1975. Effect of LH injection on plasma and follicular steroids in the chicken. *Endocrinology,* **96:**969.

5

Ovarian Follicles, Ovulation, and Corpora Lutea

Some animals (spontaneous ovulators) ovulate at regular intervals in response to an apparently spontaneous release of LH from the pituitary gland. In others (induced ovulators) copulation leads to an outpouring of LH and hence to ovulation. In neither case is it known how LH brings about the rupturing of follicles and the shedding of eggs. After ovulation the empty follicle of mammals is filled with luteal tissue, which is formed by proliferating granulosa cells. Progesterone is secreted by the follicle in small amounts even before ovulation. As the luteal tissue increases in amount, it secretes increasing quantities of progesterone until the corpus luteum begins to wane; then the rate of hormone secretion also declines. Presumably the spent corpus luteum, called the corpus albicans, secretes no progesterone.

In some species follicles frequently fail to ovulate at the proper time and continue to grow until they form cystic follicles. In some animals (dairy cattle) cystic follicles cause nymphomania; in pigs, heats become extremely irregular. The usual assumption that cystic follicles secrete more estrogen than normal follicles is not justified.

NORMAL FOLLICULAR GROWTH

We already know that follicular growth is controlled during the cycle by the hormones FSH and LH, both of which must be present if normal follicular growth and function (estrogen secretion) are expected. Follicular growth has been found to follow two patterns during the cycle. In litter-bearing animals (for example, the pig) follicular growth is slow during the luteal phase, and a definite ovulatory spurt is observed during the follicular phase, within the last few days before ovulation (Table 3-2, p. 67). In most monotocous females (for example, sheep) there is slow growth of the largest follicle during the first 14 or 15 days of the cycle and the ovulatory spurt occurs while the animal is already in heat (Figure 3-4, p. 63).

In both monotocous and polytocous animals the number of follicles that develop in the follicular phase of the cycle is much greater than the number sustained to ovulation. In the pig from three to four times as many follicles are present in the follicular phase as shortly before ovulation (Table 3-2). Follicles that do not reach ovulatory size degenerate during the follicular phase. It seems that less hormone is required to initiate follicular growth than to maintain larger follicles and bring them to ovulatory size. This is seen from the fact that when exogenous gonadotrophic hormones are injected at some time during the follicular phase the number of follicles that reach ovulatory size (and ovulate) is proportional to the amount of hormone injected. Similarly, if one ovary is surgically removed the remaining ovary produces about as many follicles and eggs as the two ovaries would have produced together. This compensatory effect is probably due to the fact that after unilateral castration twice as much gonadotrophic hormone is available for the remaining ovary as there had been before. These observations further argue for the idea, expressed in another connection, that the rate of FSH secretion is probably quite steady throughout the cycle.

An unconfirmed observation suggests that the large number of follicles started during the follicular phase furnish the estrogen that seems essential for the growth of the follicles chosen to ovulate. This conclusion is drawn from an experiment in which all but one or two of the follicles were destroyed during the follicular phase. It was found that none of the remaining follicles reached ovulatory size and none ovulated. When this experiment was repeated and exogenous estrogen was injected, the remaining follicles were maintained, grew, and ovulated.

It has been shown that the pituitary glands of prepubertal females contain significantly more gonadotrophic hormone than those of sexually mature females. Although hormone secretion in immature females stimulates some follicular growth, it is apparent from the lagging growth of the duct system that little if any estrogen is secreted. It is probable that during prepuberty the pituitary secretes mostly FSH, which is able to cause some degree of follicular development. Throughout the prepubertal life of female mammals waves of follicles grow and become atretic without reaching ovulatory size. Each successive wave of follicular growth reaches a greater degree of development until finally, at the time of sexual maturity, endocrine conditions reach the most propitious stage for the occurrence of the first ovulation. The first ovulations quite frequently are not accompanied by heat. When the proper balance between pituitary FSH and LH is reached, normal cycles, including heat and ovulation, become established.

Effect of Temperature

It is usually stated that the primary steroids produced by the ovary are estrogen and progesterone, originating in the theca layer of the follicle. Progesterone is the primary hormone synthesized by the corpus luteum, although under certain experimental conditions the corpus luteum may also secrete estrogen. It has been known for a long time that avian follicles also secrete testosterone. Recent work makes it abundantly clear that mammalian follicles also secrete androgen in large quantities. While the roles of estrogen and progesterone in female reproduction are well established, the role of testosterone in mammalian females remains unknown and is now being investigated. While androgen may be the hormone responsible for sexual receptivity in women, this is not its role in other mammals.

Although the ovary normally secretes all three sex steroids under certain circumstances, it can be made to synthesize only androgen. R. T. Hill discovered that the kind of hormone synthesized by the ovary depends to an important degree on the temperature to which the ovary is exposed. Hill removed the ovaries from rats and transplanted them to the ears of genetically closely related castrated males, where the transplants frequently became established. After a time it was noted that the male accessory glands of the castrated males bearing ovarian grafts did not degenerate but were maintained in full secretory activity and weighed as much as the prostates and seminal vesicles of intact males. This indicates that androgen is being secreted by the graft but, unfortunately, no data are available on either the chemical nature or the rate of secretion of the male sex hormone. It is also possible to graft the ovary into the tail of a female rat belonging to a genetic strain that has a rudimentary prostate but is otherwise a normal female. If the temperature of the tail is raised to body temperature after the graft has begun to secrete androgen (registered by an increase in size of the rudimentary prostate), the graft reverts to the secretion of estrogen and causes both uterine growth and vaginal cornification. These experiments show that the kind of hormone produced by the ovary may also depend on the environmental temperature in which the ovary finds itself.

Of interest is the histology of these ovarian grafts. Follicles of ovulatory size are occasionally found in younger grafts but no ovulations occur—perhaps because of the physical confines imposed on the grafts within the ear or the tail. In older grafts many small follicles may be present but in all of them there is a striking abundance of ovarian interstitial tissue, which in the oldest grafts is the most prominent cell type found. Whether the interstitial tissue is the source of androgen is not known.

Also completely unknown is the role temperature plays in determining the kind of histological structure the ovary is to assume and the kind of hormone it is to secrete. Women afflicted with multiple ovarian cysts frequently show hirsutism and a coarsening of the skin and facial festures. This suggests but does not prove that androgen is produced by ovarian cysts. Such cysts may also secrete progesterone (as they are known to do in swine), which may be metabolized to androgen. This metabolic pathway is a common one in some animals; the pregnant cow is known to convert a good deal of its luteal and placental progesterone into androgen, which shows up in abundance in the feces. Similarly, exogenous progesterone injected into nonpregnant cattle is converted into androgen, which is eliminated in the feces.

Little is known about the extent to which certain tissues are restricted in their ability to produce only one type of hormone. Elsewhere in this book (p. 177ff.) we have raised the question of whether follicular cells are capable of secreting progesterone prior to ovulation or whether this preovulatory synthesis of progesterone occurs only in species whose ovaries have much interstitial tissue. Later in this chapter we shall present evidence that the corpus luteum of the hypophysectomized rat produces estrogen when stimulated by LH; normally the corpus luteum synthesizes progesterone when stimulated by LTH. Avian ovaries normally secrete androgen but the specific histologic structure of the ovary responsible for the synthesis of androgen has never been definitely identified; the interstitial tissue of chicken ovaries contains a cell which may (or may not!) be responsible. It is conceivable that the theca cells of the follicles may be able to synthesize estrogen and then progesterone. It should not be inferred that just because there are no obvious changes in the granulosa cells under the microscope none has occurred. A change in hormone synthesis may be due to a change of trophic hormones (from FSH to LH) or a change in hormone ratios (FSH:LH) impinging on the receptor system of the granulosa cells.

Effects of Estrogen and Progesterone

We have noted that large doses of estrogen can completely inhibit follicular growth in normal animals by suppressing the secretion of pituitary gonadotrophins, whereas small doses of estrogen may enhance follicular development during the normal estrous cycle, perhaps by promoting vascularization of the chosen follicles. Beyond these observations little is known about the role of estrogen in normal follicular growth.

Progesterone has a more clear-cut effect on follicular development. When this hormone is injected into women, sheep (25 mg daily), pigs (50 mg daily), or cattle (100 mg daily) the next expected menstruation and ovulation, or heat and ovulation, do not occur as long as the injections continue. Within two or three days after the injections are stopped normal cycles, including heat or menstruation and ovulation, return. The ability of progesterone to inhibit heats and ovulations can be used to synchronize the cycles of the individuals of a flock of domestic animals. In this way the heats or ovulations of different individuals can be made to occur at a time desired by the investigator. The new rhythm does not persist, however, and within two or three cycles the females synchronized by the progesterone treatment return to a rhythm in which heats and ovulations occur in the normal random fashion.

Some adverse effects result from this progesterone treatment. Not all females respond by suppression of cycles. In some females (particularly pigs) ovarian cysts are formed in response to progesterone injections, and in them heat and ovulation do not follow discontinuation of the injections. In females in which ovulation takes place the ovulated eggs show a significantly lower fertilizability and a higher fetal resorption rate than those observed in normal females. For these reasons synchronization of cycles by means of progesterone remains an experimental technique only. A practical method of synchronizing heats or ovulations would be of great importance not only to the animal breeder but the laboratory worker as well.

The way in which progesterone acts in producing the effects noted above is not known. It appears probable that in large doses it inhibits LH secretion, thus accounting for the inhibition of ovulation and of estrogen secretion and hence for the occurrence of heat. It also appears that the rate of FSH secretion may not be affected—which would account for the formation of the cysts observed in many of the females treated with large doses of progesterone.

In pregnant females there is a distinct species difference in the ovarian response to large doses of progesterone. If as much as 8 mg of progesterone is injected into pregnant rats there is no noticeable morphological effect on the ovaries. The corpora lutea are indistinguishable from those of untreated rats. By contrast, if pregnant pigs are injected with 300–400 mg of progesterone daily between the time of conception and the day 35 of pregnancy, the corpora lutea degenerate completely, but no ovarian cysts are formed, as they would be in nonpregnant females. In relation to body weight, the 8-milligram dose for rats is about 10 times as great as the 400-milligram dose injected into pigs.

FOLLICULAR CYSTS

The formation of cysts is one of the most interesting and least understood topics in the physiology of reproduction. Because Selye (1946) has provided a good description and classification of the types of cysts found in women, we shall confine ourselves here to the types found in domestic animals. Cysts occur in most animals studied, but they are more commonly encountered in some than in others. Only rarely are they observed in sheep and goats. They are common causes of sterility in dairy cattle (but not in beef cattle) and in swine.

Ovarian Cysts in Pigs

About half of the cases of sterility observed in pigs are caused by ovarian cysts. A thorough analysis of this defect has been made in this species; three types of cysts are commonly found (Table 5–1). *Single* (or "retention") *cysts* occur in the normal cycles of swine because one or two normal follicles fail to ovulate at the time the majority of the follicles rupture. These unovulated follicles continue to enlarge during the early luteal phase of the subsequent cycle until they either become luteinized without rupturing or become atretic. Single cysts do not seem to interfere with normal cyclic behavior and are probably minor accidents that should not be regarded as abnormal. The other two types of cysts found in swine, multiple large cysts and multiple small cysts, are definitely associated with sterility.

The *multiple large cysts* (Figure 5–1) are the more common, and frequently reach the enormous diameter of 10 cm. The number is always in the same range as the number of follicles of ovulatory size typical of the

Table 5-1 Comparison of cystic and normal follicles in swine

| CYST TYPE | AVG. NO. PER OVARY | HISTOLOGY | | EFFECT ON REPRODUCTION |
		CELL WALL	UTERINE ENDOMETRIUM	
Multiple, bilateral:				
Large (2–5 cm diameter)	5.6	Granulosa heavily luteinized, thick	Progestational	Sterile, cycles very irregular, heat intense
Small	22.5	Granulosa normal	Estrogen type	No nymphomania
Single or double (2–3 cm diameter)	1.2	Granulosa normal	Depends on stage in cycle	None; normal cycles
Normal follicles (0.7–0.9 cm diameter at ovulation)	5.9	Granulosa normal	Depends on stage in cycle	Normal cycles (21 days)

SOURCE: A. V. Nalbandov, *Fertility and Sterility*, 3:100, 1952.

Figure 5-1 Multiple large cystic ovaries of a pig with a long history of sterility. The largest cyst measured 5.5 cm in diameter, and the two ovaries together weighed more than 700 g.

individual or the breed. There may be considerable variation in the size of cysts found in the two ovaries. The endocrine causes of these cysts are not known, but it is possible that they are due to failure of LH release. It is significant that in spite of their large size these cysts contain less estrogen than normal follicles, either per milliliter of fluid or per total volume of cyst contents. The large cysts are always heavily luteinized, either along the whole periphery or along part of the cyst wall. In the latter case the cyst wall is said to have lutein patches. The remainder of the wall may be completely naked or may be covered with a very thin granulosa layer, often consisting of only one or two layers of very flattened cells.

Because of the heavy luteinization, these cysts are frequently mistaken for corpora lutea, especially before they reach their ultimate large size. Normal appearing corpora lutea may be present along with the large cysts, but it is not known whether the corpora lutea are retained from previous ovulations, whether they result from completely luteinized smaller cysts, or are formed after normal ovulations at the time of the infrequent and irregular heats observed in such cystic females. Both the granulosa and the theca folliculi participate in the luteinization of cysts, but usually only the smaller and medium-sized large cysts show complete luteinization or lutein patches on the cyst wall. The largest cysts are usually "naked," showing no signs of luteinization. In fact, they show none of the normal components of follicles, such as the granulosa or the theca folliculi; it is possible that these cell layers have degenerated because of the continuous great distention of the cavity with cystic fluid. Large luteinized cysts secrete progesterone, as can be seen from bioassays of cyst walls and cyst fluids and from the fact that the uterine endometria of sows showing these cysts are of the typical progestational type.

Animals with multiple large cysts have very irregular estrous cycles, with prolonged anestrous periods, the latter sometimes leading to a mistaken diagnosis of pregnancy. There is no nymphomania, but when heats occur their intensity may be greater than normal. Most females with large cysts remain sterile in spite of frequent breedings. Recovery within the reproductive life of such animals rarely if ever occurs spontaneously, and attempts to ovulate these cysts or cause them to regress by the use of hormones have not been uniformly successful. In about 60 percent of the animals that have had large cysts for a long time, the clitoris may reach a length of 3 cm (see Figure 12–1c, p. 308). It is possible that the clitoris enlarges as the result of continuous stimulation by progesterone, but it is more probable that the progesterone is converted into androgen and that the androgen is responsible for the enlargement.

Since enlargement of the clitoris also occurs in pregnant animals (especially older ones), it is not a good diagnostic sign of large ovarian cysts.

Less common than the large cysts are the *multiple small cysts* (see Figure 12-2a), which are distinguished by the fact that they are slightly larger than normal follicles of ovulatory size. Multiple small cysts occur in greater numbers than the number of mature follicles normally found in the ovary (Table 5-1). Because it is possible to produce such cysts experimentally by overstimulating normal females with a gonadotrophic hormone, such as PMSG, the guess seems valid that the endocrine cause of multiple small cysts lies in the temporary or continuous over-production of FSH by the pituitary gland. Both naturally occurring and experimentally induced multiple small cysts resemble normal follicles histologically. Small multiple cysts secrete and contain more estrogen than do normal follicles, but even here nymphomania is not observed in afflicted females. Unlike the case of multiple large cysts, spontaneous recovery from multiple small cysts may occur, but treatments with hormones have not been uniformly successful.

It is impossible to distinguish between small and large cysts by external symptoms. In both conditions estrous cycles and intervals between heats are very irregular and unpredictable. As a rule, animals with small cysts do not show enlargement of the clitoris. The significance of these types of cysts to the efficiency of reproduction will be discussed in Chapter 12. Pregnant swine are frequently found to have ovarian cysts of either the large or the small variety. It is not known whether these cysts arise after conception or whether some animals conceive in spite of them. It is possible that despite an endocrine imbalance normal follicles may mature and ovulate in the presence of large cysts and that mating at these times may result in conception. The litter size of most pregnant cystic females is significantly smaller than that of noncystic females.

Ovarian Cysts in Cattle

Among cattle, cystic ovaries are much commoner in dairy cattle than in most herds of beef cattle. Because nymphomania is frequently associated with cystic ovaries in cattle (as it is not in other species), the assumption has been popular that ovarian cysts in cattle secrete estrogen. Work has shown that this is not true; with respect to hormone secretion, the cystic ovaries of cattle resemble the large cysts of swine, including the fact that cysts in cattle secrete progesterone. Furthermore, cows with cysts often assume male characteristics, such as general coarsening of the features of the head, deepening of the voice, thickening of the neck, enlargement of the clitoris, and male copulatory behavior. The symptoms of nymphomania may thus be secondary manifestations of an endocrine upset not of particular ovarian origin, and the cystic condition of the ovaries may be a secondary rather than the primary cause of nymphomania.

One of the main reasons for suspecting a cause-and-effect relation between nymphomania and cystic ovaries is the intense sex drive of afflicted females; such a drive is normally associated with estrogen. However, psychological heat can also be induced in females by the injection of large doses of androgen. It is not known whether androgen itself produces psychological heat or whether it must first be converted to estrogen. These considerations lead to the possibility that nymphomania in cattle may be primarily caused by malfunction of the adrenal glands, that is, that the adrenal glands may under certain circumstances secrete either androgen or an adrenal steroid that is converted into substances having androgenic or estrogenic activity. This conjecture is supported by observed masculinization of the afflicted females.

This subject of nymphomania in cattle has received less study than it deserves, owing to an assumption that is completely unsupported by experimental work, that is, that cysts secrete estrogen, causing an intense sex drive. The problem of nymphomania in cattle is further complicated by the fact that nymphomania, at least in its earliest stages, may be controlled with some degree of success by the administration of gonadotrophic hormones rich in LH (such as HCG), which presumably luteinize the ovarian cysts. Although this fact seems to strengthen the argument that the ovaries are the primary cause of nymphomania in cattle (which may be the case in the earliest stages of the condition), it does not exclude the possibility that other endocrine glands may be affected as the disease becomes established. Garm (1949) and others have found hypertrophy of the adrenal glands and especially of the zona glomerulosa in cattle with cystic ovaries and nymphomania.

SPONTANEOUS AND INDUCED OVULATION

In most mammalian females ovulation is a cyclic event, one that occurs at regular intervals unless the female is pregnant. Females belonging to this group are called *spontaneous ovulators*. In certain other species ovulation follows stimulation of the cervix, normally by the penis during copulation. Females that normally ovulate only after copulation, such as the rabbit, cat, ferret, short-tailed shrew, and the mink, are called *induced ovulators*.

In both spontaneous and induced ovulators follicles rupture as the result of LH action, but with some differences. In the spontaneous ovulator LH action is cyclic, independent of copulation, and provoked by an interplay of the neuroendocrine system. In the induced ovulator it occurs only when the cervix or parts of the vagina are appropriately stimulated. Evidence shows that the nervous impulse travels from the cervix to the basal hypothalamic area below and perhaps overlapping with the arcuate nucleus (see Figure 3–13, p. 99), where the LRF is presumed to be produced. The LRF reaches the adenohypophysis via the portal system and causes it to release LH into the peripheral circulation. At present it is not clear what pathway is followed in spontaneous ovulators, since no neural signal is necessary for the release of LH. An LRF has been extracted from the hypothalami of spontaneous ovulators, e.g., cattle and rats, which suggests that an LRF does participate in the release of LH in them, too. Furthermore, ovulations can be caused in rats by injecting them with LRF. The signal for LRF to come into play is presumed to be neural since the LRF-producing hypothalamic site may be sensitive to one or both of the ovarian steriods. If estrogen or progesterone reach a certain titer in the blood, the hypothalamic area sensitive to that steroid concentration may respond by the release of LRF. From here the pathway is the same as that for induced ovulators. It will be recalled that in chickens the injection of minute quantities of progesterone into approximately the same area of the hypothalamus causes LH release, resulting in ovulation.

Some of the considerations that lead to the above interpretation are as follows. In induced ovulators ovulation rarely occurs in the absence of copulation; nor does it occur if the pituitary gland is removed within 60 minutes after copulation, or if the pituitary stalk is sectioned within that time. If either of these surgical interventions is performed later than the time indicated, ovulation proceeds normally. The length of time required for the copulatory stimulus to elicit the secretion of enough LH for complete ovulatory response argues for the intervention of a humoral agent and against a direct nervous connection from cervix to pituitary gland. That a nervous impulse also is involved, at least in part of this chain, is demonstrated by the finding that ovulation can be prevented by injecting a nerve-blocking agent (such as Dibenamine, or N,N-dibenzyl-chloro-ethylamine) within one minute after the end of coitus. If the injection is delayed for more than one minute, ovulation is not blocked. Finally, LH release must be the terminal reaction of copulation since ovulation can be caused in rabbits and other induced ovulators by the injection of LH or LRF without copulation. In the rabbit either copulation or injection of LH leads to ovulation in about 10 hours.

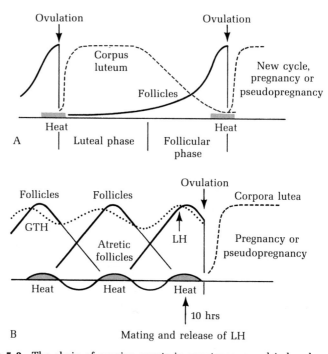

Figure 5–2 The chain of ovarian events in spontaneous and induced ovulators is basically the same. **A.** Spontaneous ovulation is cyclic, following specific periods of heat in the estrous cycle. **B.** Induced ovulation occurs as a result of stimulation of the cervix, normally copulation. Induced ovulators experience alternating estrous and anestrous periods and theoretically are sexually receptive at all times before copulation.

In induced ovulators follicles may occasionally rupture without copulation. In rabbits one to five percent (and in some strains as many as 30 percent) of the females may ovulate spontaneously. This usually happens if several females are kept in the same cage, the necessary neural stimulus for the release of LH being obtained from mounting, playing, and other physical contact. For this reason it is important to keep females used for experimental work in solitary cages. Even then as many as two percent of them may ovulate spontaneously.

Induced ovulators have no estrous cycle comparable to that of spontaneous ovulators. Theoretically, induced ovulators are sexually receptive at all times before copulation. In rabbits a period of sexual receptivity of two or three days is followed by an anestrous period of approximately equal duration. The endocrinology of these abbreviated cycles has not been worked out in detail, but it seems that the alternation of estrous and anestrous periods corresponds to the growth and atresia of follicles. During follicular growth estrogen secretion induces sexual receptivity, but rising estrogen levels eventually inhibit the pituitary gonadotrophins. A decrease in gonadotrophic hormones results in follicular atresia and the onset of the anestrous period. Follicular atresia leads to decreased estrogen flow, which is followed by an increased flow of gonadotrophins, the growth of a new crop of follicles, and a new period of sexual receptivity. Ovarian events in induced and spontaneous ovulators are compared graphically in Figure 5-2.

It is interesting to speculate on the evolution of induced and spontaneous ovulation. The former is found primarily in animals that live singly in field or forest and therefore might not be near a male at the time of the periodic and widely spaced heats. The more or less continuous state of sexual receptivity provides a good chance that at the time of occasional meetings between solitary males and females the latter will be in heat and that mating and reproduction can result. In some species, however, females that lead a solitary life nevertheless ovulate spontaneously. In them the chances of procreation are increased by a gathering of both sexes into droves during the breeding season. Rutting seasons and the endowment of females in heat with an especially pungent odor also attract distant males. It is probably futile to speculate on the advantages of one method of procreation over another or to compare the efficiency of the two systems. They both serve the purpose for which they have evolved, and the difference illustrates the enormous variation in the different genetic stocks.

THE MECHANISM OF OVULATION

Hormones Causing Ovulation

We have already raised the question of the identity of the ovulation-inducing hormone (OIH) by pointing out that although LH does cause ovulation, FSH is also known to have that ability. At present it is not known whether the role of FSH is normally facilitatory or obligatory in this event. We shall now provide additional data on this point without being able to come to any definite conclusion.

If the OIH is released normally or given by injection, a variable period elapses before ovulation occurs. This interval is 10–11 hours in the rabbit and the rat, 8–14 hours in the chicken, 40 hours in the pig, and 25 hours in the sheep. In all of these animals the release peak is very sharp and of short duration (see Fig. 3-2, p. 60; Fig. 3-3, p. 62). Further, it is known that the half-life of both FSH and LH is very short, certainly not much longer than 30 minutes. Thus, it appears that the OIH initiates metabolic changes in the follicle that culminate in ovulation. The exact nature of these changes is not known. Nor is it known whether the OIH attaches to receptor sites in the follicular cells and continues to exert its action over a period of hours or whether it simply turns on metabolic events and soon becomes metabolized. It is known that following the appearance of the OIH steroid synthesis begins almost immediately in the follicle. Whether this synthesis (of estrogen, testosterone, and progesterone) is a prerequisite for ovulation remains to be determined.

There is also the suspicion that protein synthesis may be involved in ovulation; if cyclohexamide or puromycin (both of which block protein synthesis) are injected into rabbit follicles, ovulation is blocked. If following the injection of these substances HCG is injected systemically, follicles thus treated will ovulate. Unfortunately, these experiments are not necessarily conclusive because drugs that are known to block alpha and beta adrenergic receptors, and which deplete catecholamines when injected into either the follicles of rabbits or the follicle wall of chickens, also block ovulation. This last observation implies (but does not prove) that nerves may also be involved in the ovulatory process. These results illustrate some of the complexities involved in unraveling the mechanism of ovulation.

Under normal circumstances the OIH reaches the follicle through the ovarian circulatory system. A very promising injection technique has recently been developed and may make future study of the ovulatory mechanism easier. A very fine-drawn capillary tube is carefully inserted into the ovarian stroma and directed toward a follicle; the follicle wall is pierced and LH or FSH is injected into the follicular lumen. If this is done carefully the follicle will not be damaged and will ovulate, which can be proven by demonstrating ova in the oviduct. Ovulation occurs at the normal time of 10.5 hours after injection of either hormone. It is interesting to note that the hormone is injected directly into the follicular fluid and that apparently its ovulating effect is produced simply by bathing the granulosa cell lining of the lumen. Whether the hormone penetrates to the theca layers is not known. Using this technique, one or more follicles in an ovary can be treated with various hormones or hormone doses to demonstrate the fact that each follicle acts as an individual entity without being obviously affected by activity in a neighboring follicle. In all experiments using this technique some follicles are injected with saline to test for effects of puncturing.

Experiments using this technique have resulted in the following observations. If 10 ng or more of LH is injected, ovulation occurs within 10.5 hours. If the dose is less than 10 ng the follicle luteinizes with entrapped ovum (Figure 3-7, p. 72). The degree of luteinization is proportional to the dose of LH injected until, beginning with doses around 1 ng/follicle, no solid corpus luteum is formed and both the granulosa and theca layers become transformed into lutein cells.

With FSH the story is significantly different. If 100 ng of FSH is injected, ovulation occurs at 10.5 hours; however, if less than 100 ng is injected the follicles remain normal and do not luteinize as they do with small LH doses. If FSH and LH are injected together, the minimal ovulatory doses are 50 ng and 5 ng, respectively. Thus, it appears as if the two hormones are synergistic, and one is tempted to speculate that the normally OIH is a mixture of the two gonadotrophins.

Mechanics of Ovulation

It has been shown in the hen that a follicle can ovulate normally even if the stalk that connects it to the ovary and hence to the circulatory and nervous systems is severed or clamped off in such a way as to preclude the possibility of either vascular or neural connection. For these reasons it seems most likely that LH only initiates the changes in the follicle wall and that this effect is produced within a short time after the OIH is released into the blood stream.

Ovulation in several species has been frequently seen and recorded on film, but its exact mechanism remains unknown. Several of the older theories can be definitely discounted. The follicle does not burst because its "ultimate" size has been reached or because the interior pressure of the liquor folliculi is so great that it bursts the wall. Ovulation is not an explosive process but an oozing one. Under certain experimental conditions the follicle can continue to grow long past the time when it normally should have ruptured. Cysts are much larger than follicles of ovulatory size, and the internal pressure of cysts is incomparably greater, yet they do not rupture readily. In some species (especially swine, but also cattle and sheep) follicles become very flabby a few hours before ovulation even though there is no indication of a break in the follicular wall that would permit follicular fluid to escape and thus lower the pressure. The assumption that ovulation is aided by intestinal movements or by the massaging action of the fimbria is equally unfounded, for ovulation in the pig and in the chicken proceeds normally in females in which the fimbriated ends of the oviducts have been amputated. In the chicken ovulation will even occur in vitro if the follicle is maintained at the proper temperature and humidity.

The process of ovulation in the chicken furnishes a clue that may form a basis for an explanation of the mechanism of ovulation. The complex and prominent vascular pattern of the avian follicle is one of the striking features of both the maturing and the mature follicle (Figure 2-5, p. 26). Within a few hours after LH release or administration the follicle "blanches" because of the drastically decreased blood flow

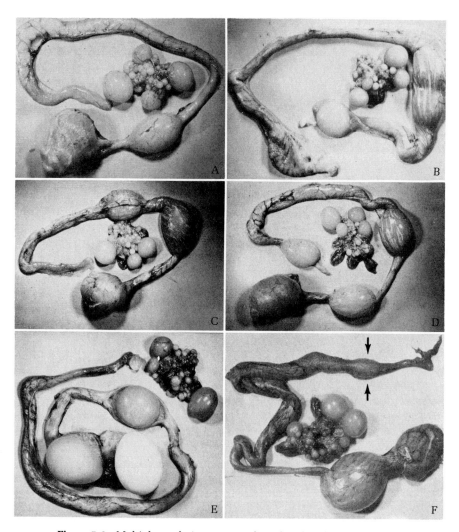

Figure 5-3 Multiple ovulations are easily induced in hypophysectomized hens injected with a mammalian LH. Note that in F (arrow) a very small ovum was caused to ovulate even though larger ova are present in the ovary. This happens only after hypophysectomy. [H. Opel and A. V. Nalbandov, 1961, *Endocrinology*, 69:1029.]

through it. The stigma (which, contrary to general impression, is not avascular) becomes wider, and many of the capillaries that extend across it become constricted and devoid of blood. Finally, a small tear appears in one of the corners of the stigma and the ovum bulges through it. The rip widens and the whole ovum slips out of the follicle, which collapses. Neither abdominal pressure nor massage by the fimbria contributes to the process of ovulation, for it is possible to block these influences in properly prepared and anesthetized birds.

Data obtained by Opel strongly support the possibility that LH may cause ovulation by initiating an ischemia in the wall of the follicle destined to ovulate. Briefly, the evidence for this is the following. By the injection of LH in intact laying hens it is possible to hasten the ovulation of the follicle destined to ovulate next, but it is not possible to ovulate any of the smaller follicles in the hierarchy. However, if a laying hen is hypophysectomized and injected with LH, not only the largest but also the second, third, and even smaller follicles can be ovulated. Thus, three or even more ova may be present in the oviduct of hypophysectomized hens (Figure 5-3), a situation that can never be duplicated in animals with intact adenohypophyses. These findings suggest that normally the hypophysis secretes a factor that inhibits ovulation of any but the largest follicle. The "inhibiting" factor in the intact hen may be FSH, which, by virtue of its folliculotrophic action, may keep follicles from becoming ischemic. This idea found confirmation in experiments with hypophysectomized hens in which FSH was administered immediately after hypophysectomy. It was found that subsequent injection of LH was able to cause the ovulation of only the largest follicle; in the presence of FSH the smaller follicles were not able to ovulate.

It appears reasonable to view ovulations as a two-stage phenomenon. During the first stage follicles reach their ultimate ovulatory size, which is determined by the total available circulating gonadotrophic hormone distributed via the follicular circulatory system in accord with the vascular capacity of individual follicles. That the amount of "available" hormone limits follicular size is supported by the finding that in both mammals and birds follicles can be caused to grow beyond their "normal" ovulatory size by the injection of exogenous gonadotrophic hormones. Preliminary observations in chickens support the possibility that as the largest follicle approaches its ovulatory size the amount of blood flowing through its vascular system becomes proportionally less than the amount of blood flowing through the smaller follicles. Hence, the amount of hormone available to each unit of follicular cell is lower in the largest follicle than it is in the smaller ones. Because of this reduction in the concentration of gonadotrophic hormones, the largest follicle can be viewed as having reached a stage of "physiological atresia"—a stage when the hormone concentration is inadequate to maintain active proliferation of the cellular components of follicles. Thus, during the hormone-adequate phase follicles are capable of rapid growth but are incapable of ovulating, while during the hormone-inadequate phase they become physiologically "atretic" and can be made to ovulate because they are essentially inactive physiologically.

It is not known, however, what the ovulatory peak of LH does to cause the conversion of a follicle that cannot ovulate into one that can. It is also unknown whether LH acts locally, on the stigma, or on the follicle as a whole. There is other supporting evidence of this view of the mechanism of ovulation, but the reader should refer to more complete summaries (Nalbandov, 1961).

The idea that a quasinecrotic process may be involved in ovulation is supported by observations of the ovulatory process in mammals. In cattle, swine, and sheep the follicle destined to ovulate loses much of its turgidity and becomes soft and pliable as the time of ovulation approaches. The outermost layer of the follicle wall slowly parts and the two or three inner layers protrude through the breach to form a papilla. Eventually the inner layers disintegrate and the follicle collapses. In cattle, swine, and sheep there is certainly nothing explosive about the ovulatory event, during which only part of the follicular fluid streams through the newly formed aperture. The ovum, loosened from the cumulus oophorus and lying unattached in the antrum, is usually washed out during the initial escape of the follicular fluid. Whether ova that are not washed out of ruptured follicles would ordinarily leave the follicle or whether they would become trapped in the forming corpus luteum is unknown. Occasionally it is possible to artificially wash out ova remaining in ruptured follicles. Because the fimbria closely envelops the ovary at the time of ovulation, and because it shows great motility at that time, it has been asserted that the fimbria plays a role in evacuating the contents of the collapsed follicle.

FORMATION OF CORPORA LUTEA

One of the most interesting morphological and physiological events during the cycle is the formation of the corpus luteum in place of the ruptured follicle. In a short time cells that secreted estrogen during the follicular phase become transformed into luteal cells that secrete progesterone, the two hormones being chemically related but physiologically completely different. Actually, the change from the secretion of estrogen to the secretion of progesterone may occur even before ovulation has taken place. Circumstantial evidence leading to that conclusion is that the majority of ovariectomized guinea pigs show psychological heat only if they are treated with progesterone before estrogen; neither hormone alone is capable of inducing heat. Since this species ovulates from six to 10 hours after the onset of heat, the assumption is that the progesterone necessary for the interaction with follicular estrogen comes from unovulated follicles. In rabbits progesterone is normally present in the blood just befor ovulation, and mouse ovaries can be caused to secrete progesterone before the follicles ovulate.

In moles luteinization of granulosa cells begins before ovulation, and in mares thecal luteinization can be seen at least in some preovulatory follicles. The African elephant shrew, *Elephantulus,* is unique in that follicles rupture as a *result* of the formation of corpora lutea—the follicular cavity is filled by proliferating, luteinizing granulosa cells, which literally push the ovum out of the antrum.

In most mammalian females formation of corpora lutea follows about the same pattern, but there is considerable variation among species. Immediately after ovulation the ruptured and puckered follicular cavity is filled with lymph and blood from broken thecal vessels. In most species postovulatory hemorrhage is minor and does not distend the collapsed follicle to its preovulatory size. In the sow, however, there is a considerable accumulation of blood and lymph; by the fourth, fifth, or sixth day after rupture the blood-filled follicle is much larger than it was before ovulation. In most species the corpus luteum is formed by hypertrophy and hyperplasia of the granulosa cells of the ovulated follicle. The theca interna cells may or may not participate in the formation of this gland; the role of these cells in the formation and maintenance of corpora is not sufficiently well understood to warrant detailed discussion here (see Harrison, 1948).

The easiest way to visualize the growth of the corpus luteum is to imagine a collapsed bag with nonelastic walls. Gradual thickening of the walls, due to hypertrophy and hyperplasia of the granulosa cells, eventually fills out the interior of the collapsed sac, obliterating its central cavity (see Figure 4-2, p. 129). In some species (rats, pigs, sheep) the central cavity is completely obliterated, leaving only a narrow central scar where the rapidly proliferating granulosa cells from the periphery meet. In other species (women) the growth of the corpus luteum stops short of filling the cavity; the corpus consists of greatly thickened walls and a substantial central cavity.

The initial increase in the size and weight of corpora lutea may be extremely rapid. In some species the corpus reaches 70 to 90 percent of its full size by day 3 after follicular rupture, but does not reach full growth until about day 13 of the 21-day cycle. In other species the rate of growth is somewhat more even. In sheep and cattle 50 to 60 percent of the full size of the corpus is reached by day 4; full size is reached in sheep on days 7–9 and in cows on about day 10. Similar rates exist in other mammals. The rate at which corpora enlarge is roughly correlated with the rate at which they secrete progesterone (see Figure 3-8, p. 77). In the pig progestational changes can be seen in the uterine epithelium and in the glands as early as day 1 after ovulation; maximal uterine changes have been induced by days 6–8 and are maintained until days 12–14 of the cycle by the progesterone from the corpora lutea.

In the pig the granulosa luteal cells of the corpus gradually increase in size until about day 13 of the cycle, when an abrupt and rapid decrease in size begins; this decrease continues through the follicular phase to the end of the cycle (days 19–21). On about day 14 or 15 the first regressive changes in the uterus also become apparent; these continue until the luteal phase ends and the follicular phase begins. In the cow the corpus is fully formed by the ninth day after ovulation and begins to regress on day 14 of the 21-day cycle.

It is probably safe to say that in all females with long cycles the luteal phase lasts a little longer than half the total cycle. Here again, however, much individual variation can be found. In some individuals the corpora lutea may be in perfect histologic condition well past the time they begin to regress in the majority; in such individuals the uterus also appears to remain under progesterone stimulation. It should be kept in mind, however, that in animals with long cycles there is a lag, probably one to three days, between the time the corpus begins to secrete its hormone and the time the effects of this hormone are observable in the uterine endometrium. Similarly, the corpus begins to wane long before the decrease in hormone secretion caused by this decline is reflected in the end organ. It is not safe to assume that waning corpora no longer secrete progesterone, but it is certain that the rate of secretion is greatly reduced.

In the great majority of animals the spent corpus luteum (called the corpus albicans) persists well into the luteal phase of the succeeding estrous cycle, during which new functional corpora lutea and much-regressed corpora albicantia coexist. As a rule, corpora albicantia are not macroscopically visible by the third cycle, but they may remain as histologically recognizable scar tissue a little longer. An interesting exception to this rule is the whale, in which degenerated corpora are said to persist in macroscopically and microscopically recognizable conditions for many years, possibly throughout the life of the animal.

During the histological decline of the corpus, luteal cells shrink, fibroblasts appear in increasing numbers, and large spindle-shaped cells are seen among the epithelial cells. Shortly before the end of the cycle the rapidly waning corpora lutea are composed largely of connective tissue, which eventually becomes hyalinized. The hyaline tissue becomes greatly reduced and finally disappears.

Corpora lutea are formed in a number of vertebrates other than mammals, e.g., in some viviparous snakes and lizards. In reptiles the formation of corpora lutea is not necessarily associated with viviparity, for it occurs also in oviparous species that retain the eggs in the oviduct.

Contrary to statements in some of the earlier literature, birds do not form corpora lutea; at least they do not form structures that are morphologically, histologically, or endocrinologically comparable to the luteal tissue of mammals or other vetebrates. It is possible, however, that the ovulated follicle, at least in chickens, plays a role in determining the time of oviposition.

Hormones Formed by Corpora Lutea

It has been shown repeatedly that corpora lutea contain estrogen, although of unknown origin. In some species the life span of corpora lutea may be prolonged significantly by injecting estrogen either systemically or directly into the corpus luteum. The mode of action of estrogen in prolonging the life span of corpora lutea is unknown; it may act by causing the pituitary gland to secrete prolactin (in rats) or a luteotrophic factor (in other species), or it may act by increasing the vascularity of corpora lutea, thus preventing a more rapid degeneration. In view of what we have said earlier concerning the possibility that LH may be a luteotrophic substance, the following observations are of considerable interest.

It has been found that either HCG or LH (but not FSH) injected into hypophysectomized rats caused vaginal cornification in the great majority of the treated rats. Even as long as $6\frac{1}{2}$ months after hypophysectomy, 10 out of 13 rats responded to HCG injection with vaginal cornification, which was maintained for 10 days. This suggests that both HCG and LH are luteotrophic—but in a surprisingly unusual way, by causing corpora lutea to secrete estrogen. Since the rats were hypophysectomized, the action of these hormones must have been directly on the corpora lutea. It is not yet possible to speculate intelligently on the mechanism that makes a corpus luteum secrete estrogen in the hypophysectomized rat and progesterone in the intact animal.

Some researchers have claimed that "LH" is luteolytic in both the cow and the rat. However, in the experiments that yielded this conclusion the LH preparations used were far from pure and, at least in the case of the rat, subsequent experiments with more highly purified LH failed to confirm this finding. In the intact rat LH is known to be responsible for the increased movement of cholesterol into the luteal tissue during formation of the corpus luteum, the amount of cholesterol stored being directly related to the amount of LH injected. At least in the rat this stored cholesterol is converted into progesterone under the action of prolactin.

The ability of corpora lutea to synthesize progesterone in vitro has been investigated in both pig and cow tissues. In both species it was demonstrated that corpora lutea do synthesize progesterone in vitro when stimulated by hormones. Significantly, in neither species was progesterone synthesized when prolactin was added to the substrate, but the addition of LH or HCG caused an increased rate of synthesis greater than that of control corpora lutea. The ability of corpora lutea to produce progesterone in vitro was found to be directly related to their age. Corpora lutea capable of synthesizing progesterone in vivo were found to be able to do so in vitro under the influence of the hormones mentioned. Those in late luteal phase, when the ability in vivo to produce progesterone is greatly diminished, were unable to produce progesterone in vitro.

Of some interest and possibly of significance is the finding that the uterus may secrete a humoral factor that, at least in vitro, has an effect on the ability of corpora lutea to synthesize progesterone. It was found that if epithelial scrapings from the uteri of pigs in the early luteal phase were added to corpora lutea cultured in vitro, the corpora's ability to produce progesterone was significantly enhanced. By contrast, scrapings from uteri in the late luteal phase depressed the rate of progesterone synthesis.

If a rabbit ovary containing only preovulatory follicles (no corpora lutea) is perfused in vivo with LH, HCG, or PMSG, progesterone is recovered in significant amounts from the ovarian vein. As in the case of cow and pig corpora cultured in vitro, the injection of prolactin does not cause an increase in progesterone synthesis. Once the infusion of hormones is discontinued, the ovary resumes production of estrogen. Even though this observation is obviously of great interest and significance, the rabbit is a somewhat unfortunate experimental animal for this demonstration. Because rabbit ovaries contain large masses of interstitial tissue, it remains unknown whether the recovered progesterone is produced by the interstitial tissue or the follicle cells. Preovulatory progesterone has been demonstrated in the peripheral blood of rabbits, guinea pigs, and women, all of whom have more or less abundant interstitial tissue in the ovaries. It was stated earlier that in many species (rat, pig, sheep, cow) psychological heat can be induced with significantly smaller doses of estrogen if the animals are previously primed with progesterone, and that in the guinea pig progesterone is essential for the manifestation of psychological heat. Thus, intuitively it seems that preovulatory progesterone is produced in all species, even in those without ovarian interstitial tissue. Whether this is the case remains to be seen.

Earlier (p. 68f.) we saw that prolactin is luteotrophic in the rat but not in other species, and thus we used the term *luteotrophic factor* in discussing all species other than the rat. In the preceding discussion we saw that LH-containing hormones can induce progesterone synthesis but that prolactin cannot. It is tempting therefore to say that LH is luteotrophic. However, at present we know only that LH has the ability to cause progesterone synthesis, and this may be its only role. Another, as yet unknown, substance may be needed to sustain corpora. Additional work is needed to determine whether LH functions as both a luteotrophin and progestenotrophin. At the same time, we should not ascribe luteotrophic effects to the humoral factor in uterine scrapings, since this factor may be only progestinotrophic.

Though all lutein tissue has basically the same structure, endocrine control, and function, there are certain special kinds, such as luteinized follicles, accessory corpora lutea, and corpora lutea of pregnancy, pseudopregnancy, and lactation. These are discussed elsewhere (Chap. 3).

ALTERNATION OF OVARIAN FUNCTION

We have already noted that in most birds only the left ovary is functional; both ovaries may be developed and functional in some of the hawks and owls. In all mammals both ovaries are developed, but each functions to different degrees. In the duck-billed platypus the right ovary is only about one-tenth as well developed as the left, and only the left ovary is functional. In Echidna the two ovaries are almost equally well developed. In some of the bats the left ovary is defective but occasionally functional. Most other mammals can be divided into right and left ovulators. This does not mean that they ovulate unilaterally, but that they normally ovulate more eggs from one ovary than from the other.

As a rule, the favored ovary is slightly heavier (even in immature females of the species), and the structures in it are somewhat larger. It has been thought that these differences may be due to differences in blood supply; in some species the right ovary is more liberally vascularized, in others the left. Experimental data on this point are not available.

In swine 55–60 percent of the corpora are found in the left ovary. In the mare 61 percent of the corpora are found in the left ovary, and the follicles in the left ovary are significantly larger than those in the right.

Sheep and cows are right ovulators; sheep ovulate from the right ovary 52–59 percent of the time, cows 60–65 percent. In rhesus monkeys 60 percent of the eggs are shed from the right ovary.

SELECTED READING

Baker, N. L., L. C. Ulberg, R. H. Grummer, and L. E. Casida. 1954. Inhibition of heat by progesterone and its effects on subsequent fertility in gilts. *J. Animal Sci.,* **13**:648.

Brambell, F. W. R. 1956. Ovarian changes. In *Marshall's Physiology of Reproduction,* Vol. I, Part 1, 3rd ed. Longmans.

Casida, L. E., and A. B. Chapman. 1951. Factors affecting the incidence of cystic ovaries in a herd of Holstein cows. *J. Dairy Sci.*, **34**:1200.

Garm, Otto. 1949. A study on bovine nymphomania. *Acta Endocrinologica*, Suppl. 3.

Hansel, H., and G. W. Trimberger. 1951. Atropine blockage of ovulation in the cow and its possible significance. *J. Animal Sci.*, **10**:719.

———. 1952. The effect of progesterone on ovulation time in dairy heifers. *J. Dairy Sci.*, **35**:65.

Harrison, R. J. 1948. The development and fate of the corpus luteum in the vertebrate series. *Biol. Rev.*, **23**:296.

———. 1948. The changes occurring in the ovary of the goat during the estrous cycle and in early pregnancy. *J. Anat.*, **82**:21.

Nalbandov, A. V. 1961. Mechanisms controlling ovulation of avian and mammalian follicles. In *Control of Ovulation,* edited by C. A. Villee. Pergamon Press.

Rowlands, I. W. 1956. The corpus luteum of the guinea pig. *Ciba Foundation Colloquium on Ageing,* **2**:69.

Seyle, H. 1946. The ovary. In *Encyclopedia of Endrocrinology,* Section IV, Vol. 7. Richardson, Bond, and Wright (Montreal).

The following color films show the processes of ovulation and ova transport. They are highly recommended.

Physiology of Reproduction in the Rat. R. J. Blandau. Available from Educational Films, University of Washington, Seattle, Washington.

The Formation of the Avian Egg. D. C. Warren and H. M. Scott. Available from the Poultry Department, Kansas State College, Manhattan, Kansas.

6

Hormones
of Reproduction

Brian Cook

Many hormones are involved in the coordination of the complex process of reproduction. Releasing factors (RFs), small peptides produced within the neurones of the hypothalamus, serve as a link between the nervous and endocrine systems. The target organ for RFs is the adenohypophysis, where glycoprotein gonadotrophic hormones originate. Prolactin, a protein, is also produced in the adenohypophysis. The neurohypophysis releases the octapeptide oxytocin, synthesized in the hypothalamus. The ovaries and testes, stimulated by the gonadotrophins, synthesize and release steroid hormones, which are lipids derived from cholesterol. Steroids are largely responsible for stimulating development of the reproductive systems and secondary sex characters. They modulate their own rate of production through the hypothalamus, which monitors their concentration in the blood. This information is one of the controlling influences on the output of hypothalamal RFs. The hypothalamal-hypophyseal-gonadal axis is a sort of amplification system—the output from one component catalyzes hormone production in another. The prostaglandins are a group of lipids derived from essential fatty acids. They exert profound effects on reproductive processes, but they do not always seem to meet the classic criterion of a hormone—a substance carried in the blood from a secretory gland to a site of action. Some prostaglandins appear to be synthesized in the tissues in which they act.

Many hormones have multiple functions. For example, prolactin influences the gonads as well as mammary tissue in females and prostatic tissue in males. Oxytocin stimulates smooth-muscle contraction in both the reproductive tract and the mammary glands. Since hormones are chemically so diverse, and since they influence so many different tissues, the biochemical mechanisms of hormonal action are equally diverse. Steroid hormones appear to control cellular activity by direct influence on the nucleus, whereas protein and peptide hormones influence cytoplasmic biochemical processes. It is still too early to claim this as an invariable rule. Scarcely anything is known about the mechanism of action of some compounds, e.g., prostaglandins.

INTRODUCTION

The hormones involved in reproduction originate in three principal structures: the hypothalamus, the pituitary, and the gonads. The hypothalamus assesses the internal hormonal state of the animal and receives input from exteroceptive stimuli. The appraisal and collation of this information results in the production of releasing factors (RFs) that are secreted into the portal blood. The pituitary gland receives these blood-borne hypothalamal signals and amplifies them by releasing appropriate trophic hormones. The ovaries or testes are stimulated by gonadotrophins to release steroid hormones, which act on the reproductive tract to control its function, on the hypothalamus, and perhaps on the pituitary gland as part of a feedback system. Thus, an endless hormone chain is established, each link influencing the next. Other mechanisms are superimposed on this basic structure to fulfil specific functions. For example, oxytocin is released from the neurohypophysis in response to neural stimuli for milk letdown; prostaglandins may be produced in uterine and other tissues to act on smooth muscle at parturition and possibly on corpora lutea at regression. Some of the characteristics of these various substances will be examined in this chapter.

RELEASING FACTORS

The history of the isolation and identification of RFs is typified by false starts and discarded theories. In the late 1950s compounds with trophic-hormone-releasing properties were isolated from areas of the brain. Early structural investigations of these compounds revealed them to be peptides that were similar to fragments of various pituitary hormones in their amino acid sequences. As the techniques for measuring releasing activity became more sophisticated it was realized that these peptides were not true RFs, but may simply have been contaminated with the active agents. Subsequently, certain amines related to the biogenic amines (putrescine, spermidine, etc.) and even some metal ions were shown to have releasing activity in various test systems, so the validity of the earlier work was seriously questioned.

The first RF to have its structure elucidated in a convincing manner was the thyrotrophin-releasing factor (TRF). Guillemin's group, working on material derived from sheep hypothalami, and Schally's group, working with pig hypothalami, showed in 1969 that TRF was the tripeptide L-pyroglutamyl-L-histidyl-L-proline (Figure 6–1). The structure of TRF has been confirmed by a successful synthesis of biologically active material. Since only 4.4 mg of porcine TRF could be isolated from 165,000 pig hypothalami, the availability of synthetic material facilitated continued physiological study.

Other hypothalamal factors with trophic-hormone releasing activity have been identified. In 1971 Schally's group announced the structure of a decapeptide that has the capacity to release both LH and FSH (see Figure 6–1). This compound, LRF-FRF, is apparently more scarce in the hypothalamus than TRF: only 830 μg was isolated from 165,000 pig hypothalami weighing 2.5 kg (compared to 4.4 mg TRF from the same quantity). The structure of LRF-FRF has also been confirmed by synthesis. Of great physiological interest is the fact that this one hypothalamal factor releases two gonadotrophic hormones. Because FSH and LH are present simultaneously, the usefulness of separating physiological responses to these two hormones has been questioned in the past. Recent investigations of hormone concentrations in plasma during estrous and menstrual cycles have shown that LH and FSH have closely similar temporal secretion patterns, so that whatever differences do exist are of contentious significance. It is still possible, however, that

184

(pyro)Glu——His——Trp——Ser——Tyr——Gly——Leu——Arg——Pro——Gly——NH₂

LH- and FSH-releasing hormone (LH-RH/FSH-RH or LRF/FRF)

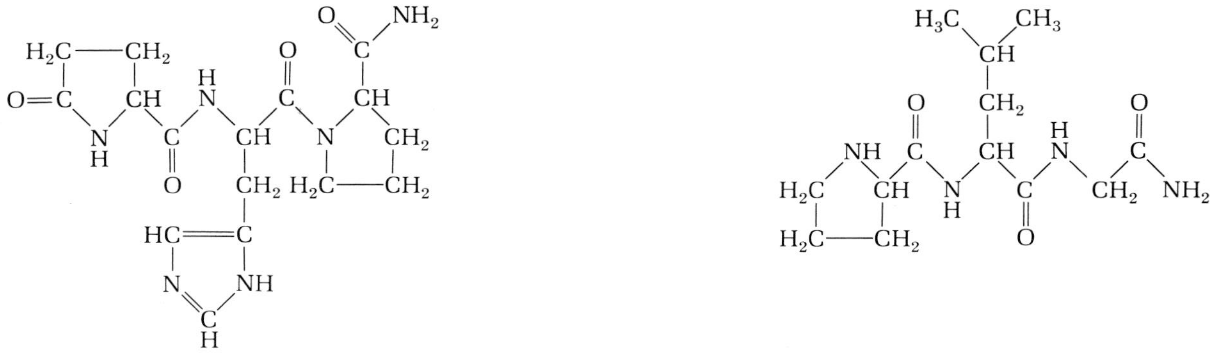

(pyro)Glu———His———Pro—NH₂
TSH-releasing hormone (TRH or TRF)

Pro———Leu———Gly——NH₂
MSH release-inhibiting hormone (MRIH or MIF)

Figure 6-1 Hypothalamal releasing and inhibiting factors.

other hypothalamal factors exist that influence the separate release of LH and, more particularly, FSH.

Melanophore-stimulating hormone (MSH) is a product of the pars intermedia of the pituitary gland. The release of MSH appears to be controlled by an inhibiting factor, MRIH or MIF, whose structure has been determined by Nair, Kastin, and Schally; like TRF, it is a tripeptide (see Figure 6-1). Prolactin secretion in mammals appears to be controlled by an inhibiting factor, the nature of which is unclear. In birds peripheral prolactin concentrations may be controlled by both releasing and inhibiting factors; the relative importance of these hypothalamal factors could differ between migratory and nonmigratory species. In general, it appears that all hypothalamal releasing and inhibiting factors will prove to be small peptides with short, unusual sequences of amino acids.

TROPHIC AND PEPTIDE HORMONES

The anterior lobe of the pituitary gland secretes the glycoproteins TSH, FSH, and LH as well as the proteins (or peptides) growth hormone, prolactin and adrenocorticotrophic hormone (ACTH). Structures other than the pituitary gland also produce gonadotrophins. The pregnant mare has specialized endometrial cups that produce a compound that is secreted into the bloodstream. This substance has principally FSH activity and is referred to as pregnant mare's serum gonadotrophin (PMSG or PMS). The fetal trophoblast in women produces a gonadotrophin that has LH activity and is generally referred to as human chorionic gonadotrophin (HCG). Various other species are thought to produce similar compounds. PMSG and HCG are widely used experimentally because they are more readily available than gonadotrophins from pituitary sources.

Much is known about the structure of the glycoproteins TSH, FSH, LH, PMSG, and HCG, which have many chemical similarities. Each consists of two chemically dissimilar subunits, designated α and β chains. The α and β chains are most dissimilar in LH and most similar in FSH. Either chain alone has little biological activity, but when the two are allowed to recombine, activity is restored. Furthermore, because the α chains of TSH, FSH, PMSG, and HCG are similar, hybrid molecules such as TSH-α/LH-β can be produced. The biological activity of such hybrids is determined by the β chain. Thus, TSH-α combined with LH-β behaves biologically like LH, and LH-α combined with TSH-β behaves like TSH. The α chain of HCG has been combined with the β chain of TSH to give a compound exhibiting TSH activity. (It has also been combined with the β chain of LH to give LH activity, but of course HCG and LH normally have similar biological effects.)

The pituitary glycoproteins all have molecular weights in the vicinity of 28,000 daltons; each subunit is approximately half the molecular weight of the parent compound. HCG seems to be a much larger glycoprotein, with a molecular weight of about 45,000 daltons, but it contains many more sugar residues than the pituitary glycoproteins. The characteristics of isolated HCG tend to be less uniform than those of the pituitary hormones because degradation, especially of the carbohydrate side chains, can take place during urine production. HCG-α appears to have a molecular weight of 18,000 daltons, and HCG-β a weight of 28,000 daltons. The two chains probably do not differ greatly, in the number of amino acid residues, although HCG-β appears to be much richer than HCG-α in sugar residues. A similar situation exists for FSH, i.e., the β chain has more sugar residues than the α chain. LH and TSH do not show profound differences with regard to the number of sugar residues on each chain. Details of the structures of the α and β chains of PMSG have yet to be elucidated.

Determining the structure of the pituitary glycoproteins is complicated by the fact that genetic polymorphisms appear to exist in the amino-acid sequences for each hormone. Consequently, in addition to the problems associated with hydrolysis of sugar side chains during isolation of the hormones, each preparation of a hormone or subunit represents a population of closely similar rather than identical amino-acid sequences. Thus, structural differences in these hormones may exist within as well as between species.

The similarities of these glycoprotein hormones raises interesting physiological and evolutionary questions. Are the different chains of one hormone produced in different pituitary cells? Do single chains circulate in plasma in normal or pathological conditions, or are they always paired? Is some of the polymorphism seen in the individual hormones due to dissociation and reassociation with hybridization during the purification process? If so, unequivocal determination of structure will be difficult.

It has been suggested for some time that the genes delineating LH, FSH, TSH, and possibly HCG evolved along similar pathways. This idea first occurred when it became known that all four hormones have a high content of both proline and cystine. Now the idea of a common origin has been strengthened by the knowledge that each hormone has one chain in common with each of the others. Perhaps the two-chain structure arose as a result of partial duplication of the genes for the α chain. Differences in this duplication process and the subsequent evolution of the duplicated chain would then give rise to different β chains.

The hormones of the pituitary gland provide several examples of the utilization of similar amino acid sequences in hormones of differing

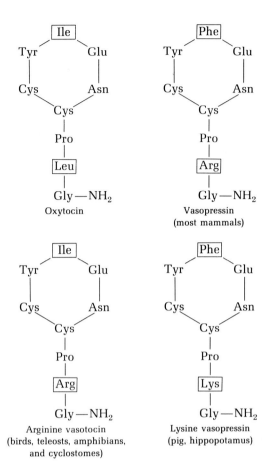

Figure 6–2 Hormones of the neurohypophysis. Boxed amino acids are points of substitution in the molecule.

function. For example, the neurohypophyseal hormones oxytocin and vasopressin differ in two amino acid substitutions in the same octapeptide (Figure 6–2). Oxytocin initiates milk let-down from the mammary gland, stimulates contractions in uterine smooth muscle at parturition, and is released during sexual stimulation, especially at orgasm (when it may assist sperm transport to the site of fertilization by stimulation of uterine muscle activity). Vasopressin is the antidiuretic hormone; it shows species differences among mammals. In other vertebrates the antidiuretic hormone is arginine vasotocin, a structure combining the oxytocin ring with the vasopressin side chain (Figure 6–2). Many synthetic analogs of these hormones have been made in an effort to understand their mechanism of action. The single-chain compounds ACTH and MSH (whose structures have been completely elucidated) share common amino acid sequences although their functions differ completely. Growth hormone and prolactin also show extensive sequential homologies in spite of their functional divergence.

Of the pituitary protein hormones prolactin is the most important for reproduction, although of the hormones participating to a major degree in reproductive processes its physiological roles are probably least well-defined. Although prolactin is named for its role in the stimulation of mammary tissue and milk secretion, it is also active in controlling ovarian function in a manner that is not yet clear. It is luteotrophic in the rat and seems to stimulate cholesterol accumulation in luteal tissue so that progesterone synthesis may ensue. In some species prolactin acts synergistically with androgens in stimulating prostatic growth. It may also stimulate other glands of the male reproductive system, e.g., seminal vesicles and preputial glands. The separate identities of prolactin and growth hormone in man were not clear for some time, and it was suspected that these two hormones were chemically the same. Human prolactin has now been separated from growth hormone; both compounds consist of a single chain of about 200 amino acids and have molecular weights in the region of 22,000 daltons. Prolactin in plasma can be determined by bioassay, by neutralizing growth hormone with a specific antibody. This technique has shown that the ratio of prolactin to growth hormone differs in various physiological conditions and, significantly, that prolactin concentrations are high in the plasma of lactating women. Growth hormone has been shown to possess lactogenic activity of its own, and this goes some way towards clarifying the earlier confusion.

Pregnancy Tests

Early diagnosis of pregnancy in women depends on the demonstration of HCG in urine. In fact, HCG can be demonstrated about the time of menstrual failure, since it is excreted very soon after conception. About one month after it is detectable in the urine it reaches a maximum titer (about 3 mg/day); a rapid decline follows, and the level remains low and constant from about the twelfth week to parturition. The function of HCG is probably to maintain the corpus luteum and its secretion of progesterone. By the time the concentration of HCG falls, placental steroid production has reached a high level.

The detection of HCG in urine was originally carried out by bioassay. In 1928 Aschheim and Zondek showed that immature mice injected with pregnancy urine would ovulate (A–Z test). In 1929 Friedman showed that virgin rabbits injected intravenously with pregnancy urine developed ovarian hyperemia. Later assays used frogs or toads, which upon treatment with pregnancy urine released sperms or ova.

These biological tests have now been replaced by immunologic techniques, which are considerably faster and more sensitive. Selye observed in 1934 that animals repeatedly exposed to HCG lose their sensitivity to the hormone. This loss of sensitivity was subsequently shown to be due to antibodies to the hormone (antihormones) in the blood of the test animals. By 1960 immunoassays became commercially feasible and widely available. The most common test is based on inhibition of agglutination. Sheep erythrocytes or latex particles are coated with HCG; when the coated agents are exposed to anti-HCG they agglutinate, that is, they form clumps and precipitate. Urine to be tested for HCG is mixed with antiserum specific for HCG, and the mixture is tested with the coated agents. If the urine contains HCG, agglutination is prevented because the antibody reacts with the HCG in the urine and therefore is not available to agglutinate the coated cells or particles. Conversely, urine lacking HCG allows a precipitate to form, indicating nonpregnancy. Immunologic tests such as this have been considerably developed and refined. Now, isotopic forms of hormones are complexed with antiserum, enabling measurements to be made by radioimmunoassay, a highly sensitive and precise technique. Radioimmunoassay permits assay of concentrations in plasma, whereas urinalysis only permits inferences of plasma hormone levels.

Radioimmunoassay

The principle of radioimmunoassay (RIA) is straightforward. The assay is based on the competition of labeled and unlabeled hormone for binding sites on antibody molecules. Labeled hormone, added in excess, can fill all binding sites. If unlabeled hormone is present, it competes with radioactive material for binding sites and less label can be taken up. Thus, the greater the mass of unlabeled hormone, the less the binding of label. When unlabeled hormone is absent, all binding sites must be occupied by radioactive hormone. When an excess of unlabeled hormone is present, few sites (essentially none) can be occupied by radioactive hormone. Thus, a curve produced by plotting the mass of unlabeled hormone against the amount of label bound has the form shown in Figure 6–3. Once such a curve has been derived for known quantities of unlabeled hormone, biological test samples can be interpolated.

The procedure for RIA involves several steps. First, antibodies specific for a hormone, e.g., HCG or LH, are obtained by repeatedly injecting the hormone into an animal, usually a guinea pig, rabbit, or sheep. The

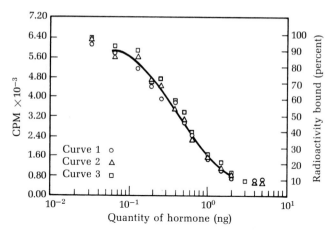

Figure 6–3 A typical radioimmunoassay standard curve showing the relationship between the amount of untreated hormone added to the reaction tube and the amount of radioactivity remaining bound to the antibody.

antibodies (known as antihormones) accumulate in the animal's blood, from which serum (known as antiserum) is prepared. A sample of hormone is then made radioactive, usually by attaching radioactive iodine (^{131}I or ^{125}I). All tubes in the assay contain the same mass of antibody and the same excess of labeled hormone, i.e. the number of binding sites is fixed and sufficient label is present to fill them all. Some tubes contain known amounts of unlabeled hormone arranged incrementally to give a standard curve as shown in Figure 6–3, whereas other tubes contain samples in which the mass of hormone is to be determined by comparison with the standards. The tubes are left to equilibrate; during this process the labeled and unlabeled hormone fill the binding sites in the ratio of their masses within individual tubes. At the end of the equilibration period labeled hormone bound to the antibody must be separated from that remaining free in solution. This separation is a major technical problem because the antiserum is generally so dilute that no precipitation occurs when the hormone–antihormone complex is formed; all radioactivity remains soluble.

Many techniques have been devised to separate bound and free radioactivity; "solid-phase" and "second-antibody" techniques are commonly used. In the *solid-phase technique* antibody is coated onto a suitable support, which may be some sort of particle or may be the assay tube itself. After the equilibration step bound label is attached to the support and the free label can be decanted. Either or both free and bound label can be determined. The *second-antibody technique* uses an antibody that reacts with the original antihormone to cause precipitation. In general, antihormones are gamma globulins, so that if an antihormone is prepared in rabbits an anti-rabbit gamma globulin prepared in a heterologous species, e.g., sheep, will react with the rabbit antihormone. Precipitation is facilitated because the antihormone-anti-gamma globulin complex is large, and the addition of a second antibody increases protein concentrations. Bound hormone remains associated with the antihormone through the precipitation step. The process is illustrated in Figure 6–4. Centrifugation facilitates separation of precipitate (bound label) and supernatant solution (free label), and either can then be determined.

For RIA to be successful, the antiserum should be highly specific for the hormone being assayed, otherwise there is a danger that more than one hormone will be estimated. This rigorous requirement can be relaxed if the labeled sample of hormone is extremely pure. All the radioactivity involved in the assay will then be associated with the hormone of interest; other antigen-antibody systems that may be present will not interfere because no radioactive label will be displaced in their reaction. The antiserum must have high avidity (strong binding) for its hormone antigen and a high titer is desirable; antisera are usually used at dilutions between 1:10,000 and 1:100,000, but some work at even greater dilutions.

The labeled hormone should behave towards its antibody in exactly the same way as unlabeled hormone. This may not happen if the iodine atom is so large relative to the hormone molecule that the configuration of the hormone is distorted. The possibility that biologic and immunologic activities are not equivalent should also be considered. The hormone sample may contain denatured hormone that is immunologically active or small hormone fragments that have no biologic activity but which can still bind with the antibody. Although caution is necessary in the use of RIA, the advantages of specificity and sensitivity have enabled great progress to be made in endocrine studies.

Radioimmunoassay for pituitary hormones is of prime importance in the study of releasing factors. The biological activity of RF preparations is tested in vivo or in vitro by assessing the amount of pituitary hormone released. Tests can be conducted in vivo by injecting the suspected RF into the carotid artery and assessing the appropriate pituitary-hormone concentration in the jugular vein. Tests can also be conducted in vitro by estimating the release of hormone from fragments of pituitary glands incubated in the presence of RF preparations. Experiments of this type enable the potency of suspected RFs to be determined, but they do not probe the *mechanism* of RF action. In particular, they shed no light on the question of whether RFs directly cause synthesis as well as the release of pituitary hormones.

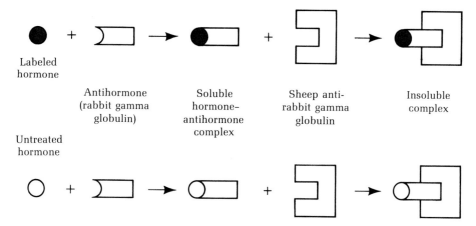

Figure 6-4   Diagram of the reaction sequence in a second-antibody radioimmuno-assay.

The specificity of an antibody for a particular pituitary hormone is a factor that has to be demonstrated. We know that the pituitary glycoprotein hormones consist of two chains, that the α chains of LH and TSH are very similar, and also that the chains of FSH are like LH-α. Thus, it is possible for an antibody to cross react with all these compounds, especially if it is directed towards the α chain. However, many antibodies are directed towards the β chain, and this enables us to discriminate between LH and TSH. The α chains seem to be weaker antigens than the β chains, but why this is so is unclear. FSH is a weak antigen— at least, antibodies against FSH are more difficult to prepare than antibodies against LH. Since the two FSH chains closely resemble one another, and since these chains resemble LH-α rather than LH-β, the same factors that make LH-α a weaker antigen than LH-β may also be responsible for the lower antigenicity of FSH.

The same features that create problems in specificity can be used to advantage in research, since antibodies to individual glycoprotein *chains* can be prepared. Such antibodies are presently being used in the study of the biology and chemistry of the individual chains constituting the intact hormone molecules. However, a second problem of specificity is that structural differences in the same hormone may exist between species. Thus, antibodies prepared against a hormone isolated from one species may not react with the corresponding hormone of a different species. For example, an antiserum against ovine LH may not react with porcine LH because the antigenic determinants in ovine LH utilized by the immunized animal to produce antibodies are lacking in porcine

LH. For convenience in RIA, cross reactions should be weak between different hormones within the same species and strong in the same hormone for different species.

Antibodies are not formed against small molecules, but such molecules can be rendered antigenic by chemically attaching them to large molecules as haptens. Antibodies can then be made against the entire complex. Typically, antibodies are made against steroids by combining the steroid molecules with bovine serum albumen (BSA) and then inject-ing the steroid-BSA complex as an antigen. Although antibodies to the entire complex are produced, the antibody also reacts with steroids not coupled to BSA because the steroids represent specific antigenic sites. Such an antibody can be used for physiological studies of steroid func-tion or for assay purposes. More antibody specificity may be obtained by using steroid derivatives in which the steroid-protein bond does not interfere with those areas of the steroid ring system that help to confer its uniqueness, that is, carbon atoms 3, 4, 11, 16, 17, 20, and 21 (see the next section).

The techniques for assaying steroids are the same as those used for protein hormones. Radioactive labeling is easier; tritiated (^{3}H) steroids with a very high specific activity can be obtained, so the problem of de-forming the molecule with the label is less severe. Separation of bound steroid from free steroid in solution is also somewhat easier. Free steroid can be adsorbed from solution with a surface-active agent such as char-coal. This leaves the antibody with its bound steroid in solution. The technique of making hapten antibodies is versatile. Antibodies against such compounds as cyclic AMP, prostaglandins, heroin, tetrahydro-cannabinol, and some vitamins have been prepared for use in immunoas-say. Now that the structures of some RFs are known and synthetic material is available, hapten antibodies against such compounds have become available for physiological use.

GONADAL HORMONES

The gonadal hormones are steroids, compounds based on a skeleton of four fused rings. Figure 6–5 shows how this ring system is conventionally illustrated; the three six-membered rings are designated A, B, and C, and the five-membered ring is designated D. Each point represents a carbon atom, numbered as shown in the figure; hydrogen atoms, which by convention are not shown, are present to satisfy the carbon valency of four. Methyl groups project from carbon atoms 10 and 13; again, by convention the hydrogens are not shown, and only a heavy projecting line indicates the carbon atom. The side chain projecting from carbon atom 17 (C-17) may be more complicated than that shown in Figure 6–5, but in gonadal steroids it does not contain more than two carbon atoms.

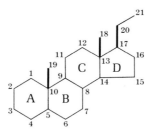

Figure 6–5 The conventional representation of the steroid ring system, showing the lettering system used to designate the rings and the numbering system used to identify the carbon atoms.

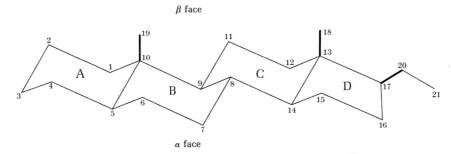

β face

α face

Figure 6–6 Perspective view of the steroid ring system, showing the β face of the molecule (above) and the α face (below).

A closer approximation to the shape of the ring system is illustrated in Figure 6-6. The molecule is a wrinkled plate, with the methyl groups at C-10 and C-13 and the side chain at C-17 projecting upwards. The upper face is called the β face; β substituents are indicated by heavy lines. The lower face is the α face; α substituents are indicated by dotted lines. It is believed that protein-steroid interactions occur at the β face.

Probably the easiest way to appreciate the structures of the different steroid hormones relative to one another and to their parent compound, cholesterol, is to examine the biosynthetic pathway outlined in Figures 6–7 and 6–8. Acetic acid, in the form of acetyl-CoA, is the precursor of cholesterol and the steroid hormones. First, three molecules of acetyl-CoA condense and are reduced to mevalonic acid, which contains six carbon atoms. Mevalonic acid is phosphorylated and decarboxylated to yield the immediate precursor of the sterols (and other terpenes), isopentenyl pyrophosphate, which has five carbon atoms. Three of these five-carbon units are condensed to give a 15-carbon unit, farnesyl pyrophosphate; the union of two of these molecules produces a 30-carbon unit, squalene. This hydrocarbon is cyclized to give lanosterol, which loses three methyl groups and is reduced to cholesterol with 27 carbon atoms. (In Figure 6–7 the carbonyl carbon of acetate is marked

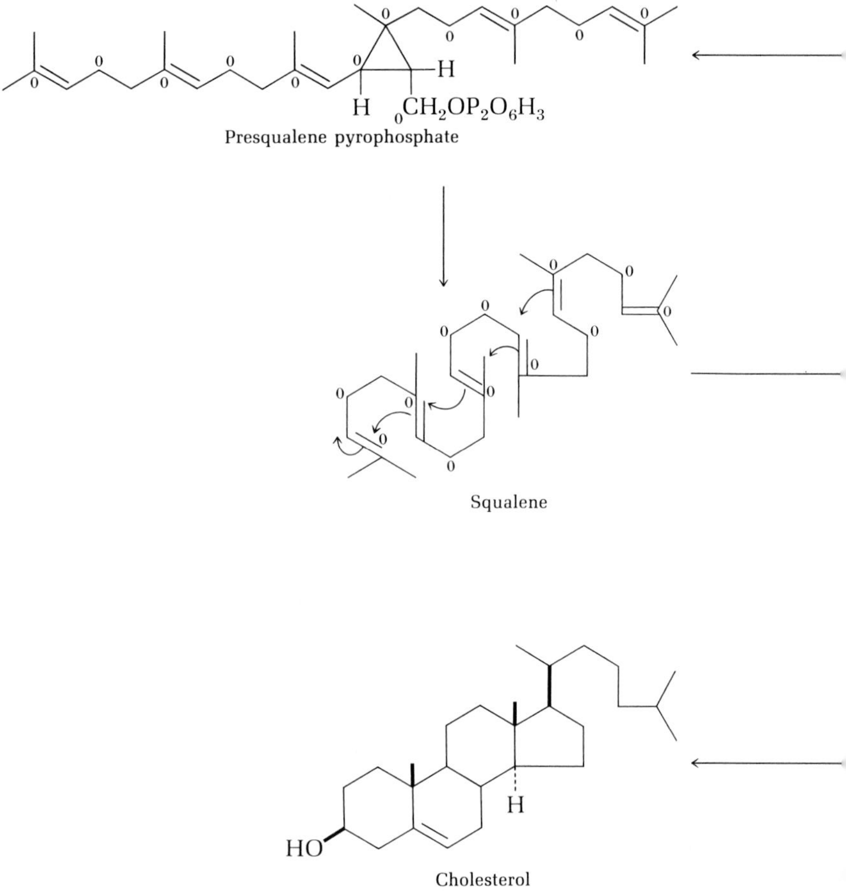

Presqualene pyrophosphate

Squalene

Cholesterol

Figure 6–7 The biosynthesis of cholesterol.

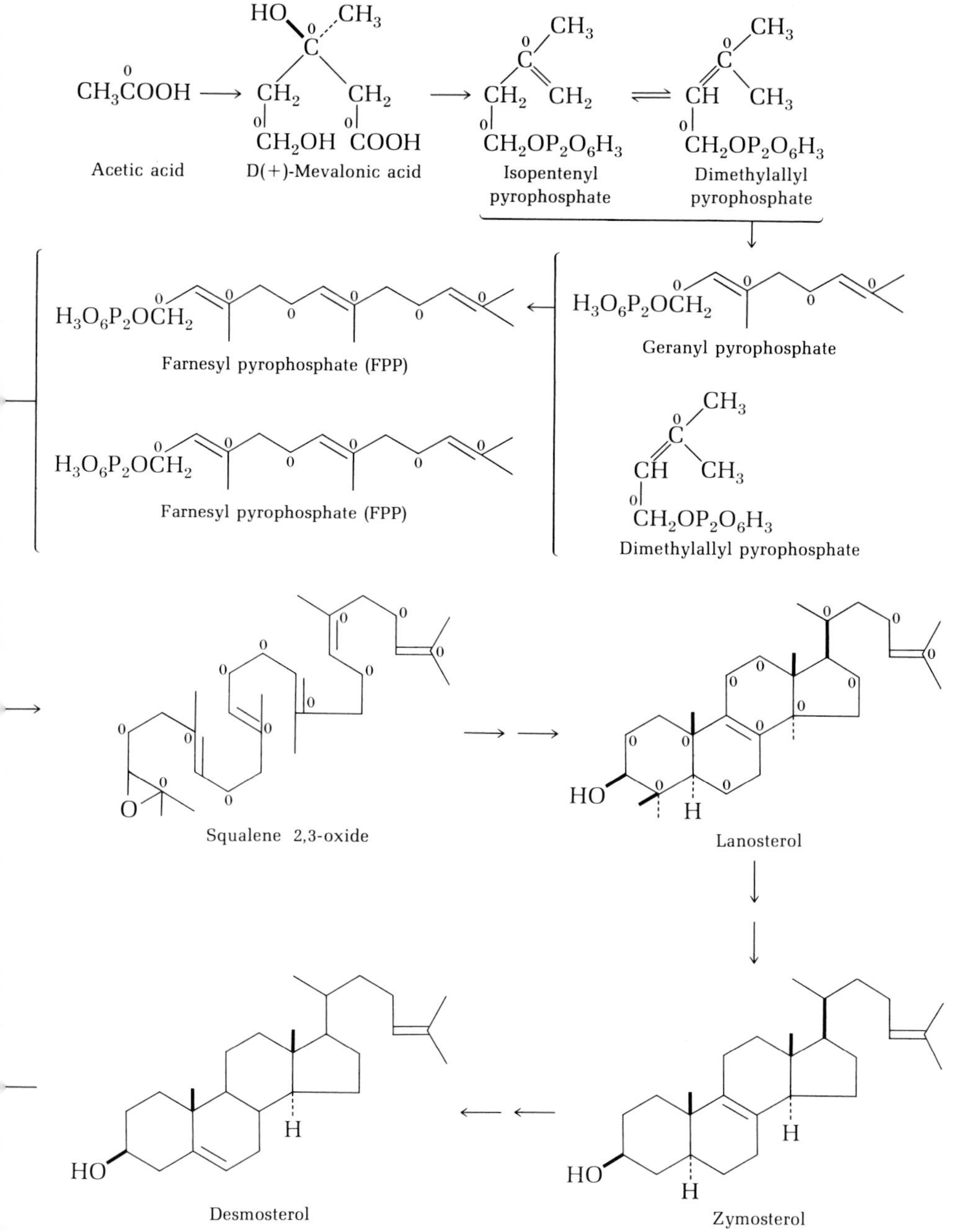

Acetic acid   D(+)-Mevalonic acid   Isopentenyl pyrophosphate   Dimethylallyl pyrophosphate

Farnesyl pyrophosphate (FPP)

Geranyl pyrophosphate

Farnesyl pyrophosphate (FPP)

Dimethylallyl pyrophosphate

Squalene 2,3-oxide

Lanosterol

Desmosterol

Zymosterol

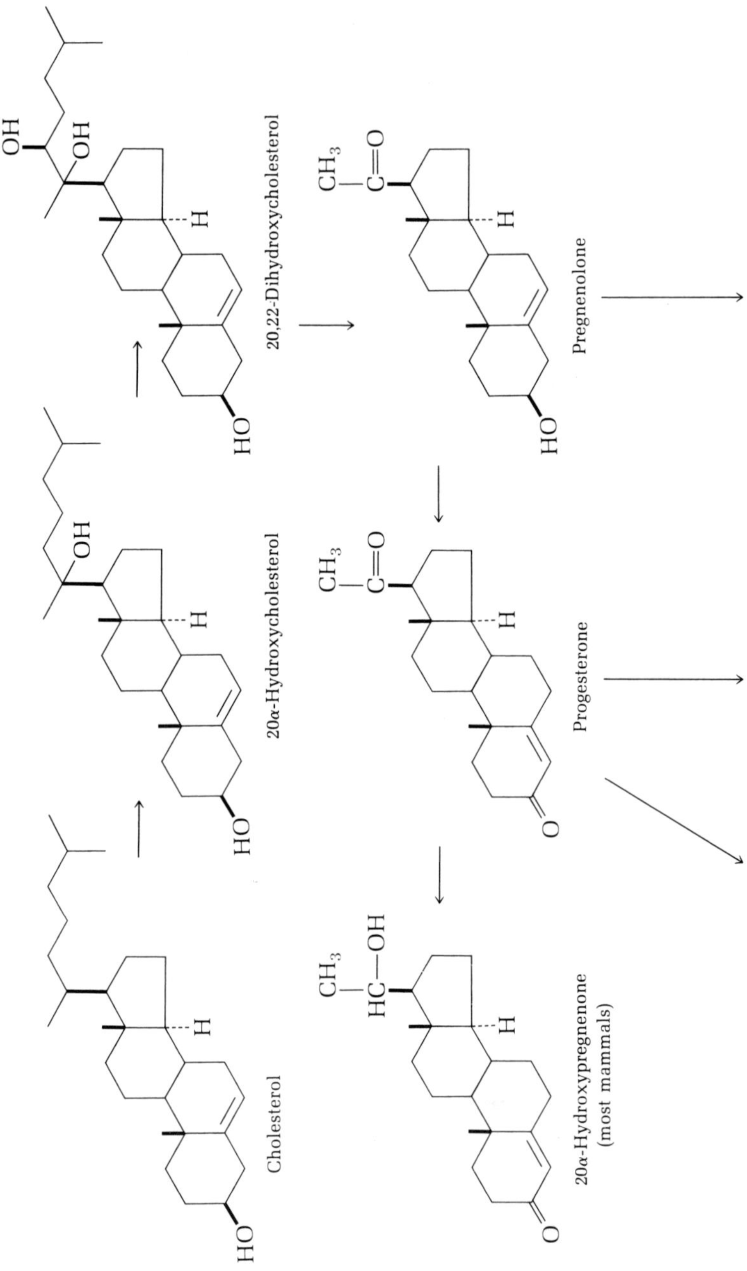

20,22-Dihydroxycholesterol

Pregnenolone

20α-Hydroxycholesterol

Progesterone

Cholesterol

20α-Hydroxypregnenone
(most mammals)

199

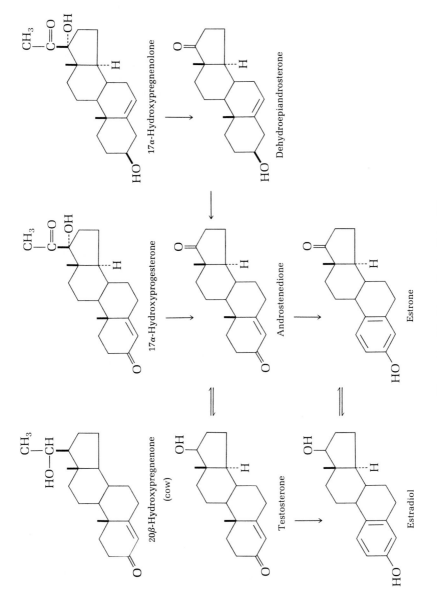

Figure 6–8 The biosynthesis of steroid hormones from cholesterol.

with an open circle so that the origin of the carbons in the steroid skeleton can be seen.) Figure 6–8 shows that hydroxylation of cholesterol precedes cleavage of the side chain to give the first C_{21} compound in the pathway, pregnenolone. Oxidation of pregnenolone gives the first hormone, progesterone, the progestational steroid. Hydroxylation of progesterone at C-17 precedes removal of the side chain from the C_{21} steroids to give the androgens, which are C_{19} compounds. Testosterone, a potent androgen, and androstenedione, a weaker androgen, are the major hormones produced by the testis; dehydroepiandrosterone, a very weak androgen, is secreted in considerable quantities by the adrenal gland in the form of a sulfate. Aromatization of the A ring with the removal of the methyl group at C-10 gives the estrogens. Estradiol is the most potent naturally occurring estrogen and is a major product of the ovary.

Thus, it is apparent that the ovary produces the male hormones testosterone and androstenedione as intermediates in estrogen synthesis, although these androgens are normally secreted in only minute quantities in the female. The testis is also able to synthesize estrogens, but normally only small quantities are released in humans. This is not true for all species. For reasons that are obscure the testes of boars and stallions produce large quantities of estrogens. In fact, stallion urine is one of the richest known sources of estrogens. The horse is further unusual in that it produces estrogens that are aromatic or unsaturated in the B ring as well as aromatic in the A ring (equilenin, equilin), especially during pregnancy.

The adrenal cortex is the most versatile steroid-producing gland. In addition to dehydroepiandrosterone and small quantities of other androgens, estrogens are produced in trace amounts, but hydroxylation of steroids at C-11, C-18, and C-21 yields the major products of the tissue, glucocorticosteroids and mineralocorticosteroids (C_{21} compounds), which the gonads normally are unable to synthesize.

The energy flow during biosynthesis of steroids is of interest. Rupture of acetyl-CoA bonds provides energy for the synthesis of hydroxymethylglutaryl-CoA, each molecule of which is reduced by 2NADPH to give mevalonic acid. Each molecule of mevalonic acid requires three molecules of ATP to give isopentenyl pyrophosphate; cleavage of these pyrophosphate bonds then provides energy for the polymerization of six five-carbon units to give squalene. The cyclization of squalene requires a mixed-function oxidase enzyme system, which utilizes oxygen from the air and NADPH to complete the reaction. In general, these reactions can be summarized as follows:

$$RH + NADPH + H^+ + O_2 \longrightarrow ROH + NADP^+ + H_2O$$

RH represents the molecule before and ROH the molecule after hydroxylation. One atom of the oxygen molecule appears in the sterol while the other is reduced to water. All the hydroxylations of the steroid ring system appear to be of this type, except that in the case of squalene cyclization a stable intermediate, squalene-2, 3-oxide, precedes the hydroxylated, cyclized product. All the methyl groups are removed oxidatively with the intervention of a mixed-function oxidase system, and this makes the pathway essentially irreversible. Estrogens cannot yield androgens; androgens cannot yield progestins. The only reactions that are reversible are those catalyzed by dehydrogenase enzymes, i.e., the conversion of androstenedione to testosterone and estrone to estradiol.

Although the gonads and the adrenal gland can synthesize steroids using acetate as the major precursor, they apparently do not always do this. Evidence is good that both the adrenal gland and the ovary utilize cholesterol derived from the blood, and produced mainly in the liver, as the starting material for hormone synthesis. The testis, on the other hand, appears to produce steroids largely from acetate. Although the reasons for this are obscure, unique features of testicular vasculature may influence the movement of blood-borne material. The pampiniform plexus serves as a thermoregulator, lowering the temperature of scrotal testes relative to the body (see p. 41ff.). In passing through the tortuous vessels of the plexus, not only is the blood cooled but also the arterial pulse is damped. How these factors influence the transfer of metabolites from blood to testicular cells is not known, but selective permeability in the testis can be illustrated by three phenomena. First, vessels of the pampiniform plexus show a particular sensitivity to blood-borne cadmium, which causes their complete destruction. Only blood vessels of the placenta show similar behaviour. Second, blood does not enter testicular tubules; the only access to these structures is through the Sertoli cells. This "blood-testis" barrier serves to protect the constituents of sperm and semen from immunologic attack. (Such protection is necessary because the constituents of sperm and semen do not develop until puberty, long after the organism's self-recognition mechanism has been established. Were they not protected, they would be regarded as foreign bodies and an immunologic attack would be mounted against them). Third, there is selective absorption of natural constituents of tubular fluid in the epididymis (see p. 48f.). Similar barriers to metabolite movement (as yet undiscovered) could limit the availability of cholesterol and favor the utilization of acetyl-CoA, produced within the Leydig cells, for androgen biosynthesis.

A curiosity of the biosynthetic pathway is that progesterone is reduced at C-20 in the ovary. Rats, rabbits, sheep, pigs, and women produce 20α-hydroxypregnenone (also called 20α-dihydroprogesterone, and incorrectly 20α-hydroxyprogesterone) but the cow appears to be the only species that produces the alternative isomer, 20β-hydroxy-pregnenone (see Figure 6–8). The role of these compounds is unknown, nevertheless their progestational activity is very weak. Some species, rabbits for example, produce large quantities of 20α-hydroxypregnenone, which may be important in feedback to the hypothalamus on the trigger-ing of the ovulatory LH peak. It has been suggested that the progesta-tional potency of the ovarian output of rats could be controlled by the modulation of progesterone reduction.

The function of dehydroepiandrosterone produced by the adrenal cortex is also unsettled; it may have a role as an anabolic steroid, that is, a steroid promoting nitrogen retention (and hence development of skeletal muscle). Activity of this type would be very important in females, in whom potent androgens, which are also anabolic, are present only in very small quantities. It should be noted that dehydroepian-drosterone output from the adrenal is controlled by ACTH, not gonado-trophins. Occasionally, enzymic derangements in the adrenal gland result in the secretion of potent androgens along with dehydroepian-drosterone. These androgens feed back to the hypothalamus to suppress gonadotrophin secretion, which in turn leads to a depression of gonadal activity. In these circumstances, whereby androgen is produced by the adrenal gland under the control of ACTH, production continues un-abated. This gonadal-adrenal interaction in its severe form, where the bulk of the steroid output is androgens rather than corticoids, is called the *adrenogenital syndrome,* and is not an uncommon defect in women. Treatment with synthetic corticosteroids suppresses ACTH secretion, which leads to a reduction of androgen output and a return of ovarian function.

Steroid Transport in Blood

Steroids are carried in the blood, associated to a large extent with specific binding proteins. The structures of these proteins are not well defined, but corticosteroid-binding globulin (CBG), or transcortin, and testosterone-binding globulin (TeBG), or sex-hormone-binding globulin (SHBG), have been investigated. The role of these blood proteins may be simply to transport steroids, which are virtually insoluble in an aqueous medium; or, the binding proteins may act as a reservoir from which steroid can be withdrawn at times of need. Another possibility is that these proteins serve to increase the half-life of the steroids, since the liver probably inactivates only free steroid. Free steroid in the blood is in equilibrium with steroid bound to the carrier; steroid bound to carrier proteins is not believed to be biologically active.

Binding globulins have been used analytically to determine steroid concentrations. Protein-binding assay follows the same procedure as the radioimmunoassay, except that the antibody used in RIA is replaced by the binding protein. Both techniques are displacement assays, which in reality are a specialized form of isotope dilution analysis.

Steroid Excretion

Because hormones constitute a control system, and because the intensity of a hormone's activity is proportional to the hormone's concentration in the blood, rapid turnover of biologically active compound is essential. Blood concentration must directly reflect the secretion rate; as the secretion rate falls, concentration must also fall. Control of this type requires a mechanism for rapid inactivation; for steroids this is achieved through chemical modification by the liver. In general, the liver changes steroid hormones in two ways: the double bonds are saturated, which renders the hormone biologically inactive, and a sulfate or glucuronic acid residue is attached, which renders the molecule water-soluble prior to its excretion in the urine. Urinary metabolites have to be water-soluble because there are no specific binding proteins to carry them in the bladder. Attachment to glucuronic acid is by a glycosidic bond at either C-3 or C-17 of the steroid ring. (The product is properly called a glucosiduronidate because it is a glucoside and a salt of glucuronic acid, but the trivial name "glucuronide" is more often used.) As an example of modification by the liver, the double bonds of progesterone are saturated to give pregnanediol, which is then coupled at C-3 to glucuronic acid to give a soluble product suitable for excretion in the urine (Figure 6–9).

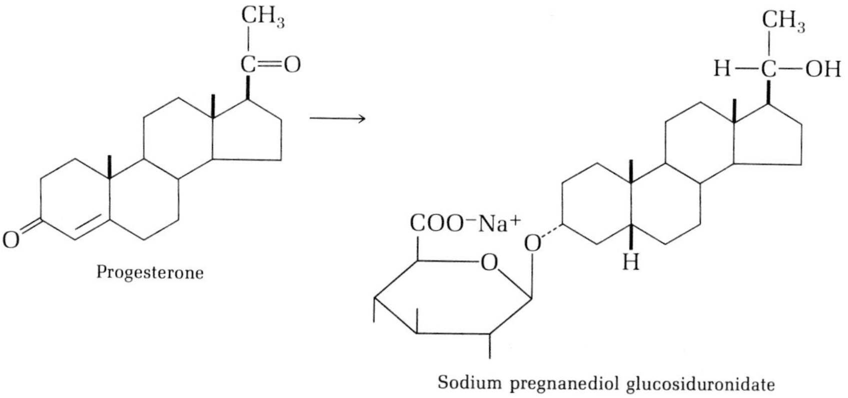

Sodium pregnanediol glucosiduronidate

Figure 6–9 Catabolism of progesterone. The major product found in the urine is 5β-pregnane-3α,20α-diol, shown here coupled in the 3 position to sodium glucuronate. (In sugar chemistry lines above and below the ring represent hydroxyl groups, not methyl groups as in steroid chemistry.)

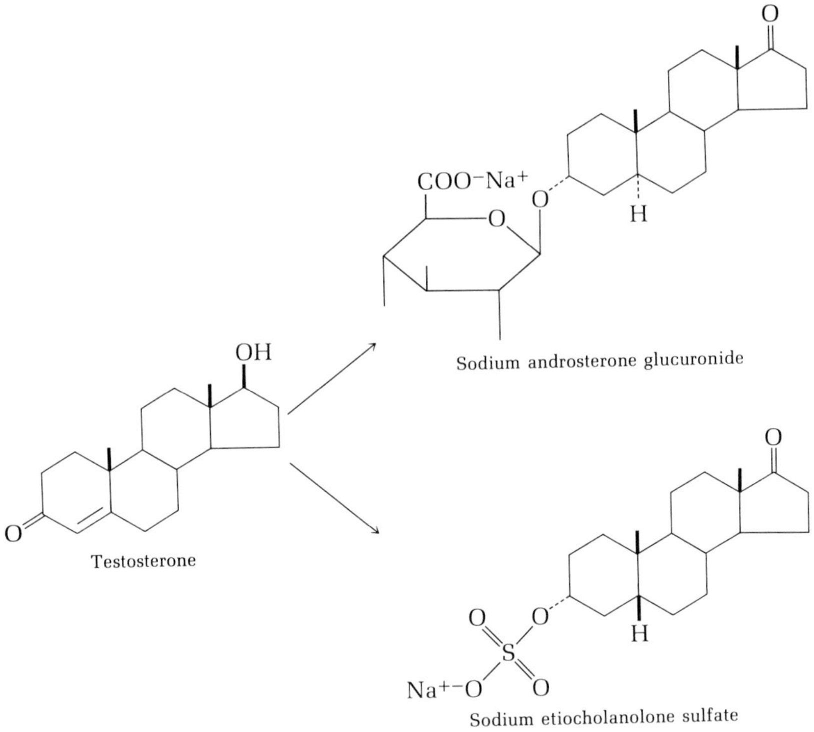

Figure 6-10 Production of 17-oxosteroids (17-ketosteroids) from testosterone. Androsterone is 3α-hydroxy-5α-androstane-17-one, whereas etiocholanolone is 3α-hydroxy-5β-androstane-17-one.

The liver is not able to saturate the aromatic A ring of the estrogens, so these compounds are inactivated only by adding sulfate or glucuronate to make them water-soluble. Testosterone appears in the urine as androsterone or etiocholanolone after inactivation by the liver. These compounds differ stereochemically in that the former has the hydrogen at C-5 introduced in the α position and the latter in the β position. These urinary metabolites are 17-oxosteroids (or 17-ketosteroids) and are shown in Figure 6-10 as a glucuronide and a sulfate. Androgens are the compounds that principally give rise to 17-oxosteroids, so these urinary excretion products result from the secretory activity of both the gonads (androstenedione and testosterone) and the adrenals (dehydroepiandrosterone). Furthermore, some corticosteroids lose their side chain at C-17 in the course of their inactivation in the liver, hence, these C_{21} adrenal products also contribute to 17-oxosteroid accumulation in the urine.

Obviously, variations in 17-oxosteroid structure can be legion. Estimation of urinary metabolites is thus not too profitable if the activity of the gonads or the adrenals is under assessment, although it can be of some value if used in conjunction with inhibitory drugs that block ACTH or gonadotrophin release. Urinary metabolites can be examined when adrenal or gonadal output is totally supressed. Nevertheless, the technique is still of limited value; androstenedione and testosterone, for example, produce the same 17-oxosteroids yet their biological potencies differ profoundly, so that if androgenicity of ovarian output was being determined, analysis of 17-oxosteroids would be inconclusive. In any case, radioimmunoassays, which are sensitive, rapid, and specific, are replacing the old urinary assays, which are unpleasant, tedious, and ambiguous.

Estriol is a steroid found in women's urine, especially during pregnancy (Figure 6–11). It is a product of the feto-placental unit and has a very low estrogenic potency. The developing fetus, especially if it is female, needs to be protected from maternal androgens because androgen determines phenotypic sex (androgen stimulates development of the male duct system, whereas lack of androgen allows the female duct system to develop). Similarly, if a mother is carrying a male fetus she needs protection from fetal androgens if masculinization is to be avoided. In women this protection is achieved partly by the intervention of the placenta and partly by the activity of the fetal liver. Maternal androgens passing through the placenta are converted to estrogens by aromatization of the A ring. The male fetus is tolerant of estrogens, but those that reach it are reduced in potency by hydroxylation at C-16 to produce estriol in the fetal liver. This estriol then returns to the maternal circulation from which it is eliminated in the urine. Conversely, fetal androgens, on passing through the fetal liver, are inactivated by hydroxylation at C-16. They then pass to the placenta where ring A is aromatized to give estriol, which is also eliminated in the maternal urine. Thus, both maternal and fetal androgens are converted to estriol, the former by aromatization followed by 16α-hydroxylation and the latter by 16α-hydroxylation followed by aromatization. The maternal liver is capable

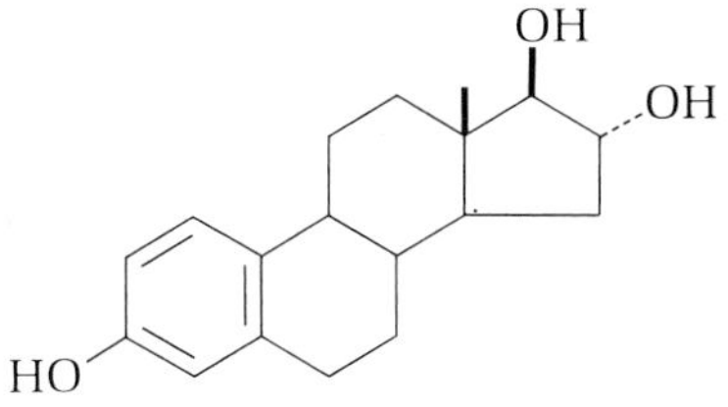

Figure 6–11   Estriol.

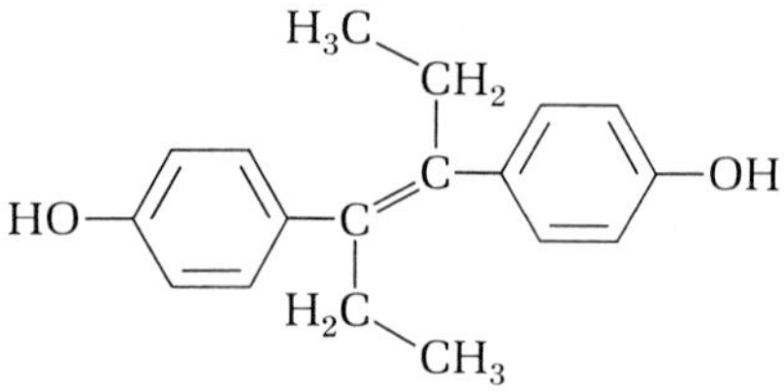

Figure 6–12   Diethylstilbestrol.

of 16α-hydroxylation, but in the adult this enzyme has relatively low activity. Thus, the fetal liver represents the major site of 16α-hydroxylation and fetal well-being can be monitored by examining maternal blood or urine for estriol.

Synthetic Hormones

Because the half-life of steroids in the body is so short, naturally occurring steroids are of little use for exogenous administration; they are degraded very quickly no matter how they are administered. To overcome this problem chemists have modified steroid structures to produce compounds that are more useful experimentally and therapeutically. The earliest modifications were rather simple: steroid esters, such as estradiol benzoate and testosterone propionate, were prepared and slowly hydrolysed within the body to release the biologically active compound. A nonsteroidal compound with estrogenic properties that has been widely used commercially and experimentally is diethylstilbestrol (Figure 6–12). Unlike steroidal estrogens, this compound is easily synthesized and is as potent as estradiol in many systems.

More sophisticated manipulations of the steroid ring system have produced compounds that are effective over long periods of time when administered orally. Such compounds are incorporated in oral contraceptives, and a selection of them is shown in Figure 6–13. A typical contraceptive pill contains two components: an estrogen and a progestin. (In the sequential contraceptive these components are separated; estrogen pills are administered before progestin pills to produce a dose regime that, it is claimed, more closely resembles the normal menstrual cycle.) Two types of progestins are commonly used: analogs of 19-nortestosterone (19-norsteroids lack C-19, the methyl group normally attached at C-10) and analogs of progesterone. Although the 19-norsteroids resemble testosterone, their action is progestational. The synthetic estrogens typically have an ethinyl group introduced in the 17α position, and sometimes the hydroxyl group at C-3 is converted to a methyl ether. The 19-norsteroids also have 17α-ethinyl groups and sometimes they are acetylated at the 17β position. The progesterone analogs are usually substituted at C-6 and acetylated in the 17α position. Dimethisterone is a hybrid, in that it is substituted at C-6 and an analog of testosterone, although it is not a 19-norsteroid.

These orally administered steroids work as contraceptives by suppressing ovulation in most cycles. Normal feedback at the hypothalamus inhibits gonadotrophin release and thus normal ovarian function. This feedback inhibition is probably initiated by the estrogenic component. The progestational component builds up the endometrium during the three weeks in which the pill is being taken so that pseudomenstruation can occur during the one week that therapy is withdrawn. Should ovulation occur, pregnancy is still most unlikely to ensue because the cervical mucus is probably not in an appropriate condition to allow the passage of sperm; tubal motility is probably inappropriate to deliver the egg to the uterus at the right time; and the uterine endometrium is probably not in exactly the right condition to allow implantation. These features of oral contraceptive therapy, forming a second barrier after ovulation inhibition, help to make these steroids among the most pharmacologically predictable compounds in use, with the result that oral contraceptives are now taken more regularly, for longer periods, and by a larger number of people than almost any other drug.

Other reproductive hormones have not been produced in an orally active form and it is unlikely that they will be. Releasing factors and pituitary hormones, composed as they are of amino-acid chains, undergo hydrolysis during digestion in the gut. Furthermore, the gonadotrophins are complex molecules, and although the amino acid sequence for some of them has been extensively investigated, little is known about the positioning of the sugar residues that appear to be necessary for biological activity. Synthetic oxytocin has been available for some time and the RFs that have been identified have been synthesized. Molecules that act biologically like ACTH are available; synthetic ACTH (for example, Synacthen) contains the N-terminal 24 amino acids, which constitute that segment of the peptide chain (which in the natural hormone contains 39 residues) necessary for full biological activity. When hormones are supplied parenterally, the length of time for which they are effective varies with the route of administration. The hormone is more accessible after intravenous injection than it is after subcutaneous or intramuscular injection because the rate of absorption is reduced by the latter routes. Frequent injections of low doses are probably more effective than fewer injections of high doses. Intravenous infusion is probably most effective of all. Recent experiments have shown that corpora lutea of sheep can

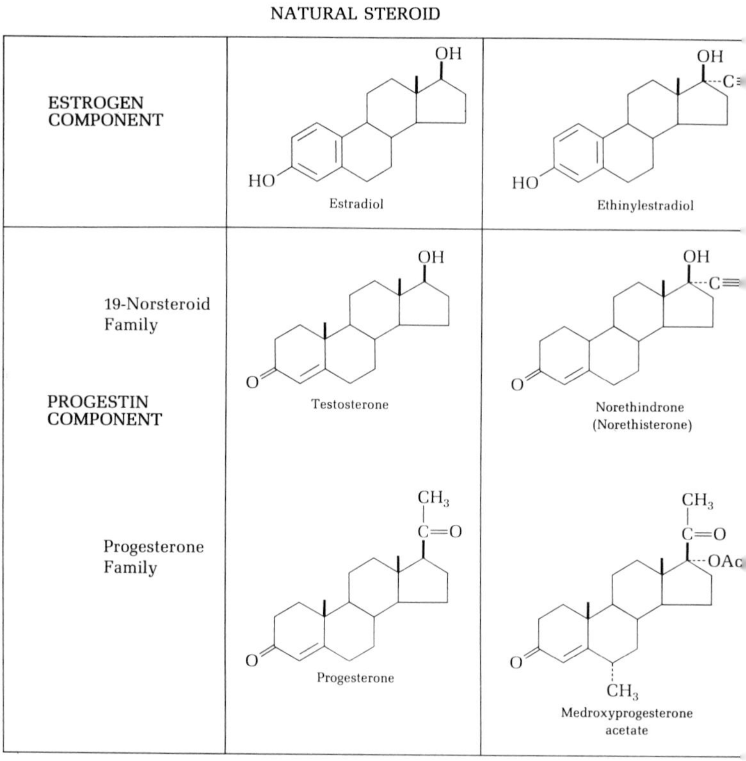

208

NATURAL STEROID

Figure 6–13 Some synthetic steroids used in oral contraceptives.

be maintained by intravenous infusion of gonadotrophins whereas daily injections of high doses, even when the dose was divided and administered at six-hour intervals, were ineffective in this test system.

PROSTAGLANDINS

The prostaglandins are a family of naturally occurring lipids that have been isolated from tissues of many species, from mammals to coral. Their diversity is as great as that of the steroids. They are C_{20} compounds derived from unsaturated fatty acids (Figure 6–14). The prostaglandins are named so that the letters denote modifications in the cyclopentane ring; the number denotes the number of double bonds in the side chains and, as in steroid nomenclature, α and β denote substituents below or above the ring. Stereochemical diversity exists; there are, for instance 64 conceivable isomers of the PGE group, but few of the possible structures are found naturally. The two main series com-

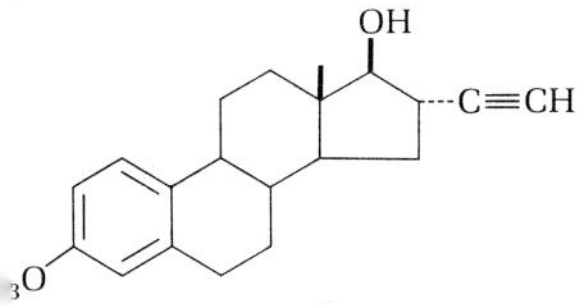

Mestranol

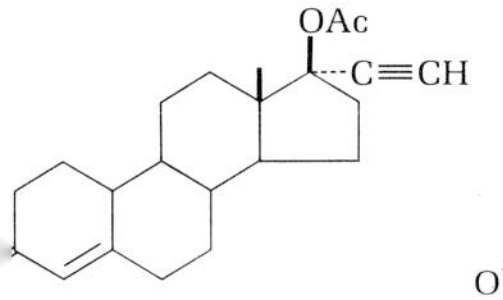

Norethindrone acetate (Norethisterone acetate)

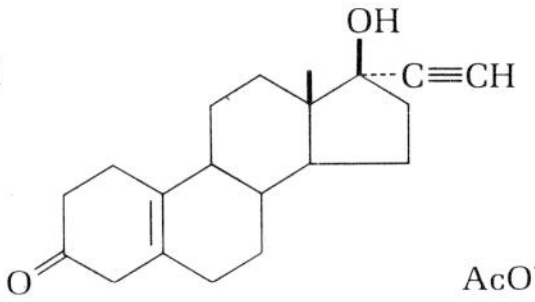

Norethynodrel

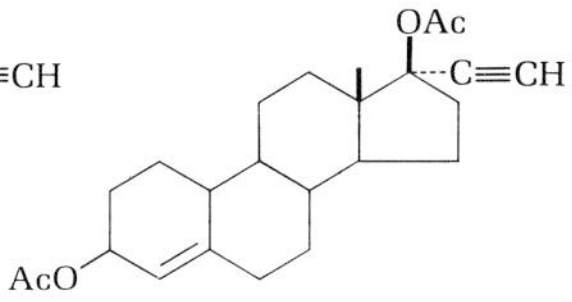

Ethynodiol diacetate

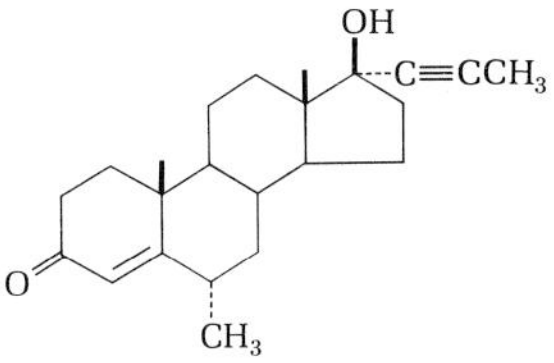

Dimethisterone

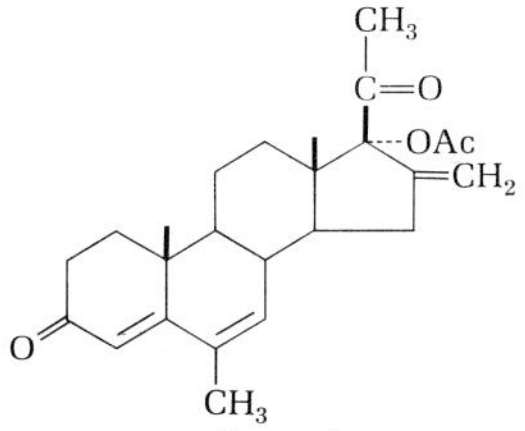

Melengestrol acetate

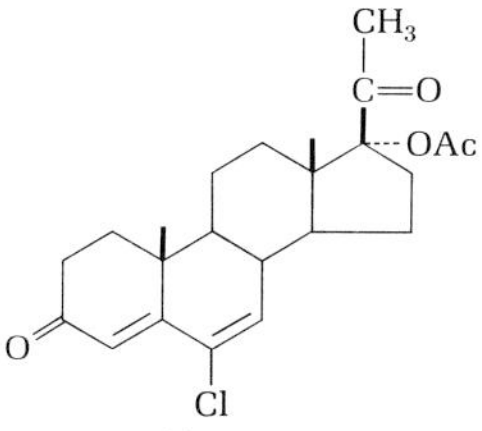

Chlormadinone acetate

monly found in mammals are the PGE and PGF compounds, which differ only in having a ketone or a hydroxyl function at C-9. Both series have hydroxyl groups at C-11 and C-15 (Figure 6–14).

Prostaglandins appear to be involved in many physiological processes, but it is not clear whether they can be considered hormones in the strict sense. In most cases they seem to modify the function of the tissue in which they are formed. Their most clearly defined role is in the stimulation of smooth-muscle contraction. They have been implicated in the process of parturition and used pharmacologically to induce abortion. PGE_2 and $PGF_{2\alpha}$ have been tested clinically and proved effective as abortifacients. Intravenous infusion of massive doses is needed to abort women less than eight weeks pregnant (25 to 100 μg/min for seven hours is a typical regimen). Because prostaglandins activate smooth muscle, the gastrointestinal side effects are prominent and unacceptable. Lower doses have proved effective when injected through a catheter inserted between the fetal membranes and the uterus. Orally administered

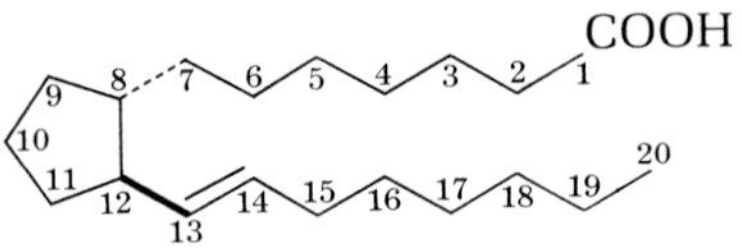

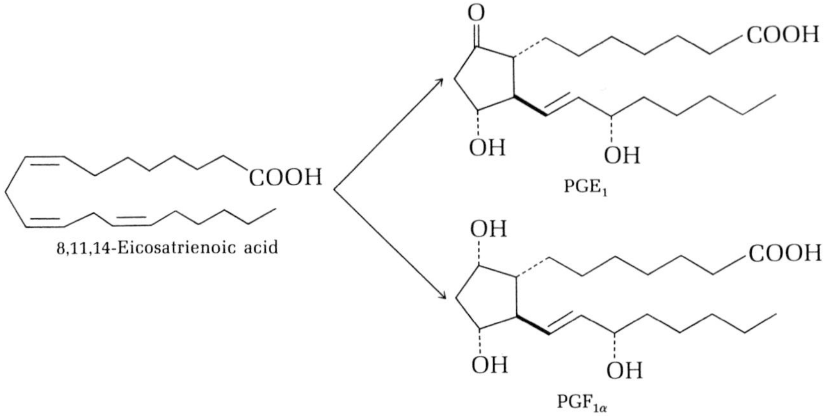

8,11,14-Eicosatrienoic acid

PGE$_1$

PGF$_{1\alpha}$

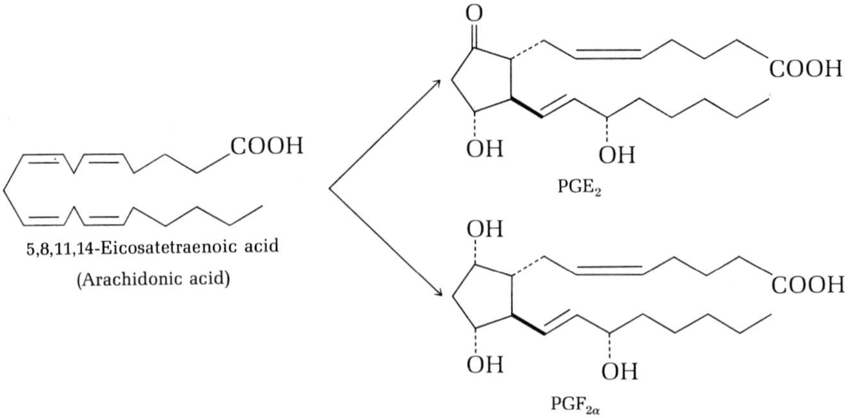

5,8,11,14-Eicosatetraenoic acid
(Arachidonic acid)

PGE$_2$

PGF$_{2\alpha}$

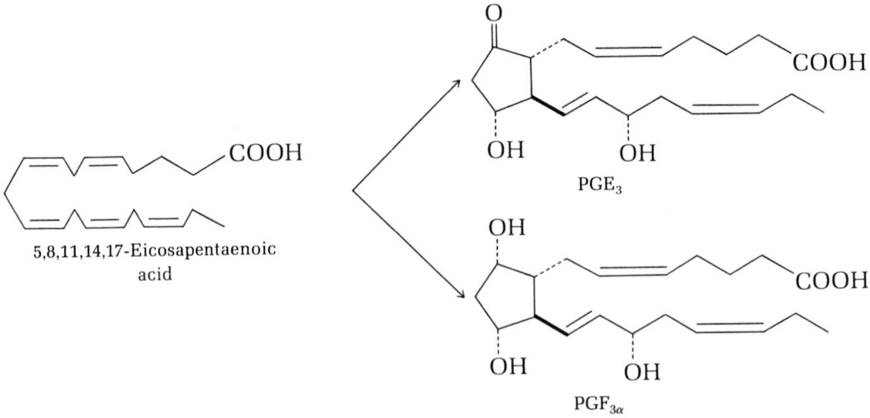

5,8,11,14,17-Eicosapentaenoic
acid

PGE$_3$

PGF$_{3\alpha}$

prostaglandins have been tested for therapeutic abortion, but they are poorly tolerated. Absorption from the vagina may lessen the severe side effects (nausea, vomiting, migraine, diarrhea, and inflammatory reactions) that administration by other routes provokes.

Although the physiological effects of prostaglandins sometimes appear paradoxical, some confusion undoubtedly results from the failure to define which particular prostaglandin is under consideration. Prostaglandin biochemistry is actively developing and the rigorous definition of structure and purity of compounds under investigation, long accepted by steroid biochemists, has not always been appreciated. Thus, in vitro the E series of prostaglandins are luteotrophic, but in vivo $PGF_{2\alpha}$ is luteolytic. In porcine granulosa cell cultures PGE_1 or PGE_2 stimulated the synthesis of cyclic AMP within minutes and the synthesis of progestins within hours. In homogenates of bovine corpora lutea PGE_1 and PGE_2 significantly stimulated cyclic AMP synthesis, but $PGF_{2\alpha}$ was less effective. The effect of LH was additive, however, meaning that prostaglandins are not the mediators of LH action, as earlier work had suggested.

Evidence is accumulating that $PGF_{2\alpha}$ is luteolytic in a variety of species, including rats, hamsters, rabbits, guinea pigs, sheep, and cows. Formerly it was suggested that the luteolytic action of $PGF_{2\alpha}$ was due to its action as a venoconstrictor, that is, its ability to reduce blood flow through the ovary. This explanation is probably not tenable, however, because rabbits in which corpora lutea were autotransplanted to the kidney capsule showed regression of both ovarian and transplanted corpora lutea when $PGF_{2\alpha}$ was injected. The effect on the transplanted corpora lutea could not be effected through venous drainage.

Evidence in sheep suggests that prostaglandins may be the long-sought uterine luteolysin. In sheep in which the ovary and uterus were transplanted to the neck, radioactive prostaglandin was found to concentrate in the ovarian artery after injection into the uterine vein. This experiment implies the existence of some sort of counter-current distribution mechanism. Larger quantities of $PGF_{2\alpha}$ have been shown to reduce progesterone secretion from the ovine ovary and to reduce ovarian blood flow. Thus, in sheep, if the uterus releases prostaglandin at the end of the cycle the material may concentrate in the ovarian vein and reduce progesterone production.

The suggestion has been made that there may be two kinds of corpora lutea. The first type regresses spontaneously at the end of the cycle unless maintained by an embryonically induced luteotrophin; this could

---

Figure 6–14 The numbering system and stereochemistry of the naturally occurring prostaglandins are shown in the top figure. Below this are the six primary prostaglandins and their fatty-acid precursors.

be the case in women, in whom HCG is produced. The second type of corpus luteum persists indefinitely unless the uterus produces a luteolysin to tear it down; this could be the case in sheep, in which prostaglandins are produced. More effort and time is needed before the hypothesis can be accepted or rejected. Even if it is shown to be true, the question of how hormones or lytic factors stimulate or depress luteal activity will still need to be answered.

MECHANISM OF HORMONE ACTION

Although the mechanism of action of any of the hormones is far from completely elucidated at the cellular level, some understanding of the events involved is emerging. It appears that responses to gonadal hormones directly involve the nucleus, whereas responses to gonadotrophins do not. For these reasons it is convenient to consider the two classes of compounds separately.

Gonadotrophins

Until recently virtually nothing was known about the suspected biochemical role of gonadotrophins in the control of gametogenesis, ovulation, and luteinization, although it is commonly believed that control of these activities is probably the major function of these hormones. In the testis, FSH has now been shown to act on Sertoli cells to induce the synthesis of a specific androgen-binding protein (ABP) that is secreted into the tubular lumen. The transfer of testosterone from Leydig cells to the germinal epithelium, which is dependent on androgen, is facilitated by ABP. Although ABP carries steroid in tubular fluid, it is totally different chemically from SHBG, which carries androgens in the blood.

In the ovary, both FSH and LH induce maturation of the ovum. LH induces synthesis of a specific protein required for the triggering of meiosis, but details of these processes remain unknown. However, the stimulation of gonadal steroid synthesis by gonadotrophins has been intensively investigated; in particular, LH has been shown to promote progesterone production in corpora lutea and testosterone production in Leydig cells from many species. Three questions can be asked about this process. Where in the cell does LH act; what step in the pathway is facilitated; and through what chain of events is the stimulation accomplished?

None of these questions has been answered unequivocally. Some of the steps of steroid synthesis occur in mitochondria, whereas others are associated with endoplasmic reticulum. LH stimulates steroid synthesis only in intact cells and no subcellular fraction has been pinpointed as the site of action. Armstrong (1968) showed that LH stimulates progesterone synthesis when cholesterol production is blocked by a metabolic inhibitor. This shows that the effect of LH is exerted beyond cholesterol in the metabolic pathway (Figure 6–8). Other experiments with radioactively labeled precursors have supported this conclusion. LH can stimulate the conversion of endogenous stores of cholesterol to progesterone even when the replenishment of those stores is blocked. The rate-limiting step in the conversion of cholesterol to pregnenolone may be the 20α-hydroxylation of cholesterol (Figure 6–8). This reaction takes place in mitochondria and is the one that should be stimulated by LH.

Whether LH acts directly or indirectly is also open to dispute. Three steps are involved in the action of a hormone: there must be an interaction with some preexisting receptor site in or on the cell; there must be a change in the biological function of the receptor molecule as a result of the interaction; and there must be an amplification of the primary response to the hormone through a cascade of secondary responses. An early consequence of LH action on luteal and Leydig cells is an increased concentration of the nucleotide adenosine 3′, 5′-cyclic monophosphate (cAMP) within the cells. Receptors for LH have been shown to exist on the surface of gonadal cells. Binding of LH to the receptor activates the enzyme adenyl cyclase, which is also membrane-bound. Adenyl cyclase catalyzes the conversion of ATP to cAMP, which is then available to act as an intermediary in the cellular response to LH and to initiate many secondary responses to the gonadotrophin. In this role cAMP is often described as a "second messenger"—LH being the first messenger, carrying information to the cell, and cAMP the second messenger, initiating responses within the cell. Four observations indicate that cAMP is a mediator of LH action. LH stimulates adenyl cyclase; cAMP itself will stimulate steroidogenesis; LH increases intracellular cAMP concentrations; and theophylline (added in vitro) potentiates the action of cAMP or LH. Theophylline inhibits the enzyme phosphodiesterase, which catalyzes cAMP breakdown, and thus allows the cyclic nucleotide to persist within the cells for longer periods.

Earlier it was said that prostaglandins of the E series stimulate progesterone synthesis in luteal cells. They have also been shown to stimulate cAMP production and glycolysis in vitro in luteal cells from rats and mice. A role for prostaglandins as an obligatory link in the cell membrane between the LH receptor and adenyl cyclase has been proposed. Such a role would make prostaglandin a second messenger and cAMP a third messenger. Although PGE$_2$ mimics the effects of LH within luteal cells, an obligatory role for prostaglandins in gonadotrophin action on the ovary has not been established. Inhibitors of prostaglandin synthesis do not block LH-stimulated cAMP production; LH

has not been shown to stimulate the conversion of exogenous arachidonic acid to prostaglandins (see Figure 6–14). These and other lines
of evidence suggest that LH and prostaglandins may initiate responses
by different mechanisms that converge on a final common path involving adenyl cyclase. Thus, for the present, the role of prostaglandins in
ovarian cells remains enigmatic.

LH-stimulated gonadal steroid production requires protein synthesis
but not RNA synthesis. These facts have been demonstrated by the
inhibition of protein synthesis with puromycin or cycloheximide and
by the inhibition of RNA synthesis with actinomycin-D. The results
suggest that LH-stimulated steroid synthesis involves translation of a
stable messenger RNA. Hence, gonadal cells contain a messenger RNA
with a long half-life that codes for a protein with a short half-life.

It has been shown that ACTH stimulates cAMP production in the
adrenal gland, and this in turn allows increased synthesis of glucocorticoids. In this system a receptor molecule for cAMP has been
identified within the cell, and the binding of the nucleotide to the
receptor activates a protein kinase. The protein kinase is believed to
phosphorylate ribosomes, which then allow protein synthesis to take
place. Presumably this protein synthesis represents translation of the
stable RNA message within the adrenal cells. It may be that the newly
synthesized protein transports cholesterol to the mitochondria, where
it can be converted to pregnenolone, the immediate precursor of the
steroid hormones (see Figure 6–8). The conversion of cholesterol to
pregnenolone has been shown to be the rate-limiting step in hormone
synthesis. It is possible that LH could initiate a similar chain of events
in luteal and Leydig cells, leading to enhanced synthesis of gonadal
steroids. Such a system would greatly amplify the response within
gonadal cells to an interaction with LH at the surface. The system is a
cascade, i.e., the product of one step is the *activator,* not the substrate,
for the next step. Thus, LH activates adenyl cyclase; cAMP activates
protein kinase; protein kinase activates ribosomes; ribosomal product
facilitates cholesterol movement to the side chain cleavage enzyme. In
this way a few molecules of LH could induce rapid maximal gonadal
hormone production.

Although the cascade model is an attractive explanation of the
mechanism of LH action, the gonadotrophin may trigger primary
responses other than adenyl cyclase activation. Mention has already
been made of the enigmatic interaction between LH and prostaglandins.
The fact that LH is a large molecule and that a receptor for it has been
identified on the cell surface does not preclude the possibility of LH
entry into gonadal cells, where interaction with another intracellular
component could trigger other, independent chains of events contributing to the cumulative cellular response to the gonadotrophin.

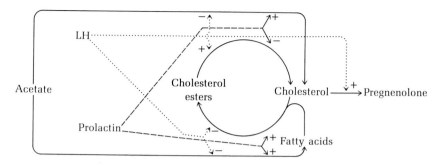

Figure 6–15 Stimulatory (+) and inhibitory (−) actions of LH and prolactin in luteal cells. [Modified from Armstrong, 1968, Rec. Prog. Hormone Res., 24:255.]

Whatever the molecular events within the cell, the gross biochemical effects of LH are fairly clear. The hydrolysis of cholesterol esters to give cholesterol and the degradation of cholesterol to yield pregnenolone are both facilitated by LH. LH also causes depletion of cellular substrate reserves in its steroidogenic activity. Prolactin appears to antagonize both of these effects, at least in luteinized rat ovaries. Prolactin stimulates cholesterol ester formation and cholesterol accretion within the cells (Figure 6–15). It also ensures the deposition of substrates on which the action of LH depends. This action of prolactin may extend to species other than rats and may explain the luteotrophic role claimed for prolactin in many experimental situations.

Gonadal Hormones

Our knowledge of the action of gonadal hormones differs considerably from what we know about the action of gonadotrophins, principally because it is relatively easy to obtain steroids tagged with radioactive atoms. A major consequence of this fact is that gonadal hormones can be localized within tissues and cells. Glasscock and Hoekstra in 1959 and Jensen and Jacobson in 1960 successfully conducted in vivo studies with normal physiological concentrations of estradiol labeled with tritium at high specific activity. This had not been possible previously; doses of steroids had to be administered in such high amounts that abnormal distribution was inevitable. Both groups of workers demonstrated that those tissues that showed growth responses to estradiol, i.e., the uterus and vagina, also concentrated and retained the hormone more than did nontarget tissues (Figure 6–16). Further work established that estradiol was not extensively metabolized by the uterus and that the binding was specific for biologically active estrogens. Estradiol-17α, androgens, or corticosteroids were not retained. These findings led to the rekindling of interest in the concept of a receptor in target cells.

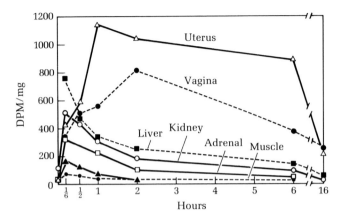

Figure 6–16 Concentration of radioactivity in rat tissues after a single subcutaneous injection of approximately 0.01 µg of estradiol (specific activity, 195
µc/µg) in saline. Liver and kidney points are mean values of four aliquots of dried,
pooled tissue; other points are median values of individual samples from six animals. Values are expressed as disintegrations per minute (DPM) of ^{3}H-estradiol
per mg of dry weight of tissue or 5 µl of blood. (From Jensen and Jacobson, 1962,
Recent Prog. Hormone Res., 18:387.)

Using a variety of experimental approaches, Gorski and his colleagues
and Jensen and his collaborators have established that an estrogen-
protein complex is formed in the cytoplasm of uterine cells and that
this complex moves into the nucleus. This transfer to the nucleus is a
temperature-dependent process; it occurs at 37°C but not at 4°C. The
binding protein has a complex subunit structure and can be isolated
in a number of forms. When isolated from cytoplasm, it has a molecular
weight of about 200,000 daltons in solutions of low ionic strength, but
in solutions of high ionic strength the molecular weight drops to about
50,000 daltons. Before the estrogen-receptor complex can enter the
nucleus it undergoes a process called "transformation," after which its
molecular weight is about 130,000 daltons. It is this transformed receptor
that is isolated from the nucleus, and the process of transformation
that is temperature-dependent. In rats, shortly after the estrogen-protein
complex has entered the nuclei of uterine cells, synthesis of RNA (presumably messenger RNA) and a specific protein can be detected. When
estradiol is added to rat uteri incubated in vitro, the appearance of the
new protein can be detected within 30 minutes. These facts support the
hypothesis advanced by Mueller in 1958 that estrogens might influence
gene expression. The role of this new estrogen-induced protein is totally
unknown at the present time.

The function of the cytoplasmic protein interacting with estrogen is
not completely clear. Binding of estrogen permits transformation to

occur, and transformation is a necessary prelude to movement of the complex into the nucleus. Within the nucleus the estrogen and the receptor protein appear to remain associated. Estrogen may therefore act as a key, allowing the steroid-receptor complex to enter the nucleus and interact with a second receptor, specific for the entire estrogen-receptor complex. This would be the site of the primary action of the hormone.

Support for these ideas has been given by O'Malley and his colleagues using a different model system. The oviduct of the chicken produces a specific protein, avidin. Avidin secretion is dependent on the action of progesterone on the estrogen-primed oviduct. A sequence of events paralleling those seen when estrogen interacts with uterine cells has been established for progesterone interaction with oviductal cells. The progesterone binds to a cytoplasmic protein and moves into the nucleus, where the protein-steroid complex becomes associated with chromatin. Studies of cell-free systems have established that the oviduct receptor protein is essential for nuclear binding of progesterone. Oviduct nuclei are needed; nuclei from other tissues scarcely react. Solutions of high ionic strength will extract the steroids still associated with the receptor protein from the nuclei. O'Malley has shown that acidic proteins of chromatin recognize the steroid receptor at "acceptor sites"; the basic histones do not appear to be involved. Thus, studies with this system have gone further than those using estrogen and uterine cells in that a nuclear acceptor site has been identified. This acceptor site could be the true receptor.

Interaction of the progesterone-receptor complex with chromatin leads to the production of a specific messenger RNA that codes for avidin. In a cell-free protein-synthesizing system derived from rabbit reticulocytes, this purified avidin message, obtained from chicken oviducts, has been shown to initiate avidin synthesis. Thus, a complete sequence of biochemical events following the action of progesterone on the chicken oviduct has been demonstrated: cytoplasmic binding of steroid is followed by transfer of a transformed steroid-receptor complex to the nucleus; interaction with chromatin leads to the production of a specific messenger RNA that codes for a specific protein. The RNA message is characterized by demonstrating its activity in a cell-free system. This method is particularly convincing in the case of avidin because a protein-synthesizing system derived from a mammal yields a product unique to birds.

This pattern for the mechanism of action of hormones on target tissues has been shown to hold for other sex steroids. The chicken oviduct responds to estradiol by producing the protein ovalbumin. Again, the complete sequence of events—cytoplasmic binding, estrogen transfer to the nucleus, stimulation of RNA production, and translation of the ovalbumin message (also confirmed in a cell-free system)—has been described by O'Malley and his colleagues.

Testosterone stimulates its target organs in a similar manner, but for some tissues there is a major difference. Many of the accessory glands of the male reproductive tract contain an enzyme that converts testosterone to 5α-androstan-17β-ol-3-one (dihydrotestosterone). It is dihydrotestosterone that binds to the cytoplasmic receptor and is transferred to the nucleus. For these tissues testosterone is said to be a *prehormone*, since testosterone 5α-reductase in the target cells converts it to the active metabolite, dihydrotestosterone. Mainwaring and his colleagues have shown that synthesis of an enzyme of general metabolism, aldolase, is stimulated in the prostatic cells of rats by androgens. A sequence of events similar to those outlined above for progesterone and estrogen action on the chicken oviduct is followed by testosterone in stimulating aldolase synthesis in rat prostatic tissue. Testosterone is reduced to dihydrotestosterone, which binds to a cytoplasmic receptor; the receptor moves to the nucleus and associates with chromatin; this association stimulates the production of messenger RNA that codes for aldolase. Thus, similar responses in appropriate target cells have been delineated for estrogens, progestins, and androgens, all involving modification of genomic expression.

The importance of testosterone reduction has not been established for all androgen target tissues. For instance, R. E. Peterson's group has recently shown that in male human fetuses development of the seminal vesicles, vas deferens, and epididymis appears to be stimulated by testosterone, whereas development of the testes, scrotum, and penis appears to be stimulated by dihydrotestosterone. Furthermore, the anabolic events of puberty in human males—the increase in muscle mass, the growth of the phallus and scrotum, and the voice change—seem to be induced by testosterone, whereas prostatic growth, development of facial hair, temporal hairline recession, and development of acne (which are also pubertal events) appear to be induced by dihydrotestosterone. The comparative biochemistry of androgen-dependent tissues, both within and between species, is yet to be fully defined.

In spite of what we know about the binding, transport, and interaction at acceptor sites of steroid hormones, the details of the primary response of the cell still elude us. It should not be long before the events occurring between the attachment of the complex at the acceptor site and the detection of new RNA are better defined. All kinds of secondary responses in target tissues for gonadal hormones have been described, and, although they are important in defining the gross response of the tissue, nothing would be gained by cataloging them here. In general, gonadal hormones promote growth; this is a complex process in which innumerable factors are involved. Growth must be coordinated, and the biochemical details of this process merely represent events secondary to the primary action of the hormone. We still have a long way to go before we will know fully how gonadotrophins or gonadal steroids regulate cellular metabolism.

SELECTED READING

Armstrong, D. T. 1968. Gonadotropins, ovarian metabolism and steroid biosynthesis. *Rec. Prog. Horm. Res.,* **24:**255–320.

Bloch, K. 1965. The biological synthesis of cholesterol. *Science,* **150:**19–28.

Canfield, R. E., F. J. Morgan, S. Kammerman, J. J. Bell, and G. M. Agosto. 1971. Human chorionic gonadotrophin. *Rec. Prog. Horm. Res.,* **27:**121–164.

Garren, L. D., G. N. Gill, H. Masui, and G. M. Walton. 1971. On the mechanism of action of ACTH. *Rec. Prog. Horm. Res.,* **24:**433–478.

Gorski, J., D. Toft, G. Shyamala, D. Smith, and A. Notides. 1968. Hormone receptors: studies of the interaction of estrogen with the uterus. *Rec. Prog. Horm. Res.,* **24:**45–80.

Grant, J. K. 1969. Actions of steroid hormones at cellular and molecular levels. *Essays in Biochemistry,* **5:**1–58.

Imperato-McGinley, J., L. Guerrero, T. Gautier, and R. E. Peterson. 1974. Steroid 5α-reductase deficiency in man; an inherited form of male pseudohermaphroditism. *Science,* **186:**1213–1215.

Jensen, E. V., and E. R. De Sombre. 1973. Estrogen-receptor interaction. *Science,* **182:**126–134.

Jensen, E. V., and H. I. Jacobson. 1962. Basic guides to the mechanism of estrogen action. *Rec. Prog. Horm. Res.,* **18:**387–414.

King, R. J. B., and W. I. P. Mainwaring. 1974. *Steroid-Cell Interactions.* Butterworth.

Klyne, W. 1960. *The Chemistry of the Steroids.* Wiley.

Lindner, H. R., A. Tsafriri, M. E. Lieberman, U. Zor, Y. Koch, S. Bauminger, and A. Barnea. 1974. Gonadotropin action on cultured graafian follicles; induction of maturation division of the mammalian oocyte and differentiation of the luteal cell. *Rec. Prog. Horm. Res.,* **30:**79–138.

Midgely, A. Rees, Jr., G. D. Niswender, V. L. Gay, and L. E. Reichert, Jr. 1971. Use of antibodies for characterization of gonadotropins and steroids. *Rec. Prog. Horm. Res.,* **27:**235–302.

O'Malley, B. W., and A. R. Means. 1974. Female steroid hormones and target cell nuclei. *Science,* **183:**610–620.

Peel, J., and M. Potts. 1969. *Textbook of Contraceptive Practice.* Cambridge University Press.

Pierce, J. G., T. Liao, S. M. Howard, B. Shome, and J. S. Cornell. 1971. Studies on the structure of bovine thyrotropin and its relationship to luteinizing hormone. *Rec. Prog. Horm. Res.,* **27:**165–212.

Ramwell, P. W., and J. E. Shaw. 1970. Biological significance of the prostaglandins. *Rec. Prog. Horm. Res.,* **26:**139–188.

Schally, A. V., A. Arimura, A. J. Kastin, H. Matsuo, Y. Baba, T. W. Redding, R. M. G. Nair, L. Debeljuk, and W. F. White. 1971. Gonadotrophin-releasing hormone: one polypeptide regulates secretion of luteinizing and follicle-stimulating hormones. *Science,* **173:**1036–1038.

7

Reproduction
in Males

The important components of the male gonad are the seminiferous tubules, which secrete sperm, and the Leydig cells of the interstitial tissue, which secrete andogren. These are discussed in turn. It can be easily demonstrated that (except in chickens) LH alone stimulates Leydig cells to secrete androgen, but for complete spermatogenesis FSH, LH, and probably androgen are needed. Androgen maintains the secondary sex characters (beard, voice, horns, comb, aggressiveness, etc.) and the accessory glands (prostate, seminal vesicles, Cowper's glands). The secretions of the accessory glands are essential components of semen. A brief outline of sperm morphology is given, and the relation of abnormal sperm to fertility is discussed. Of the many millions of sperm ejaculated, only a few hundred reach the oviduct, and only a few reach the vicinity of the ovum.

INTRODUCTION

Like females, males are classified as continuous or seasonal breeders, and in most species the breeding behavior of the male corresponds to that of the female. Among seasonal breeders, as a rule, the testes regress at the end of the season at about the same time as the ovaries of the female of the species. At the start of the season, in most species, the testis shows signs of sexual activity somewhat ahead of the awakening of the ovary. For example, the testes and duct system of some mallard drakes may be filled with mature spermatozoa at a time when none of the females show follicles larger than 0.5 mm in diameter. The majority of drakes are in full breeding condition between 10 and 20 days before the hens are ready to ovulate their first egg. The synchronization of reproductive functions in seasonal breeding males and females is not surprising, for the basic controlling mechanisms in both sexes are identical. The relatively early sexual readiness of the male in the season may be due to the greater sensitivity of male end organs to the external and internal stimuli involved.

There are exceptions to the rule that the breeding seasons of males and females coincide. In rams spermatogenesis and libido are not restricted to the breeding season of ewes. Spermatogenesis in raccoons is continuous throughout the year even though the breeding season of females is restricted to January, February, and March. A similar situation is found in domestic dogs and probably in several other species.

The endocrine control of reproductive phenomena in males is similar to that in females in that the same two pituitary hormones, FSH and LH, play the major role in the stimulation of the testes. The behavior of male rabbits even suggests that, within the breeding season of males, there is a cycle or rhythm of testicular function, although this rhythm (if it exists) is neither as accurately timed nor as well defined as the estrous cycle of the female.

Before considering the intricacies of the endocrine control of the testes, let us briefly review what happens in the normal development of the male gonad.

THE TESTES

The Seminiferous Tubules

At birth or hatching, males possess tubules that have no lumina and are lined with a single layer of small nuclei. As the male matures, the tubules gradually acquire lumina, and the germinal epithelium progresses from the one-layered state to the complex state seen in sexually mature males, in which all cell types (spermatogonia, primary and

Table 7-1. Age at which spermatogenesis occurs in some species

	AVERAGE AGE AT APPEARANCE OF		
	PRIMARY SPERMATOCYTES	SECONDARY SPERMATOCYTES	SPERMATOZOA
Man	—	—	10–15 years
Goat	—	—	110 days
Guinea pig	—	—	50–70
Rat	—	—	33–35
Boar	84 days	105 days	147
Bull	63	181	224
Ram	63	126	147
Chicken	42–56	70	84–140

secondary spermatocytes, and spermatozoa) are seen. There is considerable variation among individuals and species in the age at which spermatogenesis begins and the rate of its development. Spermatogenesis develops most rapidly in species in which the males reach sexual maturity comparatively early; the ages for spermatogenesis in some species are shown in Table 7-1. In animals with a short life (chickens) spermatogenesis continues unabated until death, and the great majority of tubules appear to be functioning normally throughout life. In mammals with a long life (man, boars, bulls) spermatogenesis may continue normally throughout life, but past middle age the tubules normally atrophy gradually, until eventually only a few show normal spermatogenic activity.

Not infrequently individuals produce viable sperm and are capable of mating long before the majority of the members of the species or population to which they belong. Precocious sexual maturity in boys and other mammals is probably commoner than is generally assumed. Complete spermatogenesis, with sperm in the epididymis, was seen in the testes of one ram at the age of 72 days, and motile sperm were obtained from a Leghorn cock at the age of 62 days. That the age of sexual maturity is genetically influenced is seen from the fact that there is considerable difference between breeds of domestic animals (among dairy cattle, Brown Swiss bulls mature from four to six months later than Jersey or Guernsey bulls). In fact, it has proved possible to select for early or late sexual maturity within practically all breeds of domestic animals in which the attempt (either consciously or unconsciously) has been made.

In seasonally breeding males the testes regress completely during the nonbreeding season, and the germinal epithelium returns to the state in which it is commonly found in young, sexually immature males. The tubules lose their lumen and are lined with a single layer of small spermatogonia. In many mammals the testes migrate from the scrotum into the body cavity, where they remain until shortly before the onset of the next breeding season. They then go through the same changes through which they had gone in reaching puberty, and do so with the onset of each new breeding season. This seasonal reawakening of the testes is commonly called "recrudescence" to differentiate it from the prepubertal changes that occur in sexually immature males.

Of considerable interest and importance is the length of time required for the completion of spermatogenesis and spermiogenesis—that is, the transformation of a spermatogonium into a finished sperm cell. It is possible to measure this time in a variety of ways. An older method was to destroy testicular and epididymal sperm by X-irradiation or heat, and then determine the time required for repair of the germinal epithelium and formation of new sperm. A more elegant and by now more common method labels the nuclei of germ cells with ^{32}P, taking advantage of the fact that the nuclei of cells in the different stages of spermatogenesis do not become radioactive. Thus it is possible to determine the interval between the uptake of ^{32}P by nuclei in the initial stages of spermato-genesis and the appearance of radioactive mature sperm cells. There is good agreement between the figures obtained by the two methods. Both show that spermatogenesis requires about 10 days in the mouse, 16–20 days in the rat, 39 days in the rabbit, 48 days in the bull, and 50 days in the ram. Injection of ^{32}P into rams has shown that the rate at which the different stages of spermatogenesis occur is extremely uneven, some being completed in a few hours and others in as long as 15 days.

That spermatogenesis is a slow process is frequently overlooked by those who are interested in studies of the rates of sperm and semen production. This fact should be remembered especially in experiments designed to study the rate of exhaustion of epididymal or testicular sperm after frequent ejaculation. Such experiments should run for 10–20 days for smaller animals and 50–60 days for all larger ones.

The Interstitial Tissue

The most important components of the intertubular tissue of the testes are the interstitial or Leydig cells. These cells are almost certainly the source of the male sex hormone, androgen. In rats, the testes of even neonatal males secrete testosterone (which apparently gets into the circu-lation because the male hypothalamus is differentiated by androgen in such a way as to become acyclic). In fact, in rats even the embryonal testes contain enough androgen to cause a local effect in adult androgen-dependent end organs, e.g., the seminal vesicles, into which they are

Table 7-2. Androgen content of bull testes in relation to age and body weight

	1½ TO 3 MONTHS	3½ TO 5½ MONTHS	10½ MONTHS TO 17½ YEARS
Number of animals	4	5	9
Body weight (kg)	<80	90–200	300–1000
Content in testes (µg)			
Testosterone	9.2 (1.5–20.2)	128 (28–405)	1,113 (155–3,240)
Androstenedione	21.5 (3.2–40.8)	83 (3–191)	67 (<25–210)
Concentration (µg/100 g)			
Testosterone	56 (10–126)	167 (25–604)	157 (32–437)
Androstenedione	124 (21–255)	138 (3–370)	10 (5–28)

Source: H. Lindner, The androgenic secretion of the testis in domestic ungulates. In *The Gonads*, edited by K. W. McKerns, Appleton-Century-Crofts, 1969.

implanted. But the urine of month-old bull calves contains no androgen. Androgens have been demonstrated in the effluent testicular blood of bulls (Table 7-2), dogs, boars, and a variety of other animals. The predominant androgen of domestic ungulates is testosterone in adults and androstenedione in juveniles. It has been shown in a variety of male animals that the infusion of an LH-containing hormone, HCG, causes an almost immediate (15–20 minutes) and very substantial increase in androgen secretion into the spermatic vein as well as an increased testosterone content of the infused testicle. These results suggest that LH(HCG) causes increased testosterone synthesis as well as release (Lindner, 1969).

There is an excellent correlation between testes weight, rate of testosterone production, and the weight of androgen-dependent end organs (Table 7-2). In chickens, for instance, there is a very high correlation across age groups between the weight of the testes and the size of the comb (r = +0.98); in rats the correlation between the weight of the testes and the size of the seminal vesicles is +0.87. Within age groups the correlation coefficient is also significant but is considerably lower than across age groups. Among 50-day-old chickens and correlation between comb size and testes weight is +0.25; among chickens 90 days of age it is +0.56. This suggests that males with comparatively larger testes may produce less androgen and hence smaller end organs than males with smaller testes.

In men the weight of the testes, the number of Leydig cells, and the rate of androgen secretion all begin to decrease at about 30 years of age (Figure 7-1). Decrease in testes weight and the rate of androgen synthesis also occurs in senile rats and probably in aging males of other species.

The feedback control mechanism of gonadotrophins is still not understood in the male and it appears possible that estrogen rather than androgen may be the main mediator in the hypophyseal control system (p.

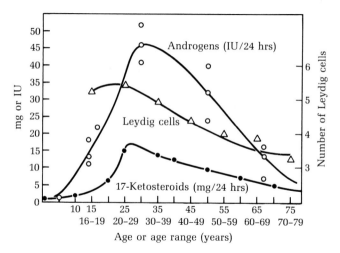

Figure 7-1 Relation between number of Leydig cells and amounts of androgens and 17-ketosteroids found in 24-hour samples of urine of normal human males ranging in age from infancy to old age. By comparison, eunuchs produce 4.0-7.7 IU and eunuchoids about 17.6 IU of androgen in 24 hours. [Androgen curve from Hamilton, 1954, *J. Clinical Endo. and Metabolism,* 14:452; 17-ketosteroid curve from Hamburger, 1948, *Acta Endo.,* 1:19; Leydig cell counts from Tillinger, 1957, *Acta Endo.,* Suppl. 30.]

230). The source of estrogen in the testes remains unknown, and both the Sertoli cells and the Leydig cells have been implicated as sites of estrogen synthesis.

ANDROGEN

Effect on Secondary Sex Characters

Androgen develops and maintains the secondary sex characters in males. The effect of androgen deficiency on the development of secondary sex characters varies, depending on the stage at which androgen is withdrawn. If, for instance, males are castrated after having reached sexual maturity, they may continue to show libido and erection. If juvenile prepubertal males are castrated, libido and secondary sex characters generally do not develop. The castration of boys causes them to retain a high voice; men castrated after puberty usually retain the pitch of their voice. The qualitative and quantitative changes produced by androgen deficiency are, in general, more intense when they occur before puberty.

One of the most frequently studied effects of androgen is growth of the comb in cocks. The comb regresses completely after castration, but it can be returned to normal size, texture, and color by the administration of androgen. Because the degree of growth is proportional to the amount of hormone injected, both immature males with small combs and castrated adults make excellent assay animals for androgen. (A few paradoxical effects of androgen in birds should be mentioned. The blackness of the beak of male English sparrows is due to androgen, for the beak lightens after castration—and during the nonbreeding season—and blackens after androgen injection. The beak of females is light but becomes black after androgen injection. This suggests that the ovary of female sparrows either does not secrete androgen or secretes it in very small quantities. In female chickens comb growth is caused by androgen, which is secreted by the ovary. Androgen produces the yellow beak of starlings and the crimson beak of the black-headed gull.)

Because of the well-known relation between libido and androgen, at one time it was expected that the hormone would correct low sex drive or impotence in males. In general, however, androgen is able to correct low sex drive only if this condition is due to testicular hypofunction and deficiency of the hormone. Many cases of male impotence seem to be psychological and thus cannot be corrected by androgen therapy. If males are experimentally used as teasers over long periods, that is, are not permitted to mate after they discover females in heat, they eventually lose their sex drive and even refuse to attempt to mate with females in heat. Such males do not regain potency even after massive doses of exogenous androgen, but they gradually recover their ability and willingness to mate if they are permitted to attempt and eventually complete matings without therapeutic interference. Such temporary loss of libido has been seen in bulls, boars, and rabbits, and some individuals become discouraged much sooner and more easily than others.

The fact that in men the sex drive is totally androgen dependent—whereas in women it apparently does not depend on hormonal stimuli of ovarian origin—makes it very difficult to devise a male contraceptive pill that depends on the inhibition of pituitary function, since this would cause a decrease in the rate of androgen production. Any male contraceptive must depend on nonhormonal substances that would prevent spermatogenesis or make sperm incapable of fertilizing the egg without in any way interfering with the ability of the Leydig cells to make adequate amounts of testosterone. One possibility is a pill containing both estrogen, to block the pituitary gland, and testosterone, to restore the libido of men taking such pills. Whether this idea has merit remains to be seen.

Effect on Accessory Sex Glands

We have noted that the effect of androgen on the secondary male sex characters, such as pitch of voice and libido, depends to some extent on factors other than the direct action of the hormone on end organs. In distinct contrast is the effect of androgen on the accessory glands, which include the seminal vesicles, the prostate, and the bulbourethral (Cowper's) glands. Morphologically and physiologically, all of these glands are totally dependent on androgen. Castration causes cessation of their secretory function and drastic reduction in their size and in the height of their epithelial linings (Figure 7–2).

Exogenous androgen administered to normal males causes an abnormal increase in the size of the accessory glands. When given to castrated males, it rehabilitates the glands both in size and in secretory function. The degree of growth in castrated males following administration of androgen is proportional to the amount of androgen injected. For this reason the accessory glands are used for androgen assays. If the concentration of androgen in an unknown sample is small, the height of the epithelium of the glands may be used as the end point. It is easier and usually more convenient, however, to use the weight of the glands. In rats the prostate is a more sensitive indicator of most androgenic compounds than the seminal vesicles. (Seminal vesicles increase somewhat in weight in response to massive doses of estrogen, but the increase is due to the action of estrogen on the fibromuscular connective tissue of the organ, and estrogen is not able to repair either the epithelium or the lost secretory activity of the seminal vesicle of castrates.)

Androgen therapy can repair the retrogressive changes that occur in the duct system of males after castration. It can also restore to normal (or raise above normal) the production of fructose, citric acid, and phosphatase by the accessory glands following castration.

Miscellaneous Effects

We have already noted that androgen plays an important role in determining the sexual behavior of the male toward the female. In addition to regulating the intensity of the male sex drive, androgen has a variety of other effects, which are only indirectly related to reproductive activity. Androgen has been shown experimentally to determine the so-called pecking order among chickens (and probably among birds in general), the social organization of fish schools, and the butting order among cows. Presumably both male and female individuals of the species assume their places in the social order according to the degree to which they are stimulated by the male sex hormone. If individuals who are normally at the top of the social order, who peck or boss all other individuals, are castrated, they cease being bossy and rapidly sink to the bottom of the order. Injection of androgen restores their previous behavior and position, which they maintain only as long as they are given the hormone. A similar social order exists in other "group-living" animals. Among mice the social position of males is determined by a series of

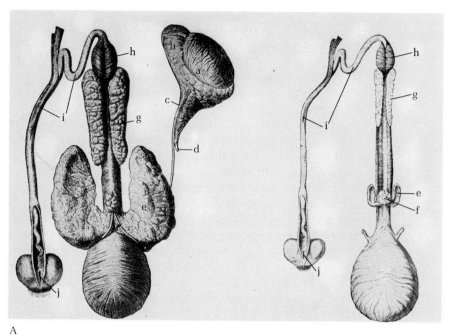

A

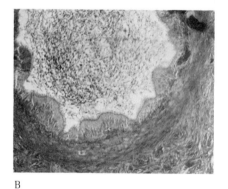

B   C

Figure 7-2   Effects of castration on the genital organs.
A. Genital organs of a normal and a castrated boar. [From *Handbuch der Ver-gleichenden Anatomie der Haustiere*, Springer Verlag, 1927.]

a testis	*f* prostate
b epididymis	*g* bulbo-urethral (Cowper's) gland
c vas deferens	*h* cavernosus muscle
d spermatic cord	*i* penis
e seminal vesicle	*j* orifice preputial pouch

B. Normal vas deferens of a bull ($\times$ 25). Note the tall pseudo-stratified epithelium and the sperm cells in the lumen.
C. Vas deferens of a steer six months after castration ($\times$25).

elimination fights, the winners of which are presumed to produce the larger amounts of androgen. Male mice that are low in the social order can be caused to rise to higher levels by the injection of androgen.

Androgen has other effects that bear no relation to reproduction. The erythrocyte count and hemoglobin concentration are lower in females than in males. Castration of males lowers both these values, and androgen administration returns them to precastration levels. Androgen stimulates protein anabolism in many of the animals in which it also causes increased nitrogen retention; this may account for the more rapid growth and greater adult weight of males. The skeletal muscles of females and castrate males grow less rapidly than those of intact males. Androgen administration increases the number and thickness of muscle fibers as well as the tensile strength and working ability of muscles.

HORMONAL CONTROL OF THE TESTES

It is almost certain that the LH of the gonadotrophic complex is responsible for the stimulation of Leydig cells, which respond by secretion of androgen. Either the LH secreted by chicken pituitary glands differs from mammalian LH or there are actually two LH-like hormones secreted by avian pituitaries. This statement is based on the fact that mammalian LH is capable of inducing only temporary comb growth in hypophysectomized cocks, whereas chicken pituitary glands normally sustain comb growth indefinitely. Furthermore, once chicken testes have lost their ability to respond to mammallian LH they can regain this ability after a few injections with whole avian pituitary glands. This ability is again lost after administration of mammalian LH, and a second interlude of treatment with avian LH is necessary before mammalian LH can cause further androgen production. Histological examination shows that mammalian LH is capable of stimulating only the already differentiated Leydig cells, which soon become exhausted. Avian LH, by contrast, not only causes androgen secretion by the differentiated Leydig cells but also causes differentiation of new Leydig cells as the old ones become pyknotic. (A similar species-specificity to gonadotrophic hormones is seen in female chickens, whose ovaries respond to mammalian substances only to a limited extent—see p. 144.)

The question of hormonal control of spermatogenesis is much more complex than that of androgen secretion. Though it is true that FSH can cause a very significant increase in the diameter of the seminiferous tubules, it appears virtually certain that both FSH *and* LH are essential to full and complete spermatogenesis. It is not clear whether LH itself is necessary or whether its effect on spermatogenesis is by its ability to

induce androgen secretion, which in turn makes complete spermatogenesis possible. The latter possibility is strongly supported by the following evidence.

If sexually mature male rats are hypophysectomized, the testes regress
very rapidly, and spermatogenesis ceases soon after the operation.
However, if androgen is injected into the rats immediately after the
operation, spermatogenesis is maintained for as long as 40 days (but
degeneration of the Leydig cells is not prevented). Still more significant is
the fact that 30–80 percent of the rats hypophysectomized *before* the
onset of spermatogenesis and immediately placed on androgen therapy
show spermatogenesis even though the injected androgen does not
stimulate the testes to increase in size. By contrast, if androgen treatment is delayed for a few days after hypophysectomy, complete testicular degeneration cannot be prevented, and spermatogenesis cannot
be revived.

The fact that androgen can maintain spermatogenesis in hypophysectomized males but cannot rehabilitate spermatogenesis raises the question of the role of androgen in the production of gametes in the testes.
Experimental evidence suggests that androgen does stimulate spermatogenesis and does not simply maintain sperm cells that have differentiated
while the testes are under gonadotrophic stimulation. It seems reasonable to postulate that gonadotrophins (FSH? both FSH and LH?) are
essential to some stages of spermatogenesis, but that the final differentiation can be accomplished by androgen. However, the intricacies of the
hormonal interrelations remain unknown.

Most puzzling is the convincing demonstration by E. and A. Stein-
berger (McKerns, 1969) that it is impossible to induce spermatogenesis
in testicular tissue cultured in vitro simply by adding all known gonadotrophic and other pituitary hormones. However, by adding vitamins
A, E, and C, and glutamine to a chemically defined medium they were
able to initiate spermatogenesis in immature rat testes. On the basis
of both their in vivo and in vitro studies they propose the following
hypothesis for the control of spermatogenesis. The transformation of
gonocytes into spermatogonia may require testosterone, but the mitotic
division of spermatogonia and the formation of spermatocytes may
occur without hormonal stimulation. The reduction–division of primary
spermatocytes may require testosterone, but the maturation of spermatids through spermiogenesis may take place without hormones. The
completion of spermiogenesis appears to need FSH.

Like estrogen, androgen can affect pituitary function. But whereas
minute, physiological quantities of estrogen can regulate pituitary function, much larger than physiological doses of androgen are required to
inhibit or significantly modify pituitary function. For this reason it is

questionable whether androgen participates in the normal feedback mechanism regulating the testicular-hypothalamo-hypophyseal axis. Since estrogen is known to be a normal product of the testes (most notably so in stallions), it appears to be a logical candidate for the role of regulator of the system. Both pituitary and plasma gonadotrophins are known to increase very significantly following castration. This is usually ascribed of course to the absence of steroidal feedback information. Administration of testosterone to castrate males, even in larger than physiological doses, does not reduce the gonadotrophin titers to their normal levels, although there is some depression. Gonadotrophins continue to be produced and released in amounts significantly higher than normal. Complete depression to normal values can be accomplished by small amounts of estrogen. Some investigators are looking for an as yet unidentified substance, referred to as "inhibin" and which may be produced by sperm, as a likely candidate for the regulation of the "push-pull" mechanism.

The question arises whether a "push-pull" mechanism, similar to the one in the female, controls the pituitary-testes relation in the male. Though no clear-cut experiments have demonstrated cyclic functioning of the male, there is a "feeling" on the part of evaluators of semen from bulls and men that the volume, density, and quality of the semen of individuals vary in a cyclic fashion. If there are indeed cyclic variations in the male, they are not as clear-cut and distinct as in the female. The effect of season on the properties of the semen and the fertility of cocks is shown in Figure 7-3.

We have already discussed in some detail the role of the releasing factors in the secretion of hypophyseal hormones and the evidence for a factor depressing the release of prolactin. Without doubt, the male has good use for releasing factors for ACTH; but what of releasing factors for LH and FSH? It has been found that LRF is present in the hypothalami of males, females, and castrated males, but no comparative studies of LRF concentration in males and females have been made. There is one bit of suggestive evidence that there may be distinct "sex difference" between the hypothalami of males and females. As we know, there is no clear-cut male gonadal cycle comparable to the cyclic behavior of female ovaries. If one transplants the adenohypophysis from a male rat into the sella turcica of a female, the male hypophysis (after becoming established) will begin to function in the same cyclic manner in which its "female" predecessor had functioned. This suggests that the pituitary gland does not know its "sex" until the hypothalamic factors direct it to function—either as an acyclic "male" or cyclic "female" pituitary gland. At present there is no reason to argue either for or against a GNRH in the normal function of the pituitary gland of males. The

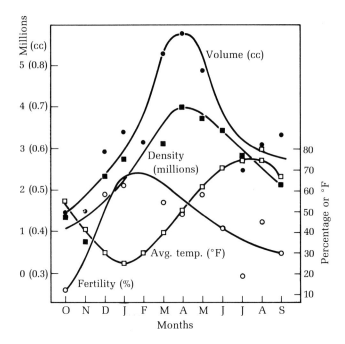

Figure 7–3 Monthly changes in the properties of the semen of cocks. Density and volume are determined by the male alone; the fertility curve is determined by the interaction of males and females. [Data from Parker and McSpadden, 1943, *Poultry Sci.*, 22:142.]

gonadotrophic RFs can be reconciled with a cyclic gonadal function in males if one assumes (as we have before) that the hormonal feedback system involves the action of an as yet unknown substance on a steroid-sensitive hypothalamic center that responds either by withholding or by releasing the RFs. While cyclicity in the male would not be as sharply defined as that of the female, it would very adequately account for the much more relaxed cyclic behavior of males mentioned earlier.

In this connection it is interesting to note that sexual excitation of the male of some species causes a dramatic increase in LH release, and this in turn causes an equally dramatic increase in testosterone in the peripheral plasma. Bulls with cannulas permanently implanted in their jugular veins were led up to cows in heat. By sniffing the vulva most bulls "recognized" that the cows were in heat, as measured by increases in plasma LH levels of the bulls; in some animals the level increased as much as twentyfold. In some bulls sight alone caused increased LH release, in others coitus was required. If the testosterone level was high prior to "recognition," the dramatic rise in plasma LH did not cause it to rise further. If it was low, the rise in plasma LH was paralleled by an equally dramatic rise in plasma testosterone (Katangole *et al.*, 1971).

That a similar effect may occur in other mammals is suggested by the fact that coitus or the presence of a receptive female has been shown to cause a striking increase in plasma testosterone in rabbits and probably men (although the data in these species are not nearly as complete as those quoted above for the bull). In a whimsical contribution to *Nature*, Anon claims that the mere anticipation of coitus causes increased beard growth, which he interprets as an indication of increased testosterone secretion.

SPERM

In an earlier section of this discussion it was pointed out that testicular and epididymal sperm are nonmotile. They become motile only when they are suspended in a fluid. This occurs when the sperm come in contact with the fluids contributed by the male accessory glands. The mixture of sperm cells and accessory gland fluids constitutes semen. The most important attributes of the seminal fluids have been briefly discussed in Chapter 2 (pp. 48–49).

Sperm Cells

The normal sperm cell consists of a head, a neck, a middle piece, and a tail (Figure 7-4). The head is covered by a protoplasmic cap (galea capitis), which at one time was thought to be present only in immature sperm cells but is now known to be a normal component of the head. It is usually dissolved when sperm are treated with fat solvents for staining. The shape of the head varies with the species. It is a flattened ovoid structure in the bull, ram, boar, and rabbit, and round in man. When motile the sperm cell swims fishlike through the suspending fluids. Only when dead does it present its flat surface to the viewer. In birds the head is an elongated cylinder; in the mouse and the rat it terminates in a distinct hook.

The neck, the middle piece, and the tail do not consist of a single solid flagellum (as frequently depicted) but are composed of several (9 or 18) strands or fibrils, which are covered by a sheath. At the very tip of the tail, where the sheath ends, these fibrils flare out into a naked brush. This anatomic detail was first seen under a simple light microscope and described in 1886 by Ballowitz, but it was not until 1941 that it was rediscovered by the use of an electron microscope (which is not so much a tribute to the power of magnification of the electronic gadget as to the thoroughness and power of observation of the earlier workers). Although the morphology of the sperm has been confirmed several times since 1943 (when the original, 1941, observation was published) sperm cells are still pictured in textbooks as having a solid flagellum.

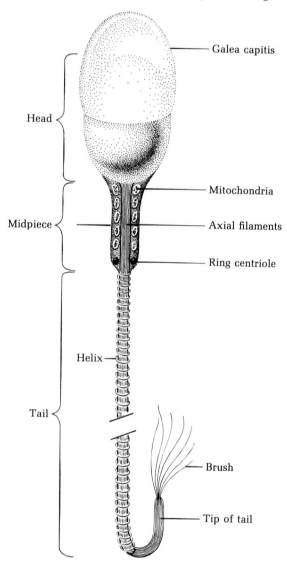

Figure 7-4 Morphology of mammalian sperm cell, seen under the electron microscope. The axial filaments, which begin in the midpiece, extend through the tail and emerge as the brush. The helix coils around the surface of the tail but does not extend to the tip.

Several deviations from the normal morphology are regarded as abnormalities. Among them are giant or dwarf heads, double heads, headless or tailless sperm (frequently caused by rough treatment in preparation for staining or preservation, but often seen in carefully treated preparations), multitailed heads, bent and coiled tails, and protoplasmic heads in the midpiece. These abnormalities are either absent or rare in normal ejaculates. When they occur in large numbers, they impair the fertility of the afflicted male. As a rule, when the number of abnormal sperm approaches 50 percent of the total sperm cells in the ejaculate, the male is sterile—even though the normal-appearing sperm cells in the ejaculate should, theoretically, be more than enough to effect fertilization (see p. 236). Abnormal sperm commonly appear in rams suffering from summer sterility, in males suffering from fever, and in males used for mating too frequently or too young. Often there is no apparent reason for the appearance of abnormal sperm cells in the ejaculate, and this defect may correct itself with time. Certain deformities of sperm cells are also known to be genetic.

As a rule, ejaculates also contain a certain number of dead (nonmotile) sperm cells. Live and dead sperm cells can be distinguished by the use of vital stains—dead cells assume the stain, live ones do not. A few dead sperm do not interfere with normal fertility, but ejaculates containing a high proportion of dead cells (approaching 50 percent of the total count) show either impaired fertility or complete sterility.

In view of the known relation between the proportion of dead or abnormal sperm in an ejaculate and the fertility of males, it seems probable either that dead or abnormal cells affect the fertilizing ability of the normal cells adversely or that the presence of abnormal cells is indicative of a constitutional weakness of the normal appearing cells.

Survival of Sperm in the Epididymis

The lifespan of ejaculated sperm in the female reproductive tract is very short (20–30 hours) and in vitro rarely exceeds a few days if stored under optimal conditions (see p. 237). By contrast, sperm survive a long time, before ejaculation, in the epididymis. This can be demonstrated in several ways; experiments involving blockage of the ducts between the head of the epididymis and the testis have been the most popular. Since this operation prevents the entry of fresh sperm from the testis into the epididymis but does not prevent ejaculation, it is possible to measure the length of time over which epididymal sperm retain their ability to fertilize eggs. In this way it was found that epididymal sperm remain motile in bulls for 60 days. By contrast, sperm isolated in the ampulla of the vas deferens lose motility in less than 72 hours. In guinea pigs epididymal sperm remain fertile for 20–35 days and motile for 59 days; in rats the corresponding intervals are 21 and 42 days. In mice epididymal sperm remain fertile for 10–14 days.

It has been observed in many species that castrated males may ejaculate fertile sperm for several weeks following castration. (This observation has been cited by ancient and even modern naturalists as evidence against the assumption that the testes are the source of sperm.) Reliable records indicate that stallions retain the ability to ejaculate fertile semen for as long as three weeks after castration, and men for as long as six weeks. In castrated rats and guinea pigs epididymal sperm may retain fertilizing ability for many months if the animals are kept on continuous androgen injection.

It is not known what properties of the epididymis or its secretions are responsible for the remarkable longevity of sperm in the epididymal lumen. Since sperm in the epididymis are nonmotile, it is possible that their longevity is attributable merely to physical and metabolic inactivity. This is improbable, however, because it has been observed that androgen plays a role in determining the lifespan of sperm by exerting an effect either directly, on the sperm cells, or indirectly, by acting on the epididymis and its secretion.

In spite of the remarkable ability of the epididymis to maintain the fertilizing ability or at least the motility of sperm for a long time, sperm eventually die even in that most favorable environment. For this reason males may ejaculate nonfertile sperm in the first two or three matings after prolonged sexual rest. In the order of their formation, and hence of their age, sperm in the tail of the epididymis may be dead while those in the head may still be motile and even capable of fertilizing ova.

Number of Sperm and Fertilization

Because of the relatively large size of the female reproductive tract, the major portion of the ejaculate never reaches the oviduct. It has been calculated that of 10–20 million sperm introduced into a rabbit only 5,000 reach the oviduct.

Only a few dozen sperm cells may reach the vicinity of the ovum; several may penetrate into the zona pellucida (polyspermy is much more common than is usually thought), but only one sperm enters the ovum proper and accomplishes fertilization. Because of this decline in the number of sperm from the time of the deposition of the semen in the vagina or cervix until the time the "sperm swarm" reaches the oviduct, it is usually thought essential that no less than 50 million living

Table 7-3. Species differences in characteristics of semen

	NORMAL VOLUME OF EJACULATE (ml)	NORMAL DENSITY OF EJACULATE (1,000 sperm/mm³)*	RANGE OF pH
Cock	0.2–1.5	50–6,000 (4,000)	6.3–7.8
Turkey	0.2–0.8	7,000 (7,000)	6.5–7.0
Boar	150–500	20–300 (100)	7.3–7.9
Bull	2–10	300–2,000 (1,000)	6.4–7.8
Ram	0.7–2	2,000–5,000 (3,000)	5.9–7.3
Stallion	30–300	30–8,000 (100)	6.2–7.8
Rabbit	0.4–6	100–2,000 (700)	6.6–7.5
Dog	2–14	1,000–9,000 (3,000)	6.7–6.8
Fox	0.2–4	30–300 (70)	6.2–6.4
Man	2–6	50–200 (100)	7.1–7.5

*The most commonly observed values are in parentheses.

sperm be present to ensure fertility. Experimentally, however, sperm concentrations even below 100,000 have resulted in good and even maximal fertility. It has been found practical to dilute bull semen for artificial insemination as much as 1:100, and with bulls of very high fertility even greater dilution has given good results. At these rates of dilution the total number of sperm inseminated varies between five and 15 million. Some of the species differences in the characteristics of semen are summarized in Table 7-3.

Preservation of Semen

The increase in the practical importance of artificial insemination is due to two factors: the development of sperm diluters, or extenders, which make it possible to inseminate several females from a single ejaculate, and the ability of semen to withstand storage for from two to four days. In general, the semen of chickens, stallions, and boars cannot be stored satisfactorily, but bull semen withstands dilution and storage very well. (Under proper conditions bull sperm may remain motile for as long as three weeks, but it loses its ability to fertilize after from seven to 10 days. Similarly, chicken, boar, and stallion sperm may, when stored, retain motility for several days, but they lose fertilizing ability within 48 hours after ejaculation.)

Much work has been done on the freezing of sperm for prolonged storage. Chicken, human, and particularly bull semen, have withstood freezing to $-79°C$ (the temperature of dry ice) and have been found, after prolonged storage and thawing out, to have retained unimpaired their ability to fertilize. This advance in semen preservation became possible through the finding that the addition of glycerol to the yolk-citrate diluter prevents crystal formation and hence the rupturing of the cells when they are frozen and thawed before use. This finding is potentially of great significance, for it makes it possible to preserve the semen of outstanding sires and to use it for artificial insemination after the genetic value of the animals has been adequately evaluated, possibly even after their death. Freezing apparently has no adverse effects on the sperm cells of their fertilizing ability, except for the sperm of boars and stallions, which does not retain fertilizing ability after freezing.

Studies at the University of Illinois have shown that cattle semen may retain its motility and fertilizing ability for seven or even 14 days at room temperature if the sperm cells (in a proper diluent) are exposed to carbon dioxide and are stored under the gas in sealed ampoules. In the presence of the gas the sperm cells become nonmotile. It is possible that the gas acts as a narcotic, immobilizing the cells and reducing their metabolic rates. This may account for their ability to survive at room temperature for a long time.

Dimorphism of Sperm

The perennial attempt to separate, mechanically, the X- and Y-bearing sperm cells, as a means of controlling sex determination, appears justifiable on the assumption that X-bearing cells are heavier than the cells bearing the smaller Y chromosome. Measurements of a large number of sperm cells of many species of mammals have provided evidence of a dichotomous distribution of the length of the heads. These measurements, however, may not be valid; for even though the dichotomy may be real it may be due to the measurement of sperm cells of varying ages and stages of maturity.

Lindahl in Sweden unsuccessfully attempted to separate sperm according to size by ultracentrifugation. He found that at lower speeds no separation is effected (as shown by an equality of the sexes in litters of young born from mothers inseminated by sperm thus treated), and that at higher speeds the sperm cells suffer so much damage that they lose their ability to fertilize eggs.

In spite of repeated claims that the two types of sperm can be separated by electrophoresis, no satisfactory demonstration has ever been made that young of predominantly one sex or the other are produced from sperm cells that aggregate at the cathode or the anode of the electric field to which they are exposed. Similar claims of separation of sperm cells by sedimentation rates thus far have not been substantiated.

SELECTED READING

Hamilton, E. B., and R. O. Greep, eds. 1975. "Male Reproductive System" (Vol. 5, Sect. 7, *Handbook of Physiology*). Williams and Wilkins.

Johnson, A. D., W. R. Gomes, and N. L. Vandemark, eds. 1970. *The Testis*, 3 vols. Academic Press.

McKerns, K. W., ed. 1969. *The Gonads*. Appleton-Century-Crofts.

Rosemberg, E., and C. A. Paulsen, ed. 1970. *The Human Testis*. Plenum Press.

Katangole, C. B., F. Naftolin, and R. V. Short. 1971. Relationship between blood levels of luteinizing hormone and testosterone in bulls and the effect of sexual stimulation. *J. of Endocrinol.* **50:**457.

8

The Germ Cells

The importance of the egg was recognized by the ancients, who asserted: "Every living thing comes from an egg." But the realization that the ovum, like all living things, goes through periods of youth, adulthood, and senility is recent. The implications of this seemingly obvious statement are important; they are discussed and documented at some length here because at least part of the high embryonal mortality discussed in Chapter 11 is due to the inability of ova fertilized during senility to complete intrauterine life normally. The male germ cell also ages, but experimental evidence for the aging of sperm is not as plentiful as that for the ovum.

Germ cells do become senile, and their age is directly related to the survival of zygotes. Ova lose their ability to survive as embryos before they lose the ability to become fertilized and to start cleaving. Data are given to show that ova fertilized by senile sperm cells have a lesser chance of completing normal embryonal development than ova fertilized by young sperm.

Sperm are transported through the female reproductive tract, not by their flagella, but by the hormone-controlled contractions of the female duct system. Sperm cells undergo a period of capacitation in the oviduct before they are able to fertilize an ovum.

Fertilization in vitro of 60–70 percent (and in some experiments even 100 percent) of rabbit ova can be accomplished by capacitated sperm. For unknown reasons, the ova of other mammals cannot be fertilized in vitro as reliably as those of rabbits, even with capacitated sperm.

AGING OF GERM CELLS

Before we present the data on aging in germ cells, a few remarks are needed to introduce the subject. Among subprimate mammals the period of sexual receptivity lasts from a few hours (rats and mice) to several days (pigs and mares) (see Table 4–3, p. 127). When left to themselves, males and the females in heat pair off and copulate frequently and repeatedly. Rats copulate almost incessantly during heat, although ejaculation does not occur at each intromission. Under domestication the breeder has taken over the task of determining when, during the sexual receptivity of the female, the single mating or artificial insemination shall take place. If insemination (either natural or artificial) occurs early in heat, the sperm arrive in the oviduct before ovulation and have to wait for the arrival of the ovum. If insemination occurs late in heat, the reverse is true: the ovum is waiting for the sperm, which may have to ripen in the oviduct before it is capable of penetrating the ovum.

Nonprimates have periods of sexual receptivity that synchronize, roughly, the arrival of the germ cells in the place of their union. Primates do not have peaks of sexual receptivity—copulation normally occurs without regard to the time of ovulation. Thus, very young eggs may be fertilized by senile sperm, or aged eggs may be penetrated by sperm of varying ages. In view of these considerations the question arises whether there is any relation between the age of germ cells and their ability to form zygotes or to complete embryonal development to parturition.

In an experiment in which data were obtained to shed light on this question, female guinea pigs were artificially inseminated at various times after ovulation. It was therefore possible to study the fate of ova that were allowed to age for various period, from eight to 32 hours after ovulation, before the sperm had an opportunity to fertilize them. The results (Table 8–1) show that guinea pig ova gradually lose their ability to be fertilized. As late as 14 hours after ovulation 56 percent of the ova could still be fertilized; even as late as 20 hours after ovulation 31 percent could still be fertilized. The important conclusion to be drawn from these data is that aged ova that had been fertilized did not develop to the stage where they could form embryos capable of completing

Table 8–1.   Effect of age of the ovum on fertility in guinea pigs

TIME OF INSEMINATION	NUMBER OF FEMALES	IMPREG- NATIONS	NORMAL PREG- NANCIES	ABNORMAL PREG- NANCIES	AVERAGE LITTER SIZE
After ovulation					
8 hours	78	67%	66%	34%	1.7
14 hours	79	56	27	73	1.6
20 hours	94	31	10	90	1.3
26 hours	86	7	0	100	0.0
32 hours	48	0	0	0	0.0
During heat (control group)	77	83	88	12	2.6

SOURCE: R. J. Blandau and W. C. Young, *Am. J. Anat.*, 64:303, 1939.

Table 8–2.   Effect of age of the ovum on fertility in cattle

HOURS FROM OVULATION TO INSEMINATION	FERTILITY OBSERVED AT 2–4 DAYS		FERTILITY OBSERVED AT 21–35 DAYS	
	TOTAL ANIMALS	ANIMALS WITH FERTILE OVA	TOTAL ANIMALS	ANIMALS WITH NORMAL EMBRYOS
2–4	4	75%	4	75%
6–8	4	75	10	30
9–12	5	60	13	31
14–16	4	25	8	0
18–20	5	40	6	17
22–28	1	0	11	0

SOURCE: G. R. Barrett, *Time of Insemination and Conception Rates in Dairy Cows,* dissertation, University of Wisconsin, 1948.

development and being born. As the age of the ova at fertilization was increased, the size of the litter decreased and the number of abnormalities increased. Similar data have been obtained for rats and cows (Table 8–2).

The fact that ova can retain the ability to be fertilized without retaining the ability to continue normal embryonal development is further emphasized by comparing the effect in cows of time of insemination on ovulation and fertility (Table 8–3).

These data demonstrate two important facts that play an important role in the reproductive efficiency of animals (a topic that will be discussed in greater detail later). First, senility in itself does not prevent an ovum from being fertilized, from undergoing cleavage, and even from becoming implanted. Second, fertilization and implantation of ova do not ensure that they will complete development to parturition. It is obvious, in fact, that the fertilization of aged ova leads at best to the very early death of the zygote, even before implantation. It is

Table 8–3. Effect of time of insemination on ovulation and fertility in cows (cows normally ovulate 14 hours after the end of heat)

TIME OF BREEDING	TOTAL COWS	COWS CONCEIVING FROM ONE SERVICE
Start of estrus	25	44.0%
Middle of estrus	40	82.5
End of estrus	40	75.0
After estrus		
6 hours	40	63.4
12 hours	25	32.0
18 hours	25	28.0
24 hours	25	12.0
36 hours	25	8.0
48 hours	25	0.0
Routine breeding	194	63.4

SOURCE: G. A. Trimberger and H. P. Davis, *Univ. Neb. Res. Bul.* 129, 1943.

probable, however, that fertilized senile ova implant frequently but temporarily, and that they are aborted sometime during the gestation period.

Nonviable embryos resulting from the fertilization of aged eggs may be the cause of the low fertility conspicuous in some animals, especially in monotocous species. The fertilization rates of both mares and cows are 80–85 percent, but only 40–50 percent of the females that conceive produce living young. In a group of women who did not practice contraception, an average of 254 copulations were required to produce living offspring. Later, more optimistic studies found that only about 30 percent of women conceive at the first opportunity after exposure.

It is obvious that most senile eggs, if fertilized, produce zygotes that die before or shortly after implantation. The fate of the exceptions—zygotes that arise from senile eggs but do not die in utero—remains unknown. They may continue to be handicapped in postnatal life by the fact that they originated from senile germ cells, but convincing data on this point are lacking.

Similar, but less overwhelming, evidence is available for male gametes. In mammals, the lifespan of sperm—and therefore the time during which they are able to fertilize eggs—is probably no longer than about 24 hours. Sperm live the longest time in the cervix and the oviducts and the shortest time in the vagina and the uterus. Some bats copulate in the fall but do not ovulate until spring. This fact was interpreted to mean that sperm remain alive in the reproductive tract of bats for several months, lying in wait for the arrival of the ovum. According to more recent evidence, however, bats copulate in the spring as well as in the fall, and fertilization is accomplished by the fresh sperm ejaculate at the spring copulation.

The only cases of very prolonged survival of sperm in the female reproductive tract have been recorded in birds, bees, and sea turtles. Sperm has been shown to survive for as long as 30 days after ejaculation in the chicken, for seven years in the queen bee, and for a few years in the turtle. In mammals spermatogenesis occurs at the lower scrotal temperature, and sperm are ejaculated into an environment of higher body temperature, where they retain fertilizing ability for some hours. In chickens, however, sperm are formed in an environment of high temperature and are ejaculated into an environment of equally high temperature, where they live for many days. Whether this difference is due to secretions of the avian oviduct or to other causes remains to be determined. As far as the ability of the two types of sperm to survive in vitro is concerned, the situation is completely reversed. Mammalian sperm may be made to live in vitro for several days, but avian sperm die in vitro within a few hours after ejaculation. One feels intuitively that a significant clue to the physiological peculiarities of sperm may lie in the differences between avian and mammalian sperm cited here, but the meaning of these differences remains obscure.

The first evidence of the deleterious effect of aged sperm on the survival of the zygote was obtained in birds. Chickens were permitted to mate at will up to a certain date, after which the males were removed. Eggs laid after the removal of the males had therefore been fertilized by sperm that were, on the average, one day older on each succeeding day. The eggs were incubated and the fertilization and hatching rates of the eggs laid each day were determined (Nalbandov and Card, 1943). The data obtained in this way showed that senile sperm retained some ability to fertilize ova, but that hatchability decreased significantly faster than fertility of the ova. There was also a strong negative correlation between the age of sperm and the age of embryos at death. These data show that senile sperm do not lose their ability to fertilize eggs, but that fertilization as such is no assurance that embryonal development will be completed. Not only do ova fertilized by senile sperm die, but they die at an increasingly earlier age as the sperm activating them become more senile. The data cited were obtained on semen aged in vivo; confirming evidence has also been obtained for avian semen aged in vitro.

Similar data have recently been obtained for bull semen that had been diluted, stored for various periods after ejaculation, and used for artificial insemination (Table 8-4). The proportion of cows that failed to remain pregnant increased with increasing age of the semen. It is con-

Table 8–4. Effect of stored diluted bull semen on pregnancy one month and five months after insemination

	AGE OF SEMEN AT INSEMINATION (days after ejaculation)				
	0	2	3	4	5+
Animals inseminated	12	726	756	970	56
Pregnant at 30 days	58%	67%	63%	54%	57%
Pregnant at 150 days	50%	57%	51%	42%	39%
Difference	(8%)	(10%)	(12%)	(13%)	(18%)

SOURCE: Salisbury et al., *J. Dairy Science,* 35:256, 1952.

cluded from these and similar data that the increasing embryonal mortality, which presumably resulted in abortion between 30 and 150 days of pregnancy, is due to the use of sperm of increasing senility.

The changes due to aging may be either physical or chemical. Frog sperm treated with methylene blue, chloral hydrate, or strychnine change their physiological constitution to such an extent that ova fertilized by them formed only pathological larvae that were unable to complete development. Though the physiology of germ-cell gerontology remains unknown and unexplored, its existence should be constantly kept in mind. It is particularly important to time matings or inseminations in domestic animals and man in such a way as to minimize the deleterious effects of aging in ova and sperm cells.

TRANSPORT OF OVA AND SPERM

In all mammals studied fertilization occurs in the oviduct (not in the uterus, as is occasionally stated). Ever since Leeuwenhoek first saw sperm under the microscope, great significance has been attached to the fact that sperm are equipped with motile flagella. It has seemed logical to assume that sperm swim from the place of their deposition to the meeting place with the egg. Despite overwhelming evidence to the contrary, this notion has not been completely dispelled even today.

Evidence shows that sperm transport from the vagina to the oviduct takes only a very few minutes. This has been demonstrated in rats, sheep, guinea pigs, and cows, and probably holds for all other species not yet investigated. The transport through the length of the female duct system is much too rapid to be accounted for by locomotion of the sperm. Furthermore, completely inert substances, such as dead sperm or particles of India ink, reach the oviduct as rapidly as live sperm. These facts have led to the assumption that uterine contractions are involved in transporting the sperm through the duct system.

Table 8-5. Time spent by the ovum in the oviduct and subsequent degree of development in some mammals

	DAYS SPENT IN OVIDUCT	STAGE OF DEVELOPMENT ON ARRIVAL IN UTERUS
Ungulates	3–4	4–16 cells
Rat, mouse, rabbit	3	Morula or blastula
Guinea pig	4	8 cells
Ferret	5	16–32 cells
Cat	8	Blastocyst
Dog	10	Blastocyst
Primates	4	16 cells

There is no complete agreement with regard to the spontaneous activity of the uterus during the cycle. In women and in certain other mammals there seems to be a peak of contractility during the follicular phase, especially shortly before ovulation; contractility gradually subsides during the luteal phase (Reynolds, 1949). By contrast, no such differences were seen in cows (Hays and VanDemark, 1953); uterine motility in cows is about the same during both heat and the luteal phase. In both cattle and sheep sperm reach the oviduct equally as fast during the luteal phase as during heat. Because it is known that oxytocin is released during the physical or psychological sexual excitation of females, it appears probable that sperm transport is accomplished by the action of oxytocin on the uterine muscle, which responds by a series of contractions greater than those normally seen in the sexually unexcited female.

Whereas sperm are transported to the oviduct extremely rapidly by muscular contractions of the duct system, the transport of the ova through the oviduct requires several days. Fertilized or unfertilized ova are propelled through the oviduct away from the ovary and toward the uterus by means of the ciliary beat. The length of time required by ova of different species to reach the uterus and the degree of development achieved during the journey are shown in Table 8-5. The rate at which ova travel through the oviduct is not uniform: three-fourths of the time is spent in the uterine half. The rate of descent may depend on the degree of estrogen stimulation since rapid descent occurs when the female is in heat and immediately after heat, when the amount of circulating estrogen is presumably greater than it is two or three days after ovulation. However, large doses of exogenous estrogen have an adverse effect on oviducal motility: the ova become arrested in the oviducts ("tube locking") and degenerate there. (By contrast, ova that are prevented from leaving the oviduct by surgical occlusion do not degenerate but proceed to form blastocysts.)

Table 8-6. Capacitation of rabbit sperm deposited in oviducts at various times before and after ovulations

TIME OF INSEMINATION RELATIVE TO OVULATION	HOURS	NO. OF RABBITS	NO. OF OVULATIONS	OVA RECOVERED	
				TOTAL	FERTILIZED
Controls					75–100%
Before	8	5	50	47	55
Before	6	4	49	41	78
Before	4	5	39	34	6
Before	2	5	42	31	0
After	1	2	14	10	0
After	2	5	58	40	0

SOURCE: M. C. Chang, *Nature*, 168:697, 1951.

FERTILIZATION

Even though it can no longer be doubted that sperm are transported through the female duct system by uterine and oviducal contractions and that some of them reach the oviduct within a few minutes after being deposited in the vagina, it remains unknown whether the first sperm to arrive can fertilize the egg. This becomes all the more problematic in view of the work of Chang (1955, 1975) and Austin (1952), who found in rabbits and rats, respectively, that sperm must be exposed for at least six hours to tubal, uterine, or vaginal secretions before they acquire the ability to fertilize eggs. This phenomenon, called "capacitation," is well illustrated in Table 8-6. In vitro fertilization was quite unsuccessful until rabbit eggs were fertilized using sperm recovered from the oviducts in which they had become capacitated. The changes that sperm undergo in acquiring fertilizing capacity involves enzymes in the sperm head.

Rabbit ova can be easily fertilized in vitro; the exposed eggs cleave normally and, if transferred into suitably prepared foster mothers, they implant and are carried to term. In an exhaustive study by Dauzier and Thibault, in vitro fertilization was achieved in 67 percent of 7,605 ova from 1,954 does. Dauzier and Thibault offer some evidence that the rabbit ovum may be coated with a substance they call "fertilizine," which repels even the capacitated sperm. If the eggs are frequently washed with Locke's solution (in which they are also cultured), fertilizine is removed from the egg surface and sperm penetration occurs within three to seven hours after exposure. They further postulate that the Fallopian tube normally contains a substance they call "antifertilizine," which counteracts fertilizine and makes ova penetrable to sperm. The evidence supporting the existence of these two substances is not completely convincing, and the question arises whether the fertilizability of an egg may not simply depend on its age or physiological maturity.

For unknown reasons, eggs from other mammals (sheep, pigs, and cattle) cannot be fertilized in vitro as easily and reliably as those of rabbits.

TRANSPLANTATION OF OVA

Ever since W. Heape in 1890 succeeded in obtaining living young by transplanting fertilized eggs from a rabbit of a small breed into a foster mother of a large breed, this technique has excited the imagination of research workers. The potential uses of the technique appear great, but the difficulties encountered have kept it from becoming as useful as artificial insemination. It was hoped that one could shorten the length of generations of slowly maturing domestic animals by superovulating sexually immature females and transplanting the eggs thus obtained into foster mothers, thus determining the genetic worth of the donor long before it loses its ability to furnish more eggs. It was also hoped that once genetically valuable females had been found one could increase enormously the number of their progeny by relieving them of the time-consuming tedium of pregnancy, using them only as egg producers, while other genetically less valuable females were used as incubators. Today, in spite of intensive work, the method of ova transplantation remains only a valuable tool for basic biological research; it has been wholly unsuccessful in the improvement of breeds.

For ova transplants to be successful in practical breeding, several basic conditions must be met:

1. The number of eggs ovulated must be greater (especially in monotocous species) than at normal ovulation.
2. Either the time of ovulation must be accurately known, or, preferably, ovulation must be made to occur at the convenience of the investigator.
3. The ova must be obtainable without major surgical invasion of the donor and must be implanted in the recipient without surgery.
4. Recipient females must be in a stage in the reproductive cycle when the transplanted eggs have a good chance of finding a uterine environment favorable for implantation.

Although each of these conditions can be met separately, difficulties have been encountered in meeting all four, especially in larger animals. Great individual variability in response to superovulating dosages of

Table 8-7. Success of transplantation of ova in sheep and rabbits

	TOTAL ANIMALS	NO. OF OVA TRANSPLANTED	NO. OF YOUNG BORN	PROPORTION OF YOUNG TO TOTAL OVA
Sheep	18	19	8	42%
Rabbits	249	1,478	415	28%

SOURCE: G. L. Hunter *et al.*, *J. Agr. Sci.*, 46:143, 1955; and O. Venge, *Acta Zoologica*, 31, 1950.

Table 8-8. Success of superovulation techniques in ruminants

	TREATMENT FOR SUPEROVULATION	TOTAL ANIMALS	AVERAGE NO. CORPORA LUTEA	OVA RECOVERED	FERTILIZED
Cattle	PMSG—luteal phase	33	6	38%	38%
	PMSG—follicular phase	15	12	22	44
	AP extract—follicular phase	10	6.5	74	92
	Controls	7	1.0	100	100
Calves*	Gonadotrophic hormones	17	15.4	27	6
Sheep	Progesterone, PMSG	10	5.1	45	84

*From 29 days to 30 weeks old.

SOURCE: D. F. Dowling, *J. Agr. Sci.*, 39:374, 1949; W. G. R. Marden, *ibid.*, 43:381, 1953; and G. L. Hunter *et al.*, *ibid.*, 46:143, 1955.

gonadotrophic hormone have been detected, so that the number of ovulations obtained in cattle may vary from none to 50 or 60. Cattle also become refractory after one or two courses of treatment with gonadotrophins (possibly because antihormone is formed), and a long rest period must intervene before cattle ovaries again respond to hormone injections. Efforts to flush ova from the duct system of cattle without surgical intrusion into the body cavity have also failed. The most reliable method of flushing eggs from cattle oviducts still requires the slaughter of the animals; calves have been obtained using this method. From smaller animals, such as sheep and rabbits, one can obtain ova by laparotomizing the female and flushing the eggs from the oviducts. Even in these smaller animals, however, only a small proportion of the transplanted eggs ever result in living young carried to term (Tables 8-7 and 8-8).

One of the most recent developments in transplantation of embryos is the newly developed technique of freezing embryos prior to transplantation into a host. As could be expected, this technique has so far been most successful with laboratory mice; mice embryos can be frozen and stored for as long as one year without serious loss of ability to develop in a host mother. As for larger animals, it has been demonstrated that cow embryos can be frozen and later transplanted into an acceptable host. However, the same difficulties beset this technique as those mentioned for transplantation of fresh embryos (unpredictability of the success of superovulation, low fertilization rates, refractoriness of females to hormone treatment, etc.). Since the transplantation of frozen embryos has so far shown such a low rate of success (less than 10 percent in cattle, about 30 percent in sheep), it could be considered practical only under the most unusual circumstances. It might become practical to import frozen embryos into countries that prohibit importation of live animals. In cases of irreparable physical injury to genetically valuable females one could superovulate them and freeze their embryos for future transplantation.

Nevertheless, it seems highly unlikely in the near future that any of the transplantation techniques developed so far will be able to compete successfully with natural mating in commercial breeding. Transplantation will probably remain an interesting laboratory tool for specialized studies. It can be used to good advantage in studies of maternal influences (see the interesting study by Venge, 1950) and the effect of the uterine environment on the survival of embryos. It has been claimed that it is now more practical to maintain mutant strains of laboratory mice in the form of frozen embryos than to follow the much more expensive and genetically vulnerable practice of maintaining numerous strains of live animals.

SELECTED READING

Austin, C. R. 1952. The "capacitation" of the mammalian sperm. *Nature*, **170**:326.

Black, W. G., L. C. Ulberg, R. E. Christian, and L. E. Casida. 1953. Ovulation and fertilization in the hormone stimulated calf. *J. Dairy Sci.*, **36**:274.

Blandau, R. J., and E. S. Jordan. 1941. The effect of delayed fertilization on the development of the rat ovum. *Am. J. Anat.*, **68**:275.

Boyd, J. D., and W. J. Hamilton. 1952. Cleavage, early development and implantation of the egg. In *Marshall's Physiology of Reproduction*, Vol. 2, 3rd ed. Longmans.

Chang, M. C. 1955. The maturation of rabbit oocytes in culture and their maturation, activation, fertilization and subsequent development in the fallopian tube. *J. Exp. Zool.*, **128**:379.

Chang, M. C., and R. M. F. Hunter. 1975. Capacitation of mammalian sperm: biological and experimental aspects. In "Male Reproductive System" (Vol. 5, Sect. 7, *Handbook of Physiology*), edited by D. W. Hamilton and R. O. Greep. Williams and Wilkins.

Dowling, D. F. 1950. Problems of the transplantation of fertilized ova. *J. Agr. Sci.*, **39**:374.

Hays, R. L., and N. L. VanDemark. 1953. Effects of oxytocin and epinephrin on uterine motility in the bovine. *Am. J. Physiol.*, **172**:557.

Hunter, G. L., C. E. Adams, and L. E. Rowson. 1955. Inter-breed ovum transfer in sheep. *J. Agr. Sci.*, **46**:143.

Marden, W. G. R. 1953. The hormone control of ovulation in the calf. *J. Agr. Sci.*, **43**:381.

Marshall, J. G. 1974. Effects of neurohypophysial hormone on the myometrium. In "The Pituitary Gland and Its Control, Part I" (Vol. 4, Sect. 7, *Handbook of Physiology*), edited by R. O. Greep and E. B. Astwood. Williams and Wilkins.

Nalbandov, A. V., and L. E. Card. 1943. Effect of stale sperm on fertility and hatchability of chicken eggs. *Poultry Sci.*, **22**:218.

Reynolds, S. R. M. 1949. *Physiology of the Uterus*, 2nd ed. Hoeber.

Trimberger, G. W., and H. P. Davis. 1943. Conception rate in dairy cattle by artificial insemination at various stages of estrus. *University of Nebraska Research Bulletin* 129.

Venge, O. 1950. Studies of the maternal influence on the birth weight in rabbits. *Acta Zool.* (Stockholm), **31**:1.

Young, W. C. 1953. Gamete age at the time of fertilization and the course of gestation in mammals. In *Reproductive Wastage*, edited by W. C. Young. Thomas.

9

The Young
Embryo

This chapter touches very briefly upon the important and complex process of the spacing of embryos in the uterus of polytocous mammals and the migration of embryos from one uterine horn to the other. The mechanisms causing the remarkably even spacing of embryos remain unknown. The processes of implantation and placentation are reviewed, and the fundamental processes of uterine nutrition of the embryo are discussed.

INTRODUCTION

After fertilization and the early cleavages of the egg are completed in the oviduct, the young embryo, consisting of from eight to 16 cells (the blastocyst stage), arrives in the uterus in search of permanent attachment. It is now only a small ball of metabolically very active cells. Because mammalian embryos do not have yolk stores such as those that supply energy for the growth of the young embryos of birds, reptiles, and other lower vertebrates, the early mammalian young must live off the environment in which it finds itself—the uterus. During the first days after its arrival in the uterus the embryo is completely dependent upon uterine secretions for its energy. The uterine glands secrete "uterine milk," which is composed of protein, fat, and traces of glycogen. It is from uterine milk and the cellular debris from the epithelial lining of the uterus that the young embryo derives its sustenance until it implants itself and forms permanent placental connection with the maternal circulatory system.

SPACING AND UTERINE MIGRATION

An interesting problem, which remains unsolved, is the mechanism in the uterine horns of polytocous mammals that governs the spacing and distribution of the embryos in the uterus. Consider the problem: embryos usually enter the uterine horn as a loose clump; as they are propelled inside the uterine lumen by irregular myometrial contractions, one of them finds a suitable spot near the oviducal end of the uterine horn and remains behind while its mates continue on their way; other embryos drop out further along the lumen until all have found suitable endometrial folds in which to implant themselves. The distances between implantation sites are approximately equal; if there are many embryos to be accommodated, the distances are shorter than if there are few embryos. If more embryos find themselves in one uterine horn than in the other, the excess migrate to the other. (This, of course, can occur only in females in which the two uterine horns are connected and not separated as they are in females that have duplex uteri.)

The phenomenon is especially well illustrated in the pig, in which the left ovary functions with significantly greater frequency than the right. The frequency of ovulations in the two ovaries and the distribution of fetuses between the two horns are shown in Table 9-1. Though we do not know why some spots in the uterine lumen are attractive for implantation, spacing can be partly explained by the finding that once a blastocyst becomes implanted adjacent areas of the endometrium are no longer receptive to other blastocysts. This of course does not explain how the uterine lumen knows how many blastocysts it is expected to accommodate, how much space to allot to each of them, and how many of them should be shunted from one horn into the other.

Uterine migration occurs not only in polytocous but also in monotocous mammals. Implantations of blastocysts in the horn opposite to the ovary from which the egg or eggs were ovulated have been recorded in sheep (3–50 percent), goats, cattle, and horses, and also in the lower primates, in which it is rare. In rabbits, which have a duplex uterus,

Table 9–1. Comparison of right and left ovulations with the distribution of fetuses in the two horns of the pig

CORPORA LUTEA		FETUSES	
RIGHT OVARY	LEFT OVARY	RIGHT HORN	LEFT HORN
2,289	2,830	1,936	1,932
44.7%	55.3%	50.0%	50.0%

SOURCE: B. L. Warwick, *Anat. Rec.*, 33:39, 1926.

transuterine migration cannot occur, but eggs ovulated by one ovary
may be picked up by the fimbria of the oviduct on the opposite side.
Such external migration occurs in a few animals that have an open
fimbria.

IMPLANTATION

If the zygote arrives in the uterus in a preblastocyst cleavage stage (Table
8–5, p. 246), it rapidly completes the development into a blastocyst. In
ungulates and carnivores the blastocyst elongates and enlarges until it
fills a good part of the uterine cavity; this is known as *central implantation*. In rodents the blastocyst remains small and becomes lodged in a
fold of the uterine lumen, where it implants itself; this is known as *eccentric implantation*. In the guinea pig, insectivores, and man the blastocyst implants itself *interstitially* by passing through the uterine
epithelium and becoming completely cut off from the uterine lumen.

During its early life in the uterine lumen the blastocyst is unattached
and free-floating. In animals in which implantation is eccentric or interstitial, the transition from the free-floating blastocyst to the definitely
implanted embryo is rather clear-cut. In man it occurs in 6–8 days; in
mice, 5 days; in guinea pigs, 6 days; in rabbits, 7 days; and in cats, 13
days after ovulation. In the mare, the cow, and the ewe the transition is
not so clear-cut. If the uterine horn of the pregnant mare is opened as
late as eight weeks after mating, the much enlarged and elongated
chorionic sac separates from the uterus quite easily and without tearing
either fetal or maternal tissue; complete attachment occurs as late as
the fourteenth week of gestation. A similar equivocal situation exists
in the cow, in which corancular attachment is said to take place as late
as day 40 after ovulation. In sheep this happens on day 18. In the sow
blastocysts float free in the uterus until between days 11 and 20 of pregnancy. By day 11 the chorion is in apposition with the uterine epithelium,
which by that time has formed irregular ridges into which the chorionic
folds mold themselves. According to Corner, the corrugation of the
uterine epithelium is helpful in anchoring the chorionic vesicle and

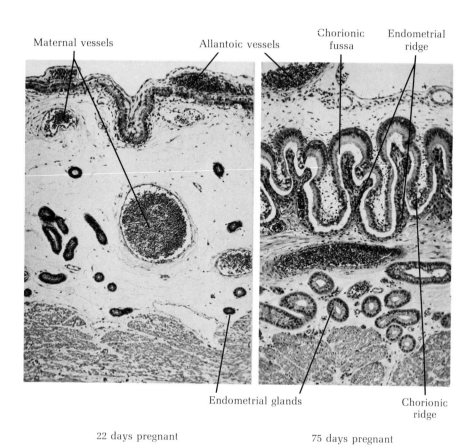

Maternal vessels   Allantoic vessels   Chorionic   Endometrial
                                                      fussa   ridge

Endometrial glands   Chorionic
                                                        ridge

22 days pregnant   75 days pregnant

Figure 9-1 Relation between the allantochorion and the endometrium in the sow early and late in pregnancy. [By permission of Professor E. C. Amoroso.]

making attachment of the embryo possible (see Figure 9–1). (For endocrine mechanisms causing embryo attachment, see Chapter 11, "Delayed Implantation," p. 290.)

PLACENTATION

As the blastocyst increases in size, it can no longer absorb enough nutritive material by diffusion, as it does during the early stages of its sojourn in the uterus. Implantation is a step toward the eventual formation of embryonic membranes that give the growing embryo access to the maternal circulatory system. Thus, the transition from embryotrophic nutrition, during which the embryo subsists on "uterine milk" (a product of the uterine glands), to hemotrophic nutrition is made. During the hemotrophic stage the placenta (literally, "flat cake") is formed. Mossman

Table 9–2. Classification of types of placentation (according to Grosser)

	GROSS SHAPE	EXAMPLES OF SPECIES
Epitheliochorial	Diffuse	Pig, horse, donkey
Syndesmochorial	Cotyledonary	Sheep, goat, cow
Endotheliochorial	Zonary or discoid	Cat, dog, ferret
Hemochorial	Discoid or zonary	Primates
Hemoendothelial	Discoid or spheroidal	Rat, rabbit, guinea pig

characterizes the formation of the placenta as an "intimate apposition or fusion of the fetal organs to the maternal tissues for physiological exchange."

It is not within the scope of this book to provide a detailed discussion of placentation, a large, complex, and fascinating subject; the reader is urged to consult the excellent summary by Amoroso (1952). In animals with centrally implanting blastocysts there is no destruction of the surface epithelium of the uterus where the apposition of epithelium and vesicle occurs. In animals with eccentric and interstitial implantation, however, the uterine epithelial cells that are in contact with the trophoblast appear to erode. Shortly after attachment the organization of placental membranes begins, and the chorion, amnion, and allantois are formed.

The different types of placentae found in mammals can be classified in several different ways, one of which is summarized in Table 9-2 (see also Figures 9-2 and 9-3). The epitheliochorial and syndesmochorial types fall into the class of "apposed," or "nondeciduate," placentae, in which there is no intimate fusion between the maternal and fetal placentae. At parturition the fetal placenta of these types separates easily from the uterine endometrium, without bleeding or major damage. The other three types of placentation are "conjoined," or "deciduate." In them an intimate connection is established between fetal and maternal tissues, leading to difficult separation, tearing of maternal tissue, and bleeding at parturition. In marsupials and in the mole the placenta is not shed at parturition but is gradually absorbed.

There appears to be no obvious correlation between the type of placentation and the readiness of females to conceive after parturition. Animals with apposed placentae, in which practically no uterine damage occurs when the placenta is sloughed off at birth, fall into two categories. Mares and sows (provided the latter do not lactate) have postpartum heats and ovulations in 3–10 days after parturition and can conceive if bred; cows have no heats or ovulations for 30–60 days after parturition, and during this interval the uterus involutes. It should be remembered that ruminants have large, specialized coruncular areas, at which attach-

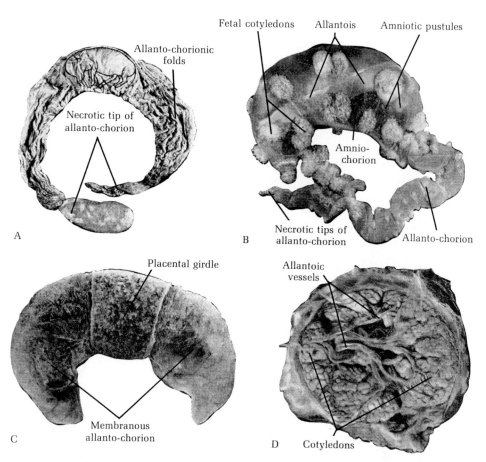

Figure 9-2 Types of placentae. **A.** Epitheliochorial-diffuse, from a sow. **B.** Syndesmochorial-cotyledonary, from a cow. **C.** Hemochorial-zonary, from a cat. **D.** Hemoendothelial-spheroidal, from a rabbit. [By permission of Professor E. C. Amoroso.]

ment of the embryo occurs. In the pregnant cow these may reach the dimensions of a hand, but in the nonpregnant they are only about 1–2 cm in diameter (Figure 9–4). The relation between the enormous corancular development in ruminants and the need for their involution, on the one hand, and the absence of heats and ovulations after parturition, on the other, is not clear. (See the discussion, pp. 109–110, of the effect of beads implanted in the uterine horns of sheep during the estrous cycle.)

By contrast, both rats and mice, which have conjoined placentae and suffer severe uterine trauma at parturition, can conceive within 24 hours after giving birth to a litter (provided they are not allowed to nurse the

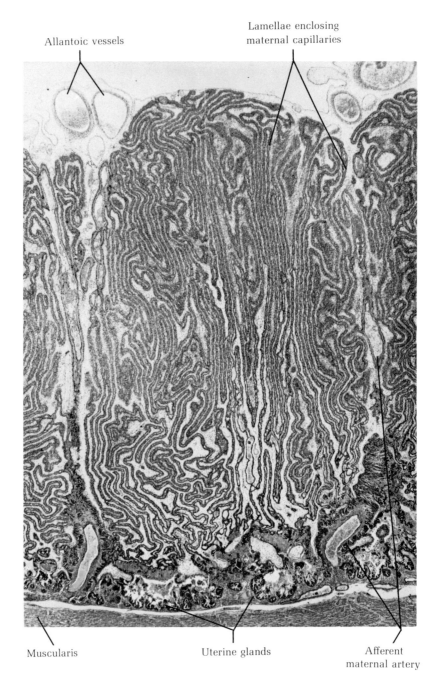

Allantoic vessels

Lamellae enclosing
maternal capillaries

Muscularis

Uterine glands

Afferent
maternal artery

Figure 9–3 Section through the chorioallantoic placenta and uterine wall of a cat near term. Compare the intricacy of this system with the simplicity of organization in the sow (Figure 9–1) and the cow (Figure 9–4). [By permission of Professor E. C. Amoroso.]

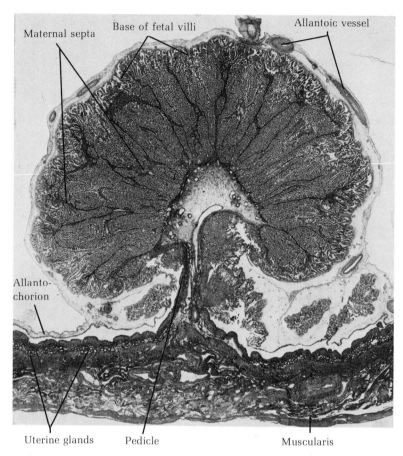

Maternal septa   Base of fetal villi   Allantoic vessel

Allanto-
chorion

Uterine glands   Pedicle   Muscularis

Figure 9–4 Vertical section through a cotyledon of a cow on day 90 of pregnancy. Note the relation between the maternal tissue and the fetal villi. Compare with Figure 9–3. [By permission of Professor E. C. Amoroso.]

young). Similarly, women not infrequently are known to conceive from one to three weeks after giving birth to a child at term. It is possible, of course, that the job of repairing the damage inflicted on the uterus by placentation is greater in imagination than in reality and that repair is easily accomplished by a vigorous organism in a short time.

NUTRITION OF THE EMBRYO

Both the mare and the pig have epitheliochorial placentation, and in both of them embryotrophe continues to play an important role in that the uterine milk secreted by uterine glands furnishes a good share of the nutrients that diffuse through the placental membranes and enter the fetal circulation. In advanced pregnancy in these species, however, and throughout pregnancy in the species with other types of placentation, the nutrients must come from the maternal circulatory system and pass through the placenta into the fetal circulation.

The sheep fetus weighs about 1,000 gm at 80 days and 8,000 gm at 140 days of gestation; the rabbit fetus weighs less than 1 gm at 16 days and 64 gm at term (32 days). In view of this extremely rapid growth of the fetus and the corresponding though less spectacular increase in weight of the fetal membranes, the question arises how the conceptus manages to divert so much nutrient material from the maternal circulation to its own use. It is possible that in all physiologic systems tissues with a higher metabolic rate have a prior claim to the available nutrients. It is known that the fetal and placental cells have a higher metabolic rate than the cells of the maternal tissues, and that fetal red cells have a greater oxygen-carrying capacity than maternal red cells.

This concept of metabolic gradients has been elaborated by Huggett and Hammond (1952) to account for a hierarchy of priorities within the growing fetus itself: (in the order of decreasing metabolic activity and rapidity of growth) the central nervous system, bone, muscle, and fat. This hierarchy is well demonstrated by experiments in which pregnant females are fed severely deficient diets; although the mother loses weight, the young increase in weight. Furthermore, the nervous tissue of young gestated under severe dietary restrictions is the least affected. Bone is the second most affected, and muscle tissue, with its lower metabolic level, is affected most adversely.

Different types of placentae show varying degrees of permeability to minerals, carbohydrates, fats, and proteins. In both pigs and cows (epitheliochorial and syndesmochorial placentations) the fetal blood shows a significantly higher concentration of blood glucose than the maternal blood. By contrast, in all other animals studied (dog, rabbit, rat, guinea pig, and man) blood sugar is always lower in the fetal blood than in the maternal blood. The significance of these differences is not known, but it is noteworthy that the babies of diabetic women are significantly heavier at birth than the babies of normal women, and that the blood-sugar level of diabetic women is higher than that of normal women. Whether the higher growth rate of the fetus and the ultimately greater size of the infant are the result of the abnormal carbohydrate metabolism of the mother or whether they are due to other causes is not known.

It is now well established that fetal size and the birth weight of the young, up to the "normal" size and weight, depend on the plane of nutrition during the second half of gestation. A low plane of nutrition during the first half of pregnancy does not affect ultimate fetal weight adversely, provided the dietary intake is adequate during the second half. An inadequate plane of nutrition during the second half of pregnancy, or competition for available nutrients between twins in monotocous animals or among very large litters in polytocous animals, reduces the size of the young at birth. Excessive food intake by the mother, however, does not cause fetal growth beyond the "normal" size.

In both rats and sheep injection of progesterone throughout or during part of gestation causes an increased size of the embryos. In this connection it should be mentioned that rats maintained on a completely protein-free diet during gestation carry gestation to normal completion if they are also injected with progesterone throughout. They lose much weight, and the young are much smaller than normal at term, but the size of the litter is not affected.

PLACENTAL PERMEABILITY
TO HORMONES AND AGGLUTININS

Placental permeability increases with advancing gestation. The placenta is permeable to insulin and to steroids. Thus, baby girls and occasionally boys may be born with developed breasts capable of lactation (witches milk). The vaginal epithelium is also proliferated and cornified, indicating that maternal steroids are responsible for these effects. Shortly after birth both the breasts and the vaginal epithelium regress. The placenta is probably permeable to ACTH but not to the gonadotrophins. Foster and his colleagues (1972) have shown that elevation of LH levels in maternal peripheral circulation by the injection of LHRH does not result in a rise of fetal LH levels. Conversely, very significant rises in fetal LH due to injection of LHRH does not cause any change in the maternal base level of LH even though the fetal rise is maintained for as long as a few hours.

In contrast to the impermeability of the placenta to LH, the placenta is apparently permeable to PMSG: a very significant stimulation of the fetal gonads coincides with the peak in PMSG secretion by the mare. Similarly, later in gestation, when maternal estrogen becomes prevalent, the growth of fetal gonads and accessory glands is presumed to result from estrogen stimulation.

Somewhat out of context, we mention here the unique case of freemartins. In about 11 out of 12 cases in which a female is gestated as a twin to a male in cattle, the reproductive system of the female is modified to a greater or lesser degree, and she is made completely sterile by intersexuality. This condition was first described in 1792, but it was not until 1917 that Lillie advanced an explanation for the intersexuality of the female. Lillie found that in all cases of cattle twins of different sex in which an anastomosis of the chorionic vessels occurs a modification of the female comes about. This is thought to occur because androgen secreted by the testes of the male twin passes directly through the anastomosed chorionic vessels into the sexually undifferentiated female twin and modifies its reproductive system to the point of intersexuality. Though this explanation is still generally accepted, there is some unhappiness over the fact that anastomoses of chorionic vessels occur in other animals, notably in cats and in marmoset monkeys, in which no intersexuality is noted in spite of careful searches for it. The work of K. Ryan may provide a possible explanation for this discrepancy. He found that the bovine placenta is relatively unable to convert androgen to estrogen, while the placenta of the marmoset monkey can do so readily. This fact may explain why no intersexes of the freemartin type are known in this species, even though anastomoses of chorionic vessels are frequently seen.

To return to placental permeability, young placentae are generally impermeable to bacteria and molecules much beyond 4–5 microns in size. The fact that young are frequently born infected with bacterial diseases is generally ascribed to placental senility or to placental lesions, which are used as a port of entry for organisms and protein molecules. There is an interesting case in which a fetal agglutinogen immunizes the mother and the maternal agglutinin crosses the placenta to the fetus. This is the celebrated case of the Rh factor or the erythroblastosis fetalis, which causes the fetus to die in utero, or shortly after birth, of anemia and jaundice. The condition is brought about by a combination of genetic and physiologic factors. The fetus inherits from the father an agglutinogen that is absent from the mother. This fetal Rh factor passes through the placenta (damaged or senile?) and causes the mother to form antibodies to Rh. Maternal iso-agglutinins for Rh return to the fetus and react with fetal Rh agglutinogens, which cause destruction of blood cells and erythroblastosis.

The immunology and genetics of this situation have been worked out only for primates, but they are potentially applicable to all proteins and to other species. Factors similar to the Rh factor may play a role in the fetal mortality of mammals other than primates (see Chapter 10).

SELECTED READING

Amoroso, E. C. 1952. Placentation. In *Marshall's Physiology of Reproduction*, Vol. 2, 3rd ed. Longmans.

Böving, B. G. 1956. Rabbit blastocyst distribution. *Am. J. Anat.*, **98**:403.

Burger, J. F. 1952. Sex physiology of pigs. *Onderstepoort J.*, Suppl. No. 2, pp. 2–217.

Foster, D. L., F. J. Karsch, and A. V. Nalbandov. 1972. Regulation of luteinizing hormone (LH) in the fetal and neonatal lamb. II. Study of placental transfer of LH in the sheep. *Endocrinology*, **90**:589.

Hafez, E. L. E., and G. Pincus. 1956. Hormonal requirements of implantation in the rabbit. *Proc. Soc. Exp. Biol. Med.*, **91**:531.

Huggett, A. St. G., and J. Hammond. 1952. Physiology of the placenta. In *Marshall's Physiology of Reproduction*, Vol. 2, 3rd ed. Longmans.

Warwick, B. L. 1926. Intra-uterine migration of ova in the sow. *Anat. Record*, **33**:29.

Winters, L. M., W. W. Green, and R. E. Comstock. 1942. Prenatal development of the bovine. *University of Minnesota Technical Bulletin* 151.

10

Efficiency of Reproduction

The rates at which reproductive phenomena occur are controlled by hormones, and the rates at which hormones are secreted are controlled by genes. All hormone-controlled phenomena, such as rate of growth, rate of milk secretion, and rate of ovulation, are therefore subject to genetic selection. This is the basis for the improvement of domestic animals by genetic selection; selection is actually for rates at which glands function, but the results of selection are measured in rate of growth, egg production, etc.

The ovulation rate also depends, to no mean degree, on age ("trajectory of reproductive performance"), previous reproductive experience (parity), and the nutritional state.

Although, in most polytocous animals the fertilization rate approaches 100 percent, very few females manage to go through gestation without embryonal mortality. About 60-70 percent of all the eggs ovulated result in live young. Embryonal mortality may be due to an "unfriendly" uterine environment, senile ova or sperm, and, in the pig at least, overfeeding.

INTRODUCTION

The number of young born depends on the number of eggs ovulated, the number of eggs fertilized, the number of fertilized eggs that are capable of cleaving and implanting themselves, and the number of implanted blastocysts that are able to survive through the whole gestation and to be born as live young. A variety of factors, which will be discussed forthwith, play a role in each of these critical events. Some of the factors are of purely maternal concern (such as ovulation rate and uterine environment). Others depend on the interaction between male and female germ cells (fertilization); and still others depend on the contribution of the parents' germ plasm to the embryo (lethals or semilethals may cause the death of the embryo). We shall consider each of these factors and present the available data.

OVULATION RATES

Genetic Control

We already know that the rate of secretion of gonadotrophic hormone determines the number of follicles ripened and the number of eggs ovulated. Later (Chapter 12) we shall consider the possibility of increasing natural ovulation rates by the use of exogenous hormones. Here only natural ovulation rates will be discussed.

As Table 10-1 shows, litter sizes vary greatly within the same species of some common domestic animals. Some breeds of domestic pigs have an average litter size of about seven (Mangalitza), but others average 11 (Norwegian Land Race). MacArthur (1949) showed what can be accomplished by genetic selection in a polytocous animal. Starting with a strain

Table 10-1. **Average litter sizes in some common domestic animals**

	LOW	HIGH
Pig	6.6*	11.2
Rat	6.1	11.1
Mouse	4.5	7.4
Rabbit	4.0	8.1
Dog	3.0	12.0

*The average litter size of the wild pig is four.

Table 10-2. Fecundity of sheep

BREED	LAMB CROP	BREED	LAMB CROP
Cheviot	89.1%	Suffolk	144.3%
Scottish Blackface	93.1	Corriedale (Canadian)	146.0
Karakul	110.0	Shropshire	162.0
Corriedale (U.S.)	118.0	Leicester	163.0
American Shropshire	126.2	East Frisian Milk Sheep	205.1
Dorset	127.4	Romanov	238.0
Lincoln	138.9		

SOURCE: S. A. Asdell *Patterns of Mammalian Reproduction,* 1946.

Table 10-3. Comparison of fecundity of Romanov ewes from single
and multiple births

EWES BORN AS	AVERAGE LITTER SIZE PRODUCED BY THEM
Singles	2.17
Twins	2.36
Triplets	2.63
Quadruplets	3.01

SOURCE: From Lopyrin, *Sovetskaia Zootechnika,* 7:88, 1940.

of mice weighing 23.2 gm, he began to select for great and small body
weight without paying attention to any other characteristic. By the
twenty-first generation he had two strains, one of which weighed 40 gm
and the other 12 gm. Since only body weight was being selected for, it
was startling to learn that by the twenty-first generation the heavy strain
ovulated 14.1 eggs and had a litter size of 10.5 while the light strain
ovulated 7.2 eggs and had a litter size of 5.3.

We can see that the ovulation rate is genetically controlled in monotocous animals by comparing the fecundity of different breeds of sheep
(Table 10–2); in some breeds multiple births are commoner than single
births (Table 10–3). Moreover, the twinning rate varies among breeds
of domestic cattle (Table 10–4). It is interesting to speculate on the way
the genes bring about these effects. The genetic differences noted are
probably due to an increased rate of secretion of gonadotrophic hormone, but this cannot be the only effect produced by genetic selection,
for a higher ovulation rate alone does not necessarily result in a greater
litter size. Harmonious interplay of all the hormones concerned with
reproduction is necessary for high rates of implantation and embryonal
survival.

Table 10-4.   Rate of twinning in domestic cattle

Swedish Red and White cattle	1.85%
Frisian cattle	3.35
Guernseys	2.78
Ayrshires	1.62
Jerseys	1.03
Beef cattle	0.44

In his experiment on mice, MacArthur simultaneously increased body weight, growth rate, and prolificacy by selecting for only one character, body size. Because this characteristic is controlled by another pituitary hormone (somatotrophin), the question arises whether by selecting for the rate of secretion of one hormone one speeds up the functioning of the pituitary gland as a whole. This possibility is supported by the data in Table 10-4, and Table 10-2. In Table 10-4 the breeds with the highest twinning rate are, in general, the greatest milk producers and are also the heaviest breeds. In Table 10-2 the two breeds showing the highest twinning rate are breeds that have been selected for high milk production. It seems, therefore, that selection proceeded simultaneously for rate of secretion of growth hormone, thyrotrophic hormone, prolactin, and gonadotrophic hormone, even though the original intention was to select for only one trait—say, high milk yield. The low twinning rate of beef cattle does not necessarily upset these speculations, for their genetic selection was probably in the direction of an entirely different endocrine environment from that, for instance, of Holstein cattle.

Similar relations roughly hold for other species, such as dogs and rabbits, in which there is a good correlation between body size and prolificacy. It would be interesting to test the possibility of combining small body size with large litter size, or vice versa, and thus demonstrate that genetic selection for more rapid secretion of one hormone does not necessarily have to be coupled with the rapid secretion of other pituitary hormones.

Effect of Age

The ovulation rate is significantly affected by age. Marshall observed many years ago that the ovulation rate rises rather rapidly from adolescent sterility to its highest point and then gradually falls with advancing age to senile sterility. This generalization appears to apply to many animals studied.

In humans less than 0.5 percent of Negro and Caucasian mothers 17 years old give birth to twins. In Negroes 37 years old the proportion of

Table 10–5. Relation between parity and prolificacy in pigs

Litter number	1	2	3	4	5	6	7	8	9
Excess of young over first litter	0.0	0.68	1.36	1.58	1.90	1.92	1.89	1.71	1.45

SOURCE: Lush and Molln, USDA Tech. Bul. 836, 1942.

twin births rises to 2.0 percent; in Caucasians of that age it is 1.3 percent. These are statistically significant differences.

Two separate effects of age are noted in pigs. At the third and fourth heats after puberty ovulation rates are significantly higher than at the first and second. After the fourth or fifth cycle the ovulation rate becomes stable. The second effect of aging shows up after the females have gone through one or more pregnancies. In the second pregnancy pigs produce 0.68 more young than in the first. The effect of the age of dams on the prolificacy of pigs is shown in Table 10–5.

It will be noted that prolificacy reaches its peak with the fifth litter, remains unchanged until the seventh litter, and then declines slightly to the ninth litter. The increase in reproductive efficiency seems to be an effect not of age but of previous reproductive experience. This is brought out by the finding that sows bred for the first time late in life produced about three pigs per litter less than sows bred at the normal time (Asdell, 1941). Similarly, rats bred very early in life produced an average of 5.8 young per litter, those bred at the normal time 6.2, and those bred for the first time late in life 4.9. These and other experiments on sheep, rabbits, and guinea pigs support the data presented in detail for the pig and emphasize the fact that the increased prolificacy of polytocous females of increasing parity is probably not a function of age (or of greater body weight) but depends primarily on previous reproductive experience. It is also a well-established fact (and a vexing one for commercial egg producers) that pullets first coming into production lay eggs significantly smaller than those laid by the same pullets two or three weeks after the initial egg. The size of the finished egg is largely determined by the size of the ovum ovulated, and the increase in the size of the ovum cannot be accounted for by increasing body size of the pullet.

It is possible that while acquiring "reproductive experience" animals find the best adjustment of the neuroendocrine niveau, which, as we emphasized earlier, involves all the glands. Thus, the proper adjustment of the rate of function of the thyroids, adrenals, pancreas, etc. may require trial-and-error periods during the time when the organism is first exposed to the new experience of producing ova, of gestating, or of lactating. These are all situations sufficiently different from each other to require a readjustment of the neurohumoral feedback system to bring all

Table 10-6.   Effect of age of first breeding on reproductive efficiency of rabbits

TIME OF FIRST BREEDING	NO. OF DOES	EGGS SHED	AVERAGE NUMBER			
			ATROPHIED BEFORE 10TH DAY, OR UNFERTILIZED	ATROPHIED AFTER 10TH DAY	NORMAL YOUNG	NORMAL YOUNG
Early age	12	10.2	3.9	0.8	5.5	55%
20 months	19	10.2	3.8	2.6	3.6	36%

SOURCE: Hammond, *Zootecnica e Veterinaria*, pp. 3–8, 1953.

Table 10-7.   The trajectory of reproductive performance of one ewe

Reproductive year	1	2	3	4	5	6	7	8	9	10	11	12	13	14	15	16	17	18	19
No. of lambs (total 36)	1	1	2	3	3	3	3	3	3	2	2	2	2	2	2	1	1	0	0

SOURCE: Pearl, *Science*, 37:226, 1913.

the glands involved into proper and optimal interrelation with each other.

The data shown in Table 10-6 further illustrate the effect of age on prolificacy and suggest that, in rabbits at least, the effect is due to increased intrauterine mortality during the second part of gestation in older mothers.

A perfect example of the phenomenon discussed above, and of the trajectory of reproductive performance, is the amazing reproductive record of Pearl's ewe (Table 10-7). The decrease in fecundity toward the end of the reproductive life of this unique female, as well as that shown by the data on pigs cited earlier, seems to apply to monotocous as well as to polytocous species.

That the nutritive state has an important relation to fecundity has been recognized for many centuries. General malnutrition, as well as specific deficiencies, such as that of certain vitamins (B, E), is known to impair or to stop reproduction completely, but mild restrictions only impair reproductive efficiency. In general, when low nutritive states are improved, the ovulation rate rises. This is probably the reason why wild animals are usually more prolific in captivity, if they reproduce at all. Experiments on domesticated animals have shown, in the majority of cases, that improved nutrition improves prolificacy. The best-known example of this is the "flushing" of sheep; the ovulation rate and the rate of twinning are significantly increased by the practice of feeding ewes more intensely shortly before the onset of the breeding season. In swine restricted energy intake (about 70 percent of normal) causes a lowered ovulation rate, but it also usually reduces embryonal mortality. For practical purposes it may be advisable to feed pigs a high-energy ration until soon after mating. Thus, the intensive feeding before breeding should increase ovulation rate, while the reduced energy intake should insure a reduced embryonal mortality during gestation.

According to experiments conducted at Cornell, dairy heifers on a low plane of nutrition reached sexual maturity (as measured by first heat and ovulation) at 65 weeks of age, those on a medium plane at 47 weeks, and those on a high plane of nutrition at 37 weeks.

Restricted feeding of males (boars and bulls), even to the point of causing distinct retardation in growth, has no significant effect on their fertility even though the total volume of semen produced by boars on 60-percent ad libitum feeding is significantly lower than the volume produced by full-fed boars.

The effects of other factors, such as light and temperature, have already been discussed in Chapter 3 ("Breeding Seasons"). The modification of ovulation rates by hormones will be discussed in Chapter 12.

FERTILIZATION RATES

The fertilization of ovulated eggs obviously depends on the meeting of germ cells in the oviduct at a time when sperm and eggs are able to fertilize and to be fertilized. We have already seen what happens when either of the germ cells is allowed to age (Chapter 8), and we shall consider here only the optimal situation, in which aging is avoided by synchronization of ovulation and insemination. This situation exists only in animals in which psychological heat is so short that neither of the germ cells has to wait for the arrival of the other. This condition is met only in animals whose frequency of mating is not restricted (most wild animals, laboratory animals, chickens, and sheep). Induced ovulators have an ideal timing device in this respect; ovulation depends on mating, and hence the sperm arrive in the oviduct at the best time in relation to ovulation.

If the timing is proper and adequate quantities of viable sperm are present, the fertilization rates of all polytocous animals approach 100 percent. In rabbits, guinea pigs, rats, and mice the rates rarely drop below 95 percent, according to most studies; the average for these species is 98.6 percent. In pigs, also, fertilization rates are very high, and, if individuals in which none of the eggs is fertilized are excluded, the rate approaches 100 percent. If all females are considered, the fertilization rate of pigs ranges from 85 to 95 percent.

At this point, a word of caution is in order. It is known that in swine and other mammals unfertilized ova frequently fragment, often showing such perfectly "normal" cleavage patterns that even an experienced observer may conclude that he is viewing normally cleaving ova. To what extent such "normally" fragmenting ova contribute to the impression that fertilizability in pigs approaches 100 percent remains unknown. The presence of sperm in the zona pellucida cannot be taken as proof that the ovum is cleaving normally because sperm have been found in the zona pellucida of ova that did not cleave at all well as in eggs that were obviously undergoing disorderly fragmentation. This observation underscores the necessary precaution that each ovum be examined carefully under the microscope, with special attention to the orderliness of cleavage planes. The presence of many disorderly cleaving ova, even if the others show orderly cleavage, should raise the suspicion that the whole lot may be undergoing fragmentation rather than normal division.

In monotocous females the story is somewhat different. The species that have been studied most intensively—cows and sheep—show fertilization rates varying from 60 to 85 percent; the majority of workers have found a figure below 80 percent. No such figures are available for primates or horses, but it is probable that the same situation exists in these monotocous species. There is no obvious explanation for this difference between monotocous and polytocous females. The low fertilization rate of monotocous animals may be due, at least in part, to lowered fertility of the males.

LOSS OF OVA AND EMBRYONAL MORTALITY

Ever since it was established that each follicle produces one ovum (leaving the possibility of multiovular follicles out of consideration for the present), comparisons have been made between the number of eggs ovulated and the number fertilized, the number of embryos implanted, and the number of young born. Such comparisons are not easily made in one individual, but they can be made statistically in groups of females, provided the experimental design makes such comparisons valid. A common method consists of sampling one group of females either by laparotomy or at autopsy within two or three days after breeding. At that time usually all the ova can be recovered by flushing of the oviduct with a few drops of saline solution. The ova can be counted and examined for signs of fertilization under low-power magnification. As a rule, and especially in the larger animals, the number of ova recovered corresponds to the number of corpora lutea counted in the ovaries. Thus it is possible to determine the average rates of ovulation and fertilization.

Table 10-8. Fate of ova shed by sows during heats at which they were mated

	IN ALL SOWS BRED	IN SOWS PRODUCING LITTERS
Loss from failure to conceive	36.4%	0.0%
Loss during gestation	19.9	31.3
Loss at parturition	2.0	3.2
Loss from birth to weaning	14.6	22.9
Live pigs at weaning	27.1	42.6

SOURCE: R. W. Phillips and J. H. Zeller, *Am. J. Vet. Res.* II, 5:439, 1941.

The time at which the second group of females should be sampled will depend on the kind of comparison desired by the investigator. If an estimate of early embryonal mortality is wanted, the sampling of the second group can be made shortly after the females have gone beyond the first expected heat after breeding. This timing is desirable because it eliminates the females that did not conceive at the previous mating. Comparison between the number of embryos found in the second group and the number of fertilized eggs found in the first group provides an estimate of early embryonal mortality. Modifications of this method can be used for estimates of reproductive efficiency during different phases of gestation. Most of the information presented here was obtained by the method described.

It has already been pointed out that the fertilization rate of litter bearers approaches 100 percent but that the implantation rate and the number of young carried to term, in the great majority, are significantly less than the number of ova fertilized. The discrepancy between the number of ova fertilized and the number of embryos found in the uterus varies greatly between the individuals of one species, but it is nearly constant for all polytocous species. Several studies of rats, rabbits, pigs, and other litter bearers have shown that 30–50 percent of the fertilized eggs are lost sometime during gestation. Most of the fetal mortality occurs during the first third of pregnancy, and probably most of the losses occur between fertilization and implantation. One illustration of the situation, taken from work on a large number of litters in pigs, includes the postnatal losses as well as those occurring during gestation (Table 10-8). The physiological factors responsible for embryonal mortality are not known, but several suggestive leads have been obtained and are worth mentioning.

It was pointed out in another connection that the aging of germ cells leads to increased embryonal mortality. It is quite possible that a large share of the average intrauterine mortality occurs at the expense of the zygotes and embryos due to the aging of either of the germ cells. It is also possible that females that show a very low embryonal mortality are

Table 10-9.   Relation between number of eggs ovulated and embryonal mortality in pigs

TOTAL FEMALES	AVERAGE NO. CORPORA IN BOTH OVARIES	AVG. NO. NORMAL EMBRYOS	EGGS THAT BECAME		
			NORMAL FETUSES	ATROPHIC FETUSES	MISSING
5	22.6	12.20	54.0%	14.2%	31.9%
5	18.8	12.20	64.9	14.9	20.2
4	16.7	11.96	71.6	16.4	11.9
5	14.8	10.99	74.3	10.8	14.9

SOURCE: J. Hammond, *J. Agr. Sci.,* 11:337, 1921.

those in which, the timing being perfect, sperm of maximal vigor fertilized ova of optimal age. Another factor that is definitely known to increase embryonal mortality is infection, such as brucellosis. (In most of the data presented here, this factor has been excluded, or at least was not present to the best knowledge of the workers.)

Data presented by Hammond in 1921 (and in later studies by others) show that embryonal mortality among pigs is higher in animals with high ovulation rates (Table 10-9). These data should not be interpreted to mean that high ovulation rates are undesirable; despite the higher mortality, the group with the highest ovulation rate produced one pig more than the group with the lowest ovulation rate and the lowest embryonal mortality.

It seems significant that in about 15 percent of the 2,800 reproductive tracts of swine obtained at slaughter the number of corpora lutea corresponded exactly to the number of embryos found in the uterus. Similarly, in about 40 percent of rat pregnancies none of the ova or embryos is lost. This observation suggests that loss of ova and embryonal mortality are not inevitable features of pregnancy in polytocous animals. The study of pigs included animals with both high and low ovulation rates, ranging from 10 to 20.5 eggs ovulated, and suggests that under certain (as yet unknown) optimal conditions all the ova can be fertilized, implanted, and carried nearly to term, regardless of the total number of embryos to be accommodated. In 90 percent of the females that lost some of the ova or embryos, there was no significant correlation between the number of eggs ovulated and the number of embryos lost. In the group with the highest ovulation rate (more than 20 corpora in the two ovaries), embryonal mortality was 33 percent; in the group with 10-20 corpora, the loss was 35 percent; in the group with less than 10 corpora, the loss was 31 percent. These findings suggest that "crowding" does not contribute to the loss of ova or to embryonal mortality. However, other workers (Perry, 1955) have suggested that pigs and other polytocous

Table 10–10. **Normal reproductive performances of two strains of mice and comparison of rates of survival of transplanted ova in crossbreeds**

STRAIN	TOTAL LITTERS	AVG. EGGS PER OVULATION	AVG. YOUNG PER LITTER	EGGS DEVELOPED INTO YOUNG
dba	220	8.2	4.8	58.3%
C57b1	236	6.7	5.6	83.9%

DONOR	RECIPIENT	EGGS TRANSFERRED TOTAL	EGGS TRANSFERRED DEVELOPED INTO YOUNG	RECIPIENTS PREGNANT NO.	RECIPIENTS PREGNANT PERCENT	EGGS DEVELOPED INTO YOUNG NO.	EGGS DEVELOPED INTO YOUNG PERCENT
dba ⟶ C57bl		1,928	18.3%	134	44.7%	352	40.3%
dba ⟶ dba		647	7.1	26	26.0	46	28.9
C57bl ⟶ C57bl		600	12.8	33	33.0	77	38.1
C57bl ⟶ dba		1,871	11.4	102	34.0	213	30.1

SOURCE: Fekete, *Anat. Rec.*, 98:409, 1947.

animals do show a positive correlation between "crowding" and embryonal mortality.

It was pointed out earlier that optimal feeding has increased ovulation rates in experimental animals (pigs and sheep). The same type of feeding, however, also increases embryonal mortality in pigs. It may become practical, eventually, to full-feed pigs in order to take advantage of the effect on the ovulation rate and then to reduce the food intake immediately after conception in order to prevent the undesirable effect of maximal food intake on embryonal mortality.

Finally, attention is called to the "unfriendly uterine environment," uncovered by Fekete (1947) in mice. A comparison of the normal reproductive performance of two strains of mice, dba and C57 black, is presented in Table 10–10. It is striking that in spite of a lower ovulation rate C57 blacks have a bigger litter size than the dba's. The two obvious interpretations of these data are: (1) that dba eggs are less viable; (2) that dba uteri are less hospitable. The latter possibility was tested in a most painstaking manner by the collection and transfer of 5,000 mouse eggs within and between the two strains (Table 10–10). Although the percentage of eggs that developed into young after transplantation was low (7–18 percent), the data obtained suggest that the dba uteri were equally inhospitable to their own eggs and to those coming from the C57bl strain. In the C57bl uteri, however, a significantly higher proportion of the young from either donor were carried to term. Because of the low rate of conception following transplantation of eggs, it appears desirable to obtain additional data of this type, but the preliminary conclusion that certain intrauterine factors may be conducive to greater embryonal mortality appears justified. What these factors are remains unknown.

Another case of probable unfavorable uterine environment has been found in certain cattle that have an embryonal mortality of 52 percent between days 16 and 34 of gestation, despite a fertilization rate of about 80 percent—a rate that indicates that there is nothing amiss with the viability and fertilizability of the eggs shortly after ovulation (Hawk et al., 1955). It is possible, however, that factors other than uterine inhospitability are responsible.

During the remainder of gestation, however, the embryonal mortality of these cattle is only 5–7 percent, and it appears to be spread at random through the remaining 200 days. Similar figures for other mammals with long gestation periods, e.g., cattle, pigs, and women, show that for most of them the peak of embryonal mortality occurs during the first third of gestation—usually during the first 20–30 days.

Efficiency of reproduction thus depends on the ovulation rate, the quality of the ovulated eggs, the ability of the eggs to become fertilized and to develop into a viable zygote, and the ability of the uterus to receive and implant the embryos and carry them through pregnancy. The fact that some females are able to turn in a perfect reproductive performance, at least once in their lifetime, is considered significant. The repeatability of such perfect performances cannot be studied, for without laparotomy there is no good and simple method of comparing ovulation, fertilization, and implantation rates with litter size in the same female over several pregnancies. It therefore remains unknown whether perfect or nearly perfect records are unique accidents in the reproductive life of females or whether some females are more able than others to produce eggs of high fertilizability and survival ability and to provide them with optimal intrauterine environments. The fact that some females show perfect reproductive efficiency presents the researcher with a most challenging and important problem, and poses the question: If some females can do it, why not all?

The embryonal mortality rate of birds appears to be comparable to that of mammals in only two respects: it depends to some extent on the fertilization rate and the aging of sperm cells. Environmental factors in the embryonal mortality of birds, such as temperature of incubation and humidity, make valid comparisons between mammalian and avian rates impossible. Two peaks of embryonal mortality occur during the 21-day incubation of chickens, the birds that have been studied most intensively. The first peak occurs during the second, third, and fourth days of incubation; the second and by far the largest peak occurs during the last two or three days of incubation. Why these particular periods are so critical is not known, but it is assumed that the physiological processes occurring then place the greatest stress on the embryo.

11

Pregnancy, Parturition, and Lactation

The important question of how the ovary knows that the uterus is pregnant is discussed, but no final and satisfactory answer is given. It seems certain that the maintenance of corpora lutea for the duration of pregnancy is initiated in the uterus by the implanted embryo and is transmitted by the nervous system or by humoral means through the hypothalamus to the anterior pituitary gland, which responds by releasing a luteotrophic substance. In some animals the ovary is the sole source of progesterone and is therefore essential throughout pregnancy; in other animals the placenta takes over secretion of progesterone sometime during pregnancy, and the ovaries can be removed from these without causing abortion. Estrogen also is secreted by the placenta.

Gonadotrophic hormones during pregnancy are produced by the endometrial tissue in Equidae and by the placenta in primates. These hormones are probably involved in the maintenance of the corpora lutea of pregnancy.

Heat occurs frequently during pregnancy in many species; ovulations during pregnancy are not nearly as common.

The size of the conceptus and the accumulation of fetal waste products are discounted as the primary causes of the onset of parturition. It appears that the onset of parturition is caused by the action of estrogen, progesterone, and oxytocin on the uterine muscle. The signal for expulsion of the conceptus may originate in the fetus in some species.

In preparation for lactation, the breast is prepared anatomically by the interaction of estrogen and progesterone; milk secretion itself is caused by the lactogenic hormone (prolactin). Let-down of milk is caused by oxytocin, which is released from the posterior pituitary gland by a variety of stimuli, such as suckling, sexual excitation, and manipulation of the genitalia.

PREPARATION OF THE UTERUS FOR PREGNANCY

Pseudopregnancy

We know that during each cycle the uterus undergoes changes in preparation for the possible arrival of a young embryo, and that these changes consist primarily of the progestational proliferation of the uterine glands and the secretion of uterine milk. In females with long cycles there is plenty of time between heats (or menstruations) for the development of the luteal tissue that secretes the progesterone that brings about the progestational uterine changes. But some mammals, such as rats and mice, have very short cycles (four or five days). During the normal cycle of such species the luteal phase of the ovary is so short that the uterus does not undergo the changes usually associated with progesterone action. Furthermore, in rats and mice eggs take about three days after ovulation to arrive in the uterus. Thus, one might think that once rat or mouse eggs are fertilized that they would arrive in a uterus that was unready for their reception, since the female would be hormonally preparing for the onset of the next cycle, with its heat and ovulation. This, however, does not happen, for copulation in rats and mice causes pseudopregnancy. Pseudopregnancy is characterized by the fact that corpora lutea from the last ovulation are maintained well beyond the time they normally would have regressed; in fact, in some species they may be maintained for an interval equal to the normal length of pregnancy. In most animals pseudopregnancy lasts about half the time of the normal pregnancy.

Pseudopregnancy in rats and mice has been shown to follow specifically stimulation of the cervix during copulation. It appears probable

Table 11-1. Comparison of animals showing pseudopregnancy

	TYPE OF OVULATION	DURATION OF PSEUDOPREGNANCY (days)	DURATION OF NORMAL GESTATION (days)
Cat	Induced	30–40	65
Dog	Spontaneous	About 60*	58–63
Fox	Spontaneous	40–50*	52
Ferret	Induced	35–40	42
Mink	Induced	Variable	Variable†
Hamster	Spontaneous	7–13	16–19
Rabbit	Induced	16–17	30–32
Rat	Spontaneous	12–14	22
Mouse	Spontaneous	10–12	19

*Pseudopregnancy follows even without copulation.
†Delayed implantation.

that cervical stimulation causes nervous impulses to act upon the hypothalamus, which releases neurohumoral substances that are carried via
the pituitary portal system to the anterior pituitary. There the substances
are instrumental in releasing luteotrophic hormone, which is essential to
the activation of corpora lutea. Another mechanism, to be discussed
in the next section, may also be responsible for the continued maintenance of the corpora.

That the mechanism activating pseudopregnancy depends on neural
transmission is supported by the following observations. Pseudopregnancy can be induced by stimulation of the cervix with a glass rod or
with an electrical current (or even by irritation of the nasal mucosa of
rats with silver nitrate). Under deep anesthesia rats do not become pseudopregnant after mechanical or electrical stimulation. Neither do they
become pseudopregnant if the cervical innervation is destroyed.

In rats and mice, then, pseudopregnancy serves the purpose of holding
the next cycle in abeyance and allowing ample time for the preparation
of the uterus for the expected implantation. If the mating is infertile and
no pregnancy results, the corpora of pseudopregnancy wane about 12
days after ovulation, and the estrous cycle is reestablished. During their
life the corpora lutea of pseudopregnancy cause an increase in uterine
size and a very significant proliferation of the mammary glands, which
in some animals, such as the bitch, may actually lactate. Nest building is
also frequently noted toward the end of pseudopregnancy. Thus, except
for the fact that there is no conceptus in the uterus of a pseudopregnant
animal, the condition is very similar to true pregnancy.

In rats, mice, and hamsters pseudopregnancy occurs only after copulation (or other cervical stimulation); in other animals it follows ovulation in the absence of mating. The bitch and the vixen, both of which

ovulate spontaneously, automatically undergo pseudopregnancy following every ovulation. In the vixen pseudopregnancy lasts as long as
gestation itself would have lasted, but in the bitch there is considerable
variation in the duration of pseudopregnancy, depending on the breed.
In the absence of accurate data it appears that in the less "perverted"
breeds of dogs (such as the German shepherd) the length of pseudopregnancy equals that of gestation. In the smaller breeds it may last as long
or may be much shorter, lasting only a week or two in some breeds
(Table 11–1). In rabbits and cats, both of which are induced ovulators,
the immediate cause of pseudopregnancy is ovulation, not mating. This
is demonstrated by the fact that in both species, in the absence of mating,
injection of LH induces pseudopregnancy; endogenous LH causes ovulation and initiates the formation of corpora lutea.

Maintenance of Corpora Lutea
Without Pseudopregnancy

As we have seen above, in some animals pseudopregnancy automatically
follows copulation or ovulation. In others, when pregnancy is probable,
there must be some mechanism that acts to maintain the uterus in progestational condition and to prevent subsequent heats or menstruations
and ovulations. What this mechanism may be is not yet clear, but the
following chain of events may participate in notifying the ovary that the
uterus has become pregnant and that the corpora lutea should be maintained. It is probable that the physical presence of embryos in the uterus
releases a neural signaling mechanism, which, either directly or via the
hypothalamus, causes the release of a luteotrophic substance from the
pituitary gland. It has already been pointed out that this luteotrophic
substance is prolactin in rats and LH in other species.

That a neural link exists between the uterine lumen and the pituitary
gland is demonstrated by the following experiment. If beads are implanted in the uterine lumen of sheep, the length of subsequent cycles is
significantly modified. That this modification is mediated neurally is seen
from the fact that the cycles remain normal in length if the uterine segment containing the bead is completely denervated. A difficulty in inter-
preting these results arises because the presence of beads in the uterine
lumen shortens the cycle (that is, has a luteolytic effect) if the beads are
implanted on about the fourth day of the cycle but prolongs the cycle
(that is, has a distinct luteotrophic effect) if the beads are implanted
during the late luteal phase (days 8–10 of the cycle). Though these observations demonstrate a neural link between the uterus and the pituitary
gland, it remains unclear just how the uterine contents control the lifespan of corpora lutea. It seems probable that the presence of a conceptus
in the uterus during the late luteal phase sets in motion the events that

cause the corpora lutea to persist much longer than they normally would. We have also mentioned that uterine contents participate in the control of the release of certain hypophyseal hormones not only in sheep but also in cattle and guinea pigs. However, neither the pig nor the rat responds to the presence of foreign bodies in its uterus by modifying the length of the estrous cycle. The work described here has been recently confirmed and extended by Hecker and others (1974) and by Wodzicka-Tomaszewska and Colleagues (1974).

THE ROLE OF HORMONES IN PREGNANCY

Though the mechanism involved in maintaining the corpora lutea in pregnant females is only incompletely understood, the general role of progesterone during gestation is well known. Progesterone is essential for keeping the blastocysts alive before implantation, during the time when they are floating free in the uterine lumen. If pregnant females are castrated before implantation, death of the blastocysts is inevitable. Castration after implantation, during the first third or first half of pregnancy, generally leads in the majority of species to resorption of fetuses or to abortion. After this crucial period some mammals can do without the ovaries and their secretions, but others must have the ovaries throughout gestation.

In rats the ovary is essential throughout the major part of pregnancy; abortion occurs if rats are castrated even as late as five or six days before expected parturition. But rats can be hypophysectomized 10 days after conception—according to some reports, even earlier—without aborting. The fact that pregnant rats tolerate hypophysectomy but not castration without aborting suggests that in this species the corpora lutea are the major—perhaps the only—source of progesterone. The pituitary gland may initiate the growth and functioning of the corpora lutea but may not be essential to their continued functioning throughout gestation.

The ovary has been found to be essential throughout pregnancy not only in the rat but also in the opossum, hamster, thirteen-striped ground squirrel, mouse, rabbit, and goat. In the cow it may not be necessary after the seventh month. Pigs abort if castration is performed anytime during the 115-day gestation period. The ewe, bitch, mare, cat, guinea pig, monkey, and women do not abort after castration if the operation is performed during the second half or, in some species, during the last trimester of gestation. Castration leads to fetal resorption in the majority of viviparous snakes, whether they are castrated early or late in gestation.

Failure of implantation, or abortion, can be prevented in all castrated animals by the administration of progesterone. Rabbits require 1 mg of progesterone daily for implantation and 2–3 mg daily for the remainder of gestation. Pregnancy can be maintained in mice with 1–1.5 mg of progesterone; sheep require 5–10 mg, goats 10 mg, and cows 50–75 mg daily. All available evidence shows that pregnancy can be maintained in castrates with progesterone alone, but that much smaller amounts of progesterone are needed if small amounts of estrogen also are injected.

In all Equidae, apparently, but certainly in mares, all corpora lutea formed after fertilization and implantation degenerate during the early part of gestation and are replaced by "accessory" corpora lutea (see Figure 11-4, p. 286). These result from the ovulation of follicles that are caused to mature by the secretion of PMSG. The accessory corpora persist until about day 180 of gestation.

The question arises why females of some species continue pregnancy in the absence of corpora lutea, whereas others abort soon after removal of the ovaries. The answer apparently lies in the fact that the placenta, as an endocrine organ, secretes estrogen and progesterone and possibly other hormones as well. It is therefore probable that in ewes, women, and others the placenta provides the progesterone that is necessary for the maintenance of pregnancy even if the ovaries are removed after implantation has been accomplished. It is inferred that in rats, goats, and other species in which abortion occurs in the absence of ovaries the placenta secretes no progesterone or secretes amounts inadequate to the maintenance of pregnancy. Progesterone has been demonstrated in the placentae of some species (women) and in the blood during pregnancy in others (ewes), but no authoritative generalizations on the relative levels of progesterone secreted by the placenta and corpora lutea are possible, since at present most determinations are made on the peripheral blood.

HORMONE LEVELS DURING PREGNANCY

Estrogen and Progesterone

Although it has been well established that both estrogens and progestins are secreted by the fetal placenta in some species but not others (such as the pig, rabbit, and rat), it is not known which placental cells are responsible for the production of these hormones. In early gestation the plasma estrogen levels are low, but show a spectacular rise shortly before parturition (see Figures 11-1 and 11-2). In cows a rise of plasma estrogens begins about 12 hours prior to parturition and reaches its peak on the day of parturition. This seems to be a phenomenon in all mammals, since similar data are available for sheep, pigs, and women.

The profile of plasma progesterone seems to be different. We have already seen that in the infertile cycle of sheep there is an abrupt and precipitous decline in both the weight of corpora lutea and in corpus

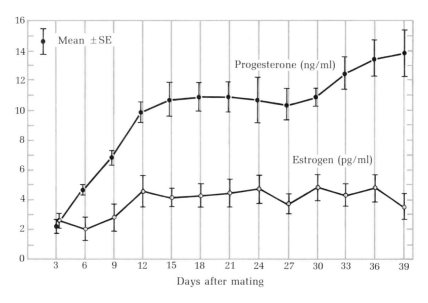

Figure 11–1 Relationship between estrogen and progesterone levels in pregnant cows. In contrast to the dramatic decline of progesterone at the end of the estrous cycle (Figure 3-8), progesterone synthesis continues in the pregnant animal (in the cow past 21 days). [From Henricks *et al.*, 1972, *Endocrinology*, 90:1336.]

luteum and plasma progesterone levels (see Figure 3-8, p. 77). If we now look at the plasma progesterone levels in bred cows (Figure 11-1) we note that there is no such abrupt decline in progesterone level, which would have been expected on about day 21 had the cow not become pregnant. Instead, the progesterone levels level off and remain high. The source of this progesterone is presumed to be the corpus luteum of pregnancy. This assumption is supported by the fact that if the ovaries are removed in the first half of pregnancy the cow aborts. The postconception level of progesterone does not change appreciably during gestation until shortly before parturition, when it declines very significantly. Note that the rise in plasma estrogen begins long before the drop in plasma progesterone takes place (Figure 11-2).

Not all animals conform to the profile of progesterone levels during gestation shown in cows. In sheep (Figure 11-3) and in women progesterone levels continue to rise throughout the duration of gestation. As in cows there is an abrupt drop in progesterone titers shortly before parturition. (We shall return to the relationship between estrogen and progesterone as one of the factors concerned with the expulsion of the young at term when we discuss parturition later in this chapter.)

While it appears generally true that in all animals studied the estrogen levels rise drastically toward the end of gestation, there may be some species in which the progesterone level does not drop prior to parturition. Women are such an exception; apparently there is no drop in plasma progesterone prior to parturition.

One other fact illustrated by Figure 11–3 is that sheep bearing twins have a higher plasma progesterone level than do sheep bearing singles. Limited data suggest that the same may be true in women. However, polytocous females with many fetuses do not always produce more progesterone than those with fewer fetuses. In pigs, for example, plasma progesterone levels bear no relation to the numbers of young carried. This difference between polytocous and monotocous species may be

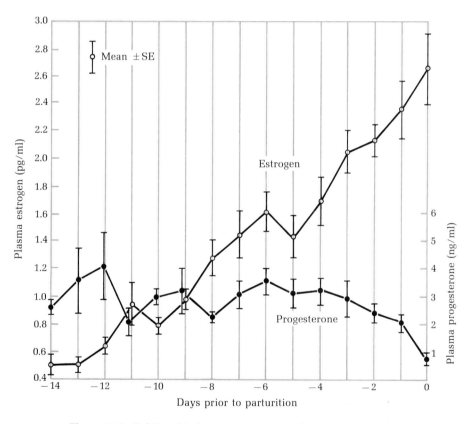

Figure 11–2 Relationship between estrogen and progesterone a few days prior to parturition. Note the decline in plasma progesterone; the dramatic increase in plasma estrogen continues until parturition. [From Henricks et al., 1972, Endocrinology, 90:1336.]

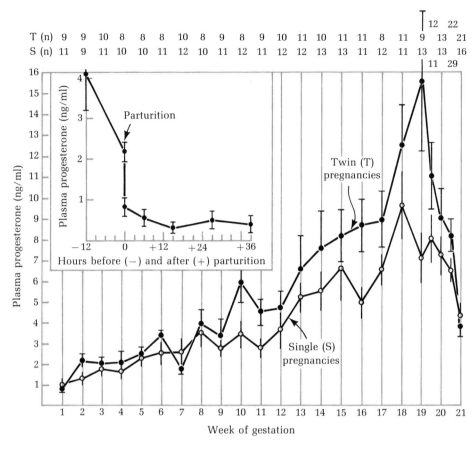

Figure 11-3 Plasma progesterone levels in sheep carrying one or two lambs. Note that progesterone rises throughout gestation and begins to decline about two weeks prior to parturition. The inset shows the drastic decline hours before delivery. [From Stabenfeldt *et al.*, 1972, *Endocrinology*, 90:144.]

due to the fact that in monotocous species the placenta is the major source of progesterone, at least toward the end of pregnancy, whereas in polytocous species probably all the plasma progesterone comes from the corpora lutea. This is also the reason why sheep and women tolerate castration without aborting in later pregnancy, whereas such animals as pigs, rabbits, and rats never do.

As a general rule the most reliable information on the hormonal status of pregnant females (or for that matter of animals in any reproductive state or sex) must be obtained from the peripheral blood, the gonadal uterine, or the umbilical blood vessels, depending on the type of information desired. Urinary or fecal hormone levels are frequently misleading and almost always difficult to interpret, since steroids are usually excreted in the urine or feces as metabolic products. An excellent example of the problem is the cow, in which very large quantities of androgen appear in the feces in late pregnancy. Although the obvious conclusion would be that androgen is a major steroid secreted by pregnant cows, fecal androgen is a conversion product of progesterone of placental and luteal origin.

Gonadotrophic Hormones

The placentae of rats is said to secrete luteotrophic hormone, which may be responsible for the maintenance of the corpora lutea of pregnancy in this animal. Evidence for the secretion of luteotrophic substances by the placenta in other species is presently restricted to primates and Equidae.

The endocrinology of the other gonadotrophic hormones—pregnant mare's serum (PMSG) and human chorionic gonadotrophin (HCG) of pregnancy—has been discussed elsewhere. We shall confine ourselves here to a discussion of the circumstances of their secretion during pregnancy.

PREGNANT MARE'S SERUM. PMSG appears in large quantities in the blood of pregnant mares at about day 40 of gestation and remains high until day 120 (Figure 11-4). This hormone is made in the so-called endometrial cups of the uterine epithelium (Figure 11-5) and is thus not a placental hormone. By the time it appears, the ovulatory corpus luteum has diminished in size and the ovaries of the mare are stimulated by the uterine gonadotrophin to form large follicles. Many of these follicles ovulate (eggs can be recovered from the oviducts) and form multiple corpora lutea, while other follicles become luteinized without ovulating. These structures, called accessory corpora lutea, persist until about day 180 of gestation; they in turn degenerate, leaving the pregnant mare without ovarian lutein tissue for the remainder of gestation.

Elephants, too, form accessory corpora lutea, but only from about the end of the sixth to the ninth month of the 24-months of gestation. It is inferred that elephants secrete a gonadotrophin similar to the one produced by pregnant mares. Accessory corpora lutea are also formed in the nilgai (Beselaphus tragocamelus), which is an antelope and a ruminant. Again, the implication is that a serum gonadotrophin is present to permit follicular development and ovulation. No serum or urinary gonadotrophic hormones have ever been found in other ruminants (cattle and sheep), nor do these species form accessory corpora lutea.

Whether the secretion of gonadotrophic hormone and the formation of accessory corpora lutea are just artifacts, without physiological significance to implantation and subsequent gestation, remains unknown. Gonadotrophic hormone in mares also acts on the gonads of the fetuses

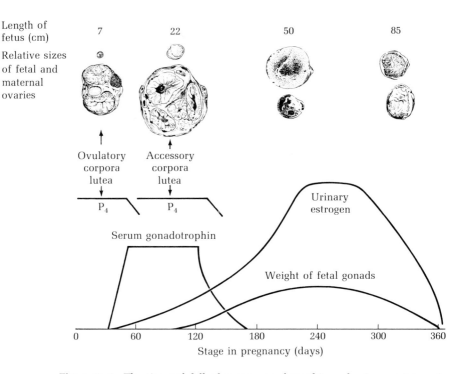

Figure 11–4 The rise and fall of serum gonadotrophin and urinary estrogen in pregnant mares, and the effect of these hormones on the maternal gonads. Note that ovulatory corpora lutea degenerate and accessory corpora are formed when the serum gonadotrophin level rises. Fetal ovaries (upper row) grow in response first to gonadotrophin and later to estrogen and become larger than the atrophic maternal ovaries. Note also that plasma progesterone (P_4) levels drop as ovulatory corpora lutea degenerate, and rise again as accessory corpora lutea are formed.

of both sexes, causing them to increase in size. In fact, the ovaries of female fetuses become larger than the ovaries of the mother, and fetal testes are larger than the testes of newborn foals (see Figure 11–4). As the influence of the gonadotrophin on the fetal gonads wears off, because of decreasing amounts of the circulating hormone, the gonads, the duct system, and the accessory glands come under the influence of estrogen and progesterone. The uterus of female fetuses becomes enlarged, and the endometrium proliferates. The connective tissue of the sex apparatus of both sexes is especially sensitive to the increasing titer of estrogen. This is particularly apparent in the seminal vesicles, which greatly enlarge because of the enormous growth of their estrogen-stimulated connective tissue.

Soon after parturition, after the duct system of the newborn has been removed from the uterine environment of high estrogen and progesterone titers, the duct system and the accessory glands are drastically reduced in size. Whereas the enlargement of the fetal gonads is typical of fetuses exposed to high levels of serum gonadotrophins, the duct systems and accessory glands in all species are affected by the high maternal titers of estrogen and progesterone. The vaginae and vulvae of newborn females are greatly enlarged and lined with stratified-squamous epithelium, but they regress to a simple form very shortly after birth.

CHORIONIC GONADOTROPHIN. HCG is not in any way comparable to PMSG. The endocrine properties of chorionic hormones were discussed earlier. Unlike PMSG, chorionic gonadotrophin is secreted by the placenta and not by the uterine endometrium. It is thought that the chorionic villi are the site of its formation, although some recent evidence implicates the trophoblast as well. We have pointed out that HCG is luteotrophic in a variety of species, including women, rats, rabbits, pigs, and others. For this reason it is tempting to ascribe to it the role of maintaining the corpora lutea during early pregnancy in women. If the hypothesis concerning the mechanism of maintenance of corpora lutea

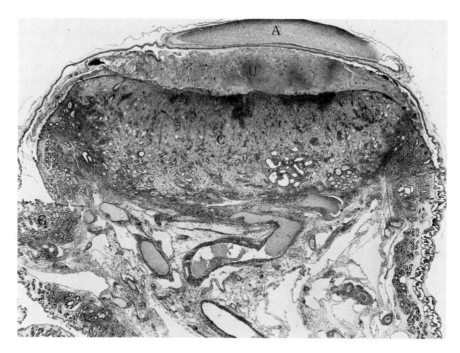

Figure 11–5 Section through an endometrial cup (C) of a mare pregnant 63 days. The space between the cup and the allantois (A) is filled with uterine milk (U). Note the uterine glands (G). (By permission of Professor E. C. Amoroso.)

proves to be approximately correct, then we may find that the corpus luteum immediately after ovulation gets enough impetus from hypophyseal luteotrophic factors to remain physiologically active until HCG begins to be formed in sufficient quantities to assume significance as a luteotrophin. Measurable amounts of HCG appear in the urine of pregnant women as early as 5–16 days after ovulation, but the urinary HCG titer does not reach a peak until 35–50 days of pregnancy (see Table 3–3, p. 75).

By the use of radioimmunoassay for HCG, it has been possible to demonstrate this glycoprotein in the plasma of women as early as eight days after known fertile intercourse. Because HCG cross-reacts with hypophyseal LH, it is not possible to be absolutely certain that what is being measured at day 8 of pregnancy is actually endometrial HCG or endogenous LH. If it is HCG then this would imply that the 8-day-old blastocyst is somehow able to turn on the necessary machinery for the synthesis of HCG, which in turn is able to counteract whatever luteolytic mechanisms may be in action and preserve the corpus luteum from destruction (which would normally occur in a nonpregnant woman).

Intermedin

The placenta contains (and probably secretes) extractable amounts of intermedin, a hormone normally secreted by the intermediate lobe (the pars intermedia) of the posterior pituitary gland. Intermedin extracted from that source is frequently contaminated with ACTH (adrenocorticotrophic hormone), and ACTH obtained from the anterior lobe usually contains intermedin. Intermedin has been extracted from the placentae of women, mares, cows, and bitches, and its concentration in the chorion increases with advancing pregnancy. In lower vertebrates this hormone causes the expansion of the chromatophores of the skin, but its physiological role in mammals remains unknown. It may be responsible for the pigmentation of the areolas, the nipples, and the linea alba of women during pregnancy.

HEATS AND OVULATION DURING PREGNANCY

It is generally thought that estrous cycles are suspended by the onset of pregnancy and that ovaries are inactive during all of gestation. However, it has been shown that pregnant females frequently have psychological heats and accept the male throughout pregnancy. This occurs much more frequently than is commonly thought. Ewes in which pregnancy was verified at autopsy showed heat indistinguishable from that observed in nonpregnant females. These heats last about 18 hours and are noted early during gestation and as late as five days before parturition. In one group of sheep checked for heat daily, it was noted that about 30 percent of the pregnant ewes showed heats, some as many as five times. It seems almost certain that no ovulations occur at these pregnancy heats, even though at the time some ewes have follicles of ovulatory size. About 10 percent of cows show one or more heats while pregnant, and this figure might be even higher if cattle presumed to be pregnant were regularly checked for heat. It is not known whether cows ovulate at pregnancy heats. The intervals between heats during gestation deviate significantly from the intervals typical of nonpregnant females. There appears to be no rhythm in the occurrence of these heats.

The fact that pregnant cows come in heat has practical significance. If they are artificially inseminated, and if the cervical seal is broken by the inseminating tube, abortion or mummification of the fetus usually results unless infection is prevented by the simultaneous injection of antibiotic drugs.

Heats during pregnancy are common in all laboratory mammals; matings have been observed in all of them throughout gestation. Infrequently some of the laboratory animals—and probably some of the larger domestic animals—ovulate during gestation. Superfetation has been recorded for rats, mice, rabbits, cattle, and sheep, two parturitions occurring a few days or weeks apart and fully formed young being born at each. Better proof of superfetation lies in the observation that the young born of parturitions that are less than a full term apart may bear color markings relating them to two different sires. It was noted in another connection that mares, as a rule, ovulate during pregnancy and that ovulations during pregnancy may also occur in women.

During the first third or even first half of pregnancy, the ovaries of all species examined show considerable follicular development; follicles of ovulatory size are frequently found in female swine, cattle, and sheep (Table 11–2) in numbers typical of the species. Little support is found in these species for the contention that during pregnancy follicles grow and become atretic at about the same rate as during the normal cycle. Though follicular atresia undoubtedly takes place during gestation, the lifetime of follicles seems to be longer during gestation than during the cycle. During the last half of gestation, and especially toward its end, the follicles diminish in size and in number, and at parturition there are no follicles larger than 1 or 2 mm in diameter. It is assumed that this reduction in ovarian activity is due to suppression of the pituitary gonadotrophins by the increasing amounts of estrogen and progesterone secreted by the placenta. This interpretation is supported by the finding that the injection of gonadotrophic hormones (PMSG) into sheep and pigs shortly before parturition causes pronounced follicular growth, ovulation, and

Table 11–2.   Relation between stage in pregnancy, number of follicles, and follicle size in ewes

STAGE IN PREGNANCY (days)	NO. OF FOLLICLES PER EWE	FOLLICLE SIZE (mm)			
		2–4	5–7	8–10	TOTAL
3–8	Average	9.5	4.0	2.0	15.5*
18–30	Average	15.1	4.6	3.0†	22.7*
45–115	Average	17.0	4.3	0.0†	21.3*

*Increase significant at one percent.
†Decrease significant at one percent.
SOURCE: Williams et al., J. Animal Science, 15:978, 1956.

even cyst formation, which show that the ovaries are capable of responding to hormone stimulation.

How heats and ovulations are held in abeyance in the majority of pregnant females is not clearly understood. The fact that follicles of ovulatory size are found in the ovaries of pregnant females shows that the pituitary-ovarian axis is not materially upset by pregnancy. It is possible that the best FSH-LH ratio for normal follicular development and ovulation does not usually exist in pregnant females but is occasionally achieved. Why females do not show psychological heat regularly and even continuously during late pregnancy, when placental estrogen reaches levels much higher than those necessary to bring nonpregnant females into heat, remains unknown. It is possible that the absence of heat is due to the well-known antagonism between estrogen and progesterone.

DELAYED IMPLANTATION

In a number of species there is a delay between the activation of the ovum by the sperm and the implantation of the blastocyst.

The best-known examples of such delayed implantation are rats and mice. Females of these species show a postpartum heat, including ovulation, within 24 hours after parturition. If copulation is permitted at this heat, the subsequent gestation is significantly longer than the normal 21 days, the young being born 30, 40, or even 50 days after mating. If the female is not permitted to nurse the first litter, the new gestation is of normal duration. Analysis of this situation shows that the eggs ovulated at the postpartum heat are fertilized normally but that the resulting blastocysts, instead of implanting themselves at the normal time, float free in the uterine lumen. The actual growth period of the embryo is 21 days, just as it is in normal pregnancies. The blastocysts can be caused to implant themselves at the proper time, in spite of concurrent lactation, by injection with estrogen. These observations are interpreted to mean that failure of implantation is caused by insufficient quantities of estrogen, which is presumed to be excreted in the milk. If the litters are large, more milk is secreted, the demands for estrogen are greater, and delay in implantation is longer than it is in females with smaller litters.

It has been found in rats that local application of estrogen to a portion of the uterine endometrium will make that particular portion hospitable to a blastocyst; only a blastocyst near the estrogen-sensitized area will implant, while the other blastocysts remain free and unattached. What specific effect estrogen has in this case is unknown, but the sensitizing of local endometrial tissue does emphasize the strictly local responses that hormones apparently can cause. A similar localized effect can be demonstrated in a bilaterally ovariectomized rat: a pellet of progesterone implanted into the lumen of only one of the horns prevents abortion of fetuses in the horn containing the hormone but not in the contralateral horn. Furthermore, if the progesterone pellet is very small, only those fetuses adjacent to the pellet survive; those further away in the same horn die.

The role of hormones in implantation is not clearly understood. As we have just noted, estrogen is obligatory for implantation in rats. In fact, samples of ovarian effluent blood from mated female rats show a very significant peak in the estrogen titer (both estradiol and estrone) in the blood four days after mating (Shaikh, 1971). This peak subsides and the levels drop to lower levels, but rise again shortly before parturition. The source of the estrogen peak on day 4 is not known, since by that time the ovary consists predominantly of corpora lutea. It is apparently this estrogen peak that triggers implantation in the rat on day 5.

In ovariectomized rabbits implantation can be caused by progesterone alone, but less progesterone is required if a few micrograms of estrogen are given concurrently. The role of estrogen in the rabbit appears to be permissive rather than obligatory as it is in the rat. In most other animals tested estrogen apparently has no effect on implantation, since implantation is brought about by progesterone alone. In none of them is there a postovulatory estrogen peak like that noted in rats. However, the question of whether estrogen is essential for implantation in nonmurine mammals remains open for further experimentation.

Totally puzzling at the present time are the implantation mechanisms in such animals as armadillos and wallabies. In the former implantation can occur after ovariectomy; in the latter it can occur after hypophysectomy, e.g., in the absence of hormonal stimuli.

In some species in which delayed implantation occurs implantation can be forced by estrogen injection (other species, e.g., the European badger, are completely nonresponsive to treatment by estrogen or any other steroid tried). Local administration of estrogen can even produce embryos of two or three distinctly different ages in the same uterine horn of rats. Although this experiment has been done in a variety of different reproductive conditions (even in castrated females), it is easiest to use postpartum lactating females that have just ovulated and in whom the resulting blastocysts will not normally implant for a period of up to two weeks after postpartum ovulation. In such females a local injection on the fifth day of as little as 0.005 μg of estradiol can cause the immediate implantation of one or two blastocysts, the others remaining free. The local injection, repeated a few days later, causes the implantation of additional blastocysts, and a still later injection can bring about the implantation of more free blastocysts. Thus, fetuses of three different ages are found in the same uterus—the oldest weighing 5 gm, those resulting from the second implantation 1 gm, and the youngest weighing only 0.25 gm. Such experiments can obviously be used in studies of the mechanisms of implantation and parturition.

Martens copulate in July or August (in America), and the eggs progress to the blastocyst stage, but implantation does not occur until January. The young are born in March about 50 days after implantation, but 250 days after fertilization. In the mink, which of all the mustelids has received the greatest attention (Hansson, 1947), the average length of gestation is about 50 days, although it varies from 39 to 74 days. In some females implantation occurs without delay. In others a significant delay is noted; the interval between mating and implantation is dependent on many factors, of which frequency of mating and environmental temperature are the most important.

Delayed implantation also occurs in the roe deer. The rutting season of this species is in July or August, although other members of the family do not mate until November or December. The ova of roe deer are fertilized at mating but lie dormant until about December, when they implant themselves; the fawns are dropped in May. The interval from mating to parturition is 40 weeks, but true gestation lasts only 20 weeks.

The physiological significance of delayed implantation in animals other than the rat is not clear. Just as in the rat, it seems to be controlled by endocrine factors, for one can shorten the gestation period of martens and mink by exposing them to prolonged light. It is assumed that this treatment works by stimulating pituitary activity, but attempts to bring about the same effect with gonadotrophic and steroid hormones have been equivocal or unsuccessful.

PARTURITION

Basically, the problem of birth is a simple one. When the period of embryonal growth is ended, the conceptuses are expelled from the uterine lumen. When the details of parturition are analyzed, however, it becomes apparent that the initiation of the birth process, if not birth itself, involves events that are complex (and at the present time incompletely understood).

First, a few facts to set the stage for subsequent discussion. Whereas only minor contractions of the uterus occur during the major part of gestation, the process of birth is precipitated by increased uterine motility and contraction. It is probable that the process is initiated by a combination of factors that lead to an increased sensitivity of the uterine muscle to hormones. An older theory held that the fetus itself might be secreting substances (accumulation of waste products) that initiated birth. It was also thought that the size of the fetus might give the signal for parturition. Two experimental observations cast serious doubt on these theories. When the fetuses of rats and rabbits were prematurely removed and the placentae left in situ, pregnancy continued and the empty placental membranes, which continued to grow, were carried for the length of time of normal pregnancy (21 days in rats and 31 days in rabbits). At term, when birth of the young would have occurred, the empty membranes were delivered. This same phenomenon has been demonstrated in rhesus monkeys.

Prolonged gestation sometimes occurs in many species (women, mares, pigs, etc.). Of greatest interest is a study of prolonged gestation in dairy cattle. Abnormally long gestation is known to be a genetic character among dairy cattle, causing calves to be carried long beyond the normal period of 278–290 days. Gestations as long as 330 days have been recorded, the calves continuing to grow in utero and reaching sizes much greater than normal (up to 200 pounds). The fetuses frequently become so large, in fact, that normal birth becomes impossible, and the calves must be delivered by surgical intervention. In normal pregnant cows blood progesterone levels are known to drop a few days before parturition, but no such drop occurs in animals with prolonged gestation. However, it is not possible to prolong gestation in normal cows with exogenous progesterone. Why the progesterone level does not drop in cows with prolonged gestation remains unclear.

Another variant of prolonged gestation has been reported. In one California herd of Jerseys, fetuses of smaller than normal or normal size are carried long beyond normal term. In this case (which appears to be due to a recessive character) the fetuses completely lack both lobes of the pituitary gland.

Prolonged gestation, the resulting large fetuses, and the delivery of membranes (after removal of the fetuses) at term all argue against the assumption that either the size of the fetus or a stimulus from the fetus has any effect on the onset of labor. It is also significant that in all monotocous species male fetuses are carried significantly longer than females. It is not quite clear whether the longer gestation of males is the cause of their larger size at birth or whether they must have a longer gestation to become physiologically ready to be born. In any event, it is quite clear that fetal size itself does not determine the time of onset of labor.

A few experimental observations may shed light on the problem of parturition. The uterine muscle contracts rhythmically when it is under the influence of estrogen. These contractions can be converted into tetanic contractions by the injection of oxytocic hormone from the posterior lobe of the pituitary gland. These facts are basic for our understanding of the birth process, but a few difficulties remain to be ironed out. If oxytocin plays a major role in initiating the uterine contractions that lead to parturition, the question arises as to how the release of oxytocin is made to coincide with the readiness of the conceptuses for delivery. The fact that birth can occur normally in totally hypophysectomized females does not necessarily argue against the role of oxytocin in parturition. In the absence of the posterior lobe the hormone may be secreted in sufficient quantities by the hypothalamus, its normal source.

Recently it has been found in women and cows that there is no demonstrable increase in plasma oxytocin at the onset of labor, i.e., uterine contractions begin without a rise in oxytocin. However, there is a considerable increase in this hormone in the plasma when the head and shoulders of the conceptuses already protrude through the vaginal canal. This suggests that factors other than oxytocin may be involved in initiating labor and that oxytocin may participate in its completion. That the factor initiating labor may be prostaglandin will be pointed out shortly.

The level of placental steroids may be of importance in the control of parturition. It has been pointed out that the estrogen level increases toward the end of gestation. It is also known that in some species pregnancy can be prolonged and parturition held in abeyance by the injection of progesterone. Once the injection is discontinued, birth occurs within a short time, indicating that the level of progesterone may be of major importance during the preamble to birth. One theory holds that a fall in the level of progesterone is a precondition of parturition. But, assuming that the placenta is the major, if not the only, source of progesterone during late pregnancy, no known mechanism could explain a decrease in the secretion of placental progesterone.

According to another theory, even if the progesterone level does not fall, the effectiveness of the hormone is reduced by the well-known antagonism between estrogen and progesterone. According to this theory, toward the end of gestation the level of placental progesterone remains unchanged while the level of estrogen continues to rise until it reaches a level sufficiently high to antagonize progesterone. At this point the uterine muscle would be under the influence of the estrogen-progesterone ratio that is most conducive to its sensitivity to oxytocin. Since the onset of labor is brought about by the normal increase of one of the hormones involved in parturition, it is not necessary to assume that oxytocin is released in response to a signal from the uterus or its contents. This theory is compatible with known physiological facts and has the advantage of being simpler than the theory that calls for juggling two hormones instead of one.

The cervix, which remains tightly closed during the whole of pregnancy, opens shortly before parturition, presumably under the influence of relaxin, which is known to be secreted by the placenta as well as by corpora lutea. Whatever the detailed mechanism of parturition may be, it is obvious that it depends on the occurrence of many synchronized events. For instance, although injection of a large dose of oxytocin shortly before the expected onset of labor may precipitate an immediate onset, unless the cervix has relaxed (to permit expulsion of the fetus), the violent uterine contraction may result in the rupture of the uterine wall rather than a normal birth process. If all the physiological events permitting parturition have taken place, the onset of labor may be hastened with oxytocin. Attempts to hasten birth by many days have not been uniformly successful.

Finally, let us return to the case of prolonged gestation in Jersey cattle in which both lobes of the pituitary gland are missing in affected fetuses. This observation has led to the following experimental protocol. If one removes (or coagulates) the pituitary or the adrenal glands of normal fetal sheep, the fetus will be carried well beyond the normal 150-day pregnancy. If ACTH is infused into fetuses lacking the pituitary or into normal ones, labor commences. Glucocorticoids can also be injected into intact fetuses or those with destroyed adrenal and cause premature labor. This suggests that the fetus itself may play a determining role in the onset of labor—forcing us to reevaluate the older theory that the fetus actively contributes to parturition. Furthermore, it is known that as parturition approaches prostaglandin $F_{2\alpha}$ (probably of placental origin) increases greatly in the peripheral blood of the mother. Whether this rise is due to the action of adrenal steroids on the placenta remains unknown (Figure 11-6; see also *Foetal Autonomy*, G. Wolstenholme and M. O'Connor, eds., Churchill, 1969, p. 218). It is known that injection of $PGF_{2\alpha}$ into such animals as pigs can hasten the delivery of living young by as much as three days prior to the expected normal parturition.

Having placed the control mechanism of parturition in the hypophyseal-adrenal axis of the fetus, we must not forget the fact stated earlier that in rats, rabbits, and monkeys the placentae are delivered at term if the fetuses are removed—which would tend to counter the idea

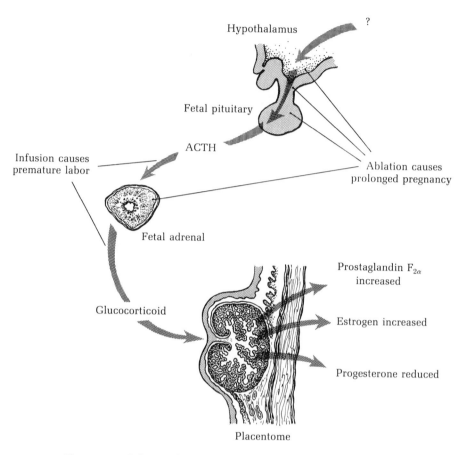

Figure 11-6 Schema of possible role of fetal glands in the precipitation of labor. [Courtesy of Dr. G. C. Liggins.]

that the fetus itself controls the time of delivery. Since the experiments on the destruction of pituitary or adrenal glands have thus far been performed only on sheep fetuses, the possibility exists that there are species in which the fetal hypophyseal-adrenal axis plays a controlling role and others (rabbits, monkeys, and rats) in which it does not.

LACTATION

Lactation may be considered the culmination of the reproductive process and is as much a part of this process as the estrous cycle or gestation. It is best discussed in two parts: (1) the basic anatomy of the mammary gland and its preparation for lactation, and (2) the initiation and maintenance of milk secretion. Although there has been much research on the physiological control of lactation, no general agreement on this subject has been reached.

Anatomy of the Mammary Gland

Mammary glands are compound glands, basically highly modified and specialized sebaceous glands that secrete milk. They are present in both sexes but as a rule become functional only in the female. The mammary glands of males are capable of becoming functional; in fact, much of the research work on the physiology of lactation has been done on males whose mammary glands had been developed by appropriate treatment. Development of the breast in males occasionally occurs spontaneously (called gynecomastia), especially in men and in male goats. There is at least one authentic case of a he-goat having produced enough milk to raise two kids.

Although the number of breasts may vary, they are invariably ventral and lateral to the midline. In some animals (elephants, sirenians, primates, etc.) they are located in the thoracic, in others (ungulates, cetaceans) in the inguinal region; in litter-bearing animals they extend from the thoracic to the inguinal region. In all species (except the monotremes) each breast has its own excurrent duct, leading to the nipple; the nipple may have a single excurrent duct, as in cows, or as many as 12–20, as in women.

In the embryo the mammary lines form on each side of the midventral line and mark the location of the future mammary glands. In animals in which only two breasts are characteristic of adults, multiple supernumerary teats (called polythelia) or even breasts may form along the mammary line; women and men with three, four, or even six teats are not rare.

Along the mammary line centers of proliferation of the Malpighian layer, called mammary buds, appear in the areas of the future breasts and in numbers characteristic of the mature individuals. Next, primary sprouts are formed as invaginations of the mammary buds. The primary sprouts give rise to side branches, which branch in turn, giving rise to secondary, tertiary, and further sprouts. As a rule, breast development in embryos stops at the stage of primary sprouts.

Between birth and sexual maturity the mammary glands continue to increase in complexity, acquiring a more extensively branched duct system together with a good amount of adipose tissue, which accounts for most of the visible increase in the size of the breast. The most extensive growth of mammary glands occurs during pregnancy or pseudopregnancy, when the elaborate duct system acquires terminal lobules, which consist of subdivisions called alveoli.

Though this account of the development of the mammary gland is by no means exhaustive and should not satisfy those who are especially interested in lactation, it is satisfactory for our purposes here. Having reviewed the basic architecture of the gland—the alveolar cells that secrete the milk, the ducts that conduct the milk toward the nipple canal and hence to the milking machine, be it living or mechanical—we are now ready to consider the endocrine control of the gland. We have deliberately ignored transformation into milk of the milk precursors conducted to the breast by the blood, not because it is not important, but because its details are part of a separate and highly complex subject.

Endocrine Control of Lactation

Because of the obvious dependence of the growth of the mammary gland on sexual development and, later on, pregnancy, it became apparent that the hormones secreted by the glands concerned with reproduction control the development of the breast. A striking demonstration of this fact is seen in newborn babies, whose mammae, stimulated by the hormones of the mother during gestation, not only show greater development than they do a few weeks after birth, but also contain at birth a secretion resembling milk (called "witches' milk").

Controlled experiments on laboratory animals have shown that two hormone systems are involved in lactation: one preparing the gland anatomically for secretory activity, the other acting on the developed gland to cause secretion of milk. We are already familiar with the concept of a dual-control system, for two hormone systems were found to be necessary to the mammalian uterus and the chicken oviduct—one for morphological development, the other for precipitation of secretory activity.

There is considerable variation among the species in the effect of estrogen and progesterone on the development of the mammary glands. We shall assume an otherwise euhormonal organism, for it is well established that both the thyroid and the adrenals play an important role in making optimal functioning of the mammary gland possible. It is true, in general, that exogenous estrogen, when given alone, produces only duct growth (although in some species this hormone, especially when given in large doses and over long periods, induces growth of the alveolar system as well). Estrogen injected into ruminants (most of the work has been done on goats) produces complete udder growth, including lobo-alveolar growth and lactation. Because most of these experiments were done on goats with intact ovaries and pituitaries, the conclusion that estrogen alone produces these effects is not justified. Progesterone alone, in the majority of species studied, produces alveolar growth without having any significant effect on duct development. As might be expected, combinations of these two steroids produce much more extensive development of the secretory tissue than either hormone alone. However, in species in which estrogen alone can cause both duct growth and alveolar development, addition of progesterone is either ineffective or only mildly synergic.

Androgen, and possibly the adrenal steroids, can also induce growth of mammary tissue, including duct and alveolar development. It is not clear whether androgen itself or a conversion product of that hormone is gynecogenic.

There is a theory that estrogen and progesterone produce their stimulatory effect on the breast indirectly, by inducing the release from the pituitary gland of a mammogenic substance (or substances), which acts on the mammary gland, causing duct and alveolar growth. However, this theory finds little support in view of the fact that complete mammary growth can be induced by estrogen and progesterone in hypophysectomized animals. Furthermore, estrogen applied by inunction to only one breast of intact males causes the proliferation of only the treated breast and not of the contralateral breast. If the estrogen effect were mediated by the pituitary gland, both glands should develop in response to the hypophyseal mammogenic agent.

Secretion of Milk

Since hypophysectomy of lactating animals always causes rapid and complete cessation of lactation, it is evident that the pituitary gland elaborates a hormone (or hormones) essential for continued milk secretion. After estrogen and progesterone have prepared the mammary gland anatomically, pituitary hormones acting on the ready gland cause it to secrete milk. It is now well established that the optimal rate of lactation is produced by the interaction of several hypophyseal hormones, of which prolactin is most important. When prolactin is injected in large doses into hypophysectomized females, lactation is induced. However, when growth hormone and ACTH are added to prolactin, lactation is more copious and the doses of these hormones required can be substantially reduced. Prolactin has its galactogenic effect by acting directly on the mammary gland. The mode of action of growth hormone is not definitely known. Growth hormone is known to be diabetogenic in many mammals and it appears possible that its galactopoetic effect is due to its ability to raise the blood glucose level and thus increase one of the substrates known to be essential in milk formation. The mode of action of ACTH appears to lie in its ability to cause the adrenal to secrete mineralo- and glucocorticoids which in some as yet unknown way affect lactation. Here the story is somewhat complicated in that if either ACTH or adrenal steroids are administered to lactating animals (except the rat), milk yield is depressed. However, the same hormones appear to be essential in the *initiation* of lactation and we will return to this point shortly.

The roles of TSH and thyroxin in lactation are at best equivocal and seem to differ among species. If just the right dose of thyroxin is chosen, it is possible to increase the milk yield of dairy cattle for short periods of time. Because of the sensitivity of cows to overdoses of thyroxine its use is not economical; a dose that is optimal for some cows may be too low for others and may overdose still others. Thus, it is practically impossible to find a dose of thyroxine that will be optimal for all cows in a herd. Only a few cases in which thyroxine preparations have been used in practice have shown a commercially significant increase in the total milk yield, and not infrequently the yield has been lowered.

Prolactin, as we have already stated, is controlled by a dual hypothalamic mechanism, a releasing factor (PRF) and an inhibiting factor (PIF). (Whether PRF is the same factor that releases TSH (TRF) or is a separate entity is presently not established.) In rats all experimental evidence shows that suckling causes the release of prolactin. Furthermore, the amount of prolactin released into the maternal circulation is proportional to the number of nipples suckled. The implication of this finding is that the prolactin released as a result of suckling is essential for the formation of new milk. As young rats grow they suckle less frequently, less prolactin is released, and the rate of lactation decreases and eventually stops. If a growing litter is successively replaced with a younger litter that maintains vigorous and frequent suckling, the period of lactation can be significantly extended beyond the normal 21-day period. That the relationship between suckling and prolactin release may be different in such animals as cows, goats, or women, is suggested by the fact that in spite of daily milking or suckling the quantity of milk synthesized in these species steadily decreases over a period of months. Women, however, are known to be able to produce some milk for as long as two to three years after parturition. In dairy cows the frequency of milking is also known to be related to prolactin release. The plasma prolactin concentration increases up to 100 percent with the onset of each milking. But this increase appears to control subsequent milk yields rather than the yield of the milking in progress. If a scheduled milking is omitted, the next milk yield will be significantly lower than if the missed milking episode had not been skipped.

As we have already stated, the mammary glands develop under the influence of estrogen and progesterone after puberty is reached. The breasts continue to respond to the cyclic ebb and flow of these two hormones during the reproductive cycles. In laboratory or domestic animals the changes in the mammary glands related to the estrous cycle are not grossly noticeable but in women they are measurable. In fact,

some women can estimate the time of ovulation from both the size and turgidity of their breasts. The mammary glands of animals in pseudo-pregnancy resemble those of pregnant animals, even forming milk toward the end of pseudopregnancy. However, pseudopregnant females, except bitches, do not lactate.

In animals in which the placenta produces estrogen and progesterone these placental steroids play the major role in building the mammary gland. Toward the end of gestation the secretory activity of the mammary gland begins with secretion of the so-called colostrum milk, which is chemically different than the milk excreted after parturition. The question why pregnant females do not begin to lactate prior to parturition has been answered by the finding that lactation can be precipitated midway through gestation and maintained by the injection of cortico-steroids. The mode of action of this steroid is not known but it may somehow activate the hypothalamo-hypophyseal system involving release of prolactin and growth hormone. That this is probably the mechanism involved in initiating lactation at parturition is supported by the finding that at the onset of labor in such females as cows, rats, and women the level of cortisol rises significantly from its base level, to which it returns after parturition.

We stated earlier that suckling causes the release of prolactin, and in the next section we shall discuss "milk let-down." In both instances a neural stimulus seems to be elicited by stimulation of the nipple, causing the release of prolactin in the one instance and oxytocin in the other. But although the role of neural stimuli appears to be important in rats (and perhaps in rabbits), it is much less clear in larger mammals. In goats in which one-half of the udder has been denervated or in which the whole udder has been transplanted to the neck, the denervated half of the udder or the whole transplanted gland produce amounts of milk comparable to intact glands. These observations raise the important question of whether in large animals there are feedback mechanisms other than those which control lactation in laboratory animals. It is abundantly clear that important species differences exist with regard to the hormone requirements for the preparation of the gland for lactation, for initiation of lactation, and for the maintenance of milk synthesis. Generalizations are therefore not possible.

Milk Let-Down

If one cannulates the teat canal of such large animals as lactating goats or cows, only the milk stored in the cistern and the larger ducts is obtained. This is a very minor portion of the total amount of milk present in the whole gland. If, however, such a stimulus as suckling or milking is

provided, milk begins to flow freely. At present the theory formulated by Gains and by Petersen and his co-workers seems to explain this phenomenon of milk "let-down" best. It is postulated that a neurohumoral mechanism is ultimately responsible for the contraction of myoepithelial cells around the alveoli, squeezing out the milk contained in the alveolar cells. The stimulus obtained at suckling (or any mechanical manipulation of the breast or teat, such as milking) causes a neural reflex to go to the hypothalamus and thence to the posterior lobe of the pituitary, which responds by releasing oxytocin; this, in turn, causes contraction of myoepithelial tissue in the breast. The assumption can be verified in a variety of ways. Injection of oxytocin into a female, without physical stimulation of the mammary gland, or into females whose mammary glands have been denervated, causes milk let-down. That this is mediated through the hypothalamus is seen from the fact that in females whose pituitary stalks have been cut suckling does not lead to let-down. Electric stimulation of the hypothalamus in goats, without manipulation of the gland, also causes milk let-down, further supporting the theory of a neurohumoral pathway between the mammary gland and the posterior lobe of the pituitary.

Stimulation of the mammary gland is not the only phenomenon that induces milk let-down. In cattle such events as the rattling of milk pails, the presence of calves, and the arrival of the scheduled time for milking are enough to initiate the neurohumoral chain leading to milk let-down. Furthermore, an event completely unrelated to milking or to nursing, such as manipulation of the external genitalia, artificial insemination, copulation, or even the presence of the male is sufficient to induce the phenomenon. Because in all these cases uterine motility and contractility increase, it is postulated that sexual excitation, whether it is connected with nursing or with mating, leads to the release of oxytocin. For centuries men have taken advantage of this chain of events by manipulating the vulva of cows to cause milk let-down. Even today, a street milk vendor in India inserts a stick into the vagina of his cow and rotates it a few times after he has found a buyer for the milk and before he milks the cow.

SELECTED READING

Austin, C. R., and A. W. H. Braden. 1953. An investigation of polyspermy in the rat and rabbit. *Australian J. Biol. Sci.*, **6:**674.

Bearden, H. J., W. Hansel, and R. W. Bratton. 1956. Fertilization and embryonic mortality rates of bulls with histories of either low or high fertility in artificial breeding. *J. Dairy Sci.*, **39:**312.

Braden, A. W. H. 1953. Distribution of sperms in the genital tract of the female rabbit after coitus. *Australian J. Biol. Sci.*, **6:**693.

Cole, H. H., G. H. Hart, W. R. Lyons, and H. R. Catchpole. 1933. The development and hormonal content of fetal horse gonads. *Anat. Record,* **56:**275.

Cowie, A. T., and S. J. Folley. 1955. Physiology of gonadotropins and the lactogenic hormone. In *The Hormones: Physiology, Chemistry and Applications,* Vol. 3, Academic Press.

Dauzier, L., R. Ortavant, C. Thibault, and S. Winterberger. 1954. Résultats nouveaux sur la gestation à contre-saison chez la brebis et chez la chèvre, possibilité d'utilisation pratique. *Ann. Zootechnie,* **2:**89.

Folley, S. J. 1955. *The Physiology and Biochemistry of Lactation,* Oliver and Boyd.

Folley, S. J., and F. H. Malpress. 1948. Hormonal control of mammary growth and Hormonal control of lactation. Two articles in *The Hormones: Physiology, Chemistry and Applications,* Vol. 1, Academic Press.

Fosgate, O. T., and V. R. Smith. 1954. Prenatal mortality in the bovine between pregnancy diagnosis at 34–40 days post insemination and parturition. *J. Dairy Sci.,* **37:**1071.

Hansson, A. 1947. The physiology of reproduction in mink, with special reference to delayed implantation. *Acta Zool.,* **28:**1–136.

Hawk, H. W., J. N. Wiltbank, H. E. Kidder, and L. E. Casida. 1955. Embryonic mortality between 16 and 34 days post-breeding in cows of low fertility. *J. Dairy Sci.,* **38:**673.

Hecker, J. F., T. Strakosch, M. Wodzicka-Tomaszewska, and A. R. Bray. 1974. Effects of size of intrauterine devices on luteal function in ewes. *Biol. of Reprod.,* **11:**73.

Jacobson, A., H. A. Salhanick, and M. X. Zarrow. 1950. Induction of pseudopregnancy and its inhibition by various drugs. *Am. J. Physiol.,* **161:**522.

MacArthur, J. H. 1949. Selection for small and large body size in the house mouse. *Genetics,* **34:**194.

Murphree, R. L., W. G. Black, G. Otto, and L. E. Casida. 1951. Effect of site of insemination upon the fertility of gonadotrophin-treated rabbits of different reproductive stages. *Endocrinol.,* **49:**474.

Nelson, M. M., and H. M. Evans. 1954. Maintenance of pregnancy in the absence of dietary protein with estrone and progesterone. *Endocrinol.,* **55:**543.

Perry, J. S. 1955. Reproductive wastage: prenatal loss. In *Collected Papers, Vol. 3: The Breeding of Laboratory Animals.* Laboratory Animals Bureau, M.R.C. Laboratories, London.

Robinson, T. J. 1950. The control of fertility in sheep, Part I: Hormonal therapy in the induction of pregnancy in the anestrous ewe. *J. Agr. Sci.,* **40:**275.

———. 1950. The control of fertility in sheep, Part II: The augmentation of fertility of gonadotrophin treatment of the ewe in the normal breeding season. *J. Agr. Sci.,* **41:**6.

———. 1957. Pregnancy. In *Progress in the Physiology of Farm Animals,* Vol. 3, Butterworth.

Rowlands, I. W. 1949. Serum gonadotrophin and ovarian activity in the pregnant mare. *J. Endocrinol.,* **6:**184.

Shaikh, A. A. 1971. Estrone and estradiol levels in ovarian venous blood from rats during the estrous cycle and pregnancy. *Biol. of Reprod.,* **5:**297.

Wiggins, E. L., R. H. Grummer, and L. E. Casida. 1951. Minimal volume of semen and number of sperm for fertility in artificial insemination of swine. *J. Animal Sci.,* **10:**138.

Wodzicka-Tomaszewska, M., J. F. Hecker, and A. R. Bray. 1974. Effects of day of insertion of intrauterine devices on luteal function in ewes. *Biol. of Reprod.,* **11:**79.

12

Fertility and Sterility

The majority of cases of sterility are due to (1) anatomical aberrations of the reproductive systems of males and females and (2) endocrine upsets. It is probable that both causes are hereditary, so that, at least in domestic animals, no attempt should be made to correct the anatomic or endocrine causes of sterility lest the condition be spread through a large part of the population.

More important than complete sterility is impaired fertility. In these cases the ovulation rate, ovulation time, and fertilization rate are normal, but embryonal mortality is far in excess of the "normal" 30 percent. The reasons why some members of a population show impaired fertility are obscure, since occasionally even such females reproduce normally if they are bred often enough.

Attempts to increase normal fertility rates by lowering "normal" embryonal mortality with the use of hormones (such as

progesterone) have not been uniformly successful, but they show some promise. Likewise, attempts to increase ovulation rates with hormones are also promising. In both instances much additional work is required.

Attention is called once again to the role of the little-understood psychosomatic factors in the efficiency of reproduction. Data are presented to show that the intensity of psychological heat in pigs is inversely related to fertility and perhaps even fecundity. It is obvious that we need more intensive study of pheromones and of animal behavior as it affects reproductive phenomena.

INTRODUCTION

In Chapter 11 we discussed some conditions under which animals show impaired fertility, and we have noted that deviations from maximal efficiency of reproduction are common and may be ascribed to a variety of causes. A variable proportion of male and female mammals and birds show complete sterility. It is the purpose of this chapter to discuss some of the common causes of sterility, other than infection by microorganisms, and to consider the possibilities of correction. The question of improving the average levels of fertility of normal animals will also be given consideration.

Sterility due to infection by microorganisms is excluded not because it is unimportant, but because this cause of sterility can be detected and treated with relative ease. Though such conditions as vaginitis, brucellosis, leptospirosis, and vibriofetus profoundly affect the reproductive performance of most mammals, they do not fall within the sphere of reproductive physiology, but are a matter of concern for veterinary medicine. It is important to be aware of the role played by infectious diseases in the physiology of reproduction and not to confound sterility and impaired fertility caused by bacteria with those that are due to endocrine or anatomic aberrations.

STERILITY

All cases of total sterility of both sexes can be divided into two categories: those due to anatomical defects, and those caused by endocrine imbalance or malfunction.

Anatomical Defects in Females

There is considerable variation among the species in the frequency with which anatomical defects of the reproductive system can be blamed for sterility. Various surveys involving large numbers of animals of known reproductive history indicate that anatomical abnormalities are rare in goats, sheep, mares, cows, rabbits, and rats, and that they are common in pigs, in which they account for about half of the cases of total sterility or greatly impaired fertility. Though no exact figures are available, anatomical abnormalities seem to be an important cause of sterility in women.

The pig has been analyzed most completely, both in the United States and in Europe, and permits some rather interesting comparisons. Several populations totaling about 10,000 animals, including both slaughter-house animals and animals whose reproductive histories were accurately known, have been studied in the United States. Results obtained at the University of Illinois are in close agreement with those from the University of Wisconsin, and so will serve to illustrate the problem.

Of 79 "sterile" females purchased from farms known to be free of infectious diseases affecting reproduction, 53 percent conceived at the first opportunity after they were brought to the university farm. The females that conceived had been bred by their owners from three to 10 times before being culled as "sterile." At autopsy only 7.1 percent of these females showed abnormalities, none of which was sufficiently serious to account for their prior failure to conceive. Neither is there an explanation for the failure of the other conceiving females to conceive in their home farms, for no improvement in their nutritive state of their general well-being was brought about by the move from their original environment. From the point of view of the breeder, these females were of course sterile, since to rebreed them as often as necessary to cause them to conceive was neither economical nor practical. Neither is it usually practical to provide them with a change of scenery in order to determine which are truly sterile and which only require repeated mat- ings (see Chapter 4, "Bizarre Phenomena Related to the Estrous Cycle").

The remaining 36 females (45.6 percent of the total) did not become pregnant. The reasons were determined at autopsy and are summarized in Tables 12-1 and 12-2 (see also Figure 12-1). Almost half of the females failed to conceive because occlusion and distention of their oviducts made the passage of ova or sperm impossible. Other abnormalities, such as unilaterally blind uterine horns, brought the total incidence of sterility due to anatomical aberrations to 64 percent. Theoretically, there is no reason why an animal having a unilaterally blind horn should not con- ceive. Nevertheless, though fertilized ova are almost invariably recovered from the patent side, pregnancy in such animals is extremely rare. Those in which half of the uterus is missing are fertile, but their litter size is of course much smaller than that of normal females.

A similar study conducted by European workers (Perry and Pomeroy, 1956; and Goethals, 1951) revealed anatomical abnormalities to be a minor cause of sterility in the populations they studied. Goethals in

Table 12-1. Anatomical findings in 36 gilts and sows that failed to become pregnant (bold numbers are subtotals)

ABNORMALITIES			PERCENT OF GILTS	PERCENT OF SOWS
TYPE	NUMBER	PERCENT		
Endocrine aberrations	**9**	**25.1%**	**9.8%**	**14.2%**
Cystic follicles (corpora lutea)	6	16.7	7.8	7.1
Cystic follicles (no corpora lutea)	2	5.6	0.0	7.1
Infantalism	1	2.8	2.0	0.0
Anatomical abnormalities	**23**	**63.9**	**43.1**	**10.8**
Hydrosalpinx, pyosalpinx	17	47.2	31.3	3.6
Unilateral blind horn	5	13.9	7.8	3.6
Unilateral missing segment*	(2)	(5.6)	2.0	3.6
Blind uterine body	1	2.8	2.0	0.0
Miscellaneous abnormalities	**2**	**5.6**	**0.0**	**7.1**
	36	94.6%	52.9%	32.1%

*This abnormality does not preclude pregnancy.

SOURCE: A. V. Nalbandov, *Fertility and Sterility*, 3:100, 1952.

Table 12-2. Anatomical findings in 79 gilts and sows that were bred from three to 10 times with boars of known fertility (bold numbers are totals)

RESULT OF BREEDING	TOTAL		GILTS		SOWS	
Pregnancy	**42**	**53.2%**	**23**	**45.1%**	**19**	**67.8%**
No abnormalities	39	49.4	21	41.1	18	64.3
Abnormalities	3	3.8	2	4.0	1	3.6
No pregnancy	**37**	**46.8**	**28**	**54.9**	**9**	**32.2**
No abnormalities	4	5.0	3	5.9	1	3.6
Abnormalities	33	41.8	25	49.0	8	28.6

SOURCE: A. V. Nalbandov, *Fertility and Sterility*, 3:100, 1952.

Belgium did not find a single case of occlusion of the oviduct in the 1,000 animals included in his sample. Perry in England found among 83 animals only three that had abnormalities—presumably including anatomical defects—other than cystic ovaries. (His findings with regard to cystic ovaries will be discussed later.)

All animals with anatomical defects have cycles of normal length and normal heats, and their condition cannot be diagnosed externally. By injecting a starch suspension intraperitoneally and staining vaginal smears with iodine to show the presence of starch granules between 12 and 24 hours after the injection, it is possible to discover oviducal occlusion, but the method does not work on females with unilaterally blind horns.

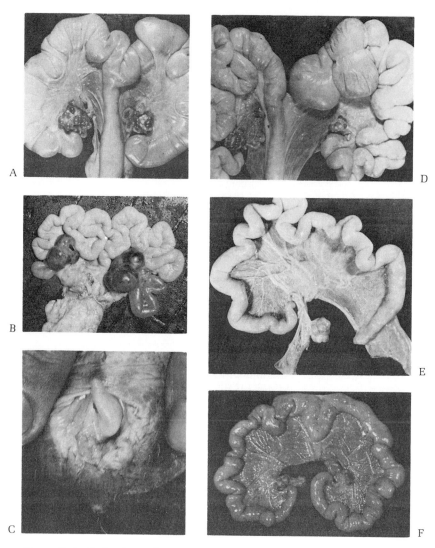

Figure 12-1 Endocrine and anatomical causes of sterility in swine. **A.** Multiple small cyst; there are 32 cysts measuring 1.5 cm in diameter in the left ovary; the right ovary has a few small cysts and corpora albicantia. **B.** This pig was sterile because she had large ovarian cysts and bilateral hydrosalpinges (compare their size with the normal oviduct in *A*). **C.** Enlarged clitoris (3.5 cm long, about 20 times the normal size), frequently found in cystic females. **D.** Blind uterine horn; this usually causes sterility. **E.** Right horn and ovary are missing; such females are fertile. **F.** Blind uterine body; the cervix and most of the vagina are missing.

Anatomical defects of the reproductive system are much less common in mammals other than pigs. Cattle frequently have a complete or partial duplication of the cervix, but it does not interfere with fertility. Complete separation of the two uterine horns seems to be very rare in cattle but not infrequent in women, in whom each horn may also have a separate cervix. There are even records of several women who conceived separately in the two horns at two ovulations about 28 days apart, with two babies being delivered about 30 days apart. (Such cases are interesting not only because they show that ovulation is not completely suppressed in pregnancy but also because they show that the readiness of the fetus to be born plays some role in the initiation of the mechanism of parturition.)

Occlusion of the oviducts is a common cause of sterility in women. The condition in women is usually ascribed to infection, but attempts to produce hydro- or pyosalpinges in pigs by deliberate local infection of the oviducts were not successful. The fact that occlusion of the oviducts seems to occur in some families of pigs more often than in others suggests an underlying genetic mechanism, and the fact that it is found almost exclusively in gilts argues further against infection as the cause. These arguments are further supported by the finding that the condition is rare in European populations of swine.

Anatomical Defects in Males

Sterility in males is also frequently caused by anatomical defects of the duct system. It occurs in all the species studied but is especially noted in men and bulls. The blockage may be found at any level of the duct system and may be unilateral or bilateral. Missing segments also are frequently noted.

Bilateral blockage of the duct system is easy to diagnose because no sperm are present in the ejaculate. Unilateral blockage is not so easy to diagnose, for the sperm count may be normal, and the rate of fertility may range from none to normal.

Inflammations or degenerative changes of the accessory glands may also cause complete or partial sterility even though the sperm count and other characteristics of the semen may be normal. In three boars, of which two were completely sterile and one produced an occasional litter of greatly reduced size, one or both seminal vesicles were found to be greatly enlarged, distended with atypical-appearing fluid, and distinctly inflamed. Sperm obtained from the vas, the epididymis, and the testes appeared normal in all three.

Endocrine Causes of Sterility

CYSTIC OVARIES IN SWINE. In the University of Illinois study discussed earlier, 21 percent of the females that did not conceive in spite of frequent breedings were found to have cystic ovaries (Figure 12–2). We know already that two kinds of cysts were found in the pigs examined

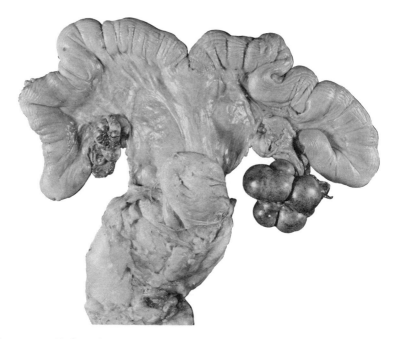

Figure 12-2 Unilateral cystic ovary of a pig with a long history of sterility. The left ovary is normal and contains fresh corpora lutea; the right ovary has eight heavily luteinized cysts 3–4 cm in diameter (compare Figures 5–1 and 12–1a and b).

and that both large and small cysts are associated with sterility (Table 5–1, p. 163). It is noteworthy that cystic ovaries were much more common in the European samples than in the American. The frequency with which cysts were found and the apparently significant seasonal fluctuation in the occurrence of cysts in sows in England are shown in Table 12–3. This tabulation includes all types of cysts, including those that may have no effect on the fertility of the females examined (for example, retention cysts). Even if single cysts are omitted, the cystic degeneration of the ovaries of swine was significantly more frequent in England than in the United States.

CYSTIC OVARIES IN CATTLE. Estimates of the frequency with which cystic ovaries occur in cattle vary by conformation (rare in beef cattle, common in dairy cattle), by breed (Holsteins have more cystic ovaries than Jerseys), and by other factors (which will be mentioned). The cystic ovaries of cattle seem to differ significantly in causes and effects from those of swine. The only similarity between the two species is that both slight and great cystic degeneration of follicles is noted in both. In cows ovarian cysts frequently lead to nymphomania, which causes afflicted

Table 12–3. Seasonal changes in the occurrence of cystic follicles in sows (arranged by months in which samples were available)

	NO. OF SOWS	PERCENT CYSTIC		NO. OF SOWS	PERCENT CYSTIC
February	22	32%	September	49	20%
March	34	35	October	147	14
April	68	35	November	150	14
May	115	36	December	85	18
June	147	30	January	5	20
	386	33.1%		436	13.3%

SOURCE: J. S. Perry and R. W. Pomeroy, *J. Agr. Sci.* 47:238–248, 1956.

females to assume several male secondary characters, such as coarseness of head and neck, male voice, and male sex behavior toward other cows. If it is assumed that ovarian cysts secrete estrogen, then it must also be postulated that estrogen in large quantities has an andromimetic effect. Since this is not supported by experimental evidence, it is possible that ovarian cysts in cattle secrete progesterone (just as they do in pigs) or that the cystic degeneration of the ovaries is a secondary effect. There is a strong suspicion that the primary difficulty may lie in an adrenal malfunction, which may lead to an upset in the pituitary-ovarian axis, culminating in cystic degeneration of the follicles. This possibility is supported by the finding that urinary 17-ketosteroids increase significantly in cows afflicted with nymphomania.

In a study involving 341 cows and 1,280 cow-service periods it was found that 18.8 percent of the cows had cystic ovaries (Casida and Chapman, 1951). Confirming results obtained in Sweden and elsewhere, this study established the fact that the condition was inherited (in this particular herd the heritability was 0.43). Of great interest is the fact that the frequency with which cystic ovaries occurred was significantly associated ($p = 0.01$) with the milk production (Table 12–4). In Table 12–5, with the exception of Group 1 (which includes heifers prior to their first parturition, in which age may be a factor), the groups are arranged in the order of the potential or actual ability of the cows to secrete milk; Group 2 was least productive and Group 4 most productive. It is tempting to ascribe the higher incidence of cystic ovaries in the animals of greater production to a rise in the rate of the pituitary function, which includes not only hormones responsible for the higher rate of milk secretion but most of the other trophic pituitary hormones. Though it is often stated that a high level of nutrition (high protein rations) is conducive to cyst formation, there is no good evidence for this contention.

Table 12–4. **Relation between milk production and incidence of cystic ovaries**

GROUP	NO. OF COWS	MANAGEMENT CONDITION	CYSTIC COWS
1	358	Not lactating	3.4%
2	457	Lactating (in general herd)	6.8
3	106	Started on test but discontinued	8.5
4	359	On official test	10.6

SOURCE: L. E. Casida and A. B. Chapman, *J. Dairy Science*, 34:1200, 1951.

The available data indicate that cysts are rare in other species, but they do occur normally or can be caused to occur by experimental manipulation of females.

OTHER ENDOCRINE DEFECTS.   In most mammalian species infantilism of the reproductive tract of males or females is not uncommon. In females reaching the normal age of puberty the ovaries remain very small and fail to produce follicles approaching ovulatory size. The duct system also remains immature, indicating an absence of trophic hormones. Infantile duct systems respond to injections of estrogen by enlarging. The ovaries, as a rule, can be caused to enlarge and to form ovulable follicles by the injection of gonadotrophic hormones. These observations indicate that infantilism is due to hypopituitarism, which is usually restricted to the secretion of the gonadotrophic hormones only, since such animals usually appear somatotypically normal. In human females infantilism is frequently caused by hypothyroidism and can be corrected by appropriate medication.

Quiet (physiological) heats present a problem in many domestic animals in that the heat is not detected and the females affected are not bred. Females having quiet heats usually ovulate normally at the expected time. It is probable that quiet heats are much commoner than is thought. If sheep are checked for heat daily with vasectomized males, it is found that certain females have intervals between heats that are multiples of 16 days, the normal duration of cycles. That ovulations occur in these females has been determined by laparotomy. Quiet heats occur more frequently in some females than in others, and the propensity for quiet heats appears in the same females in successive breeding seasons. There is no indication that ewes in quiet heat are less fertile than those not having quiet heats, but it should be kept in mind that such behavior restricts the control of the breeder over the time females conceive and is potentially dangerous if allowed to affect too many individuals. Similar observations have been made on cows, mares, and swine. No good data are available on the repeatability of the condition in individuals of any species, primarily because experiments have not been designed to differentiate between quiet heat and failure to show heat because pregnancy has been initiated but terminated (through resorption or abortion) shortly after breeding.

Correction of Sterility

It is difficult to estimate the proportion of animals that are partially or completely sterile. Estimates of the occurrence of cystic ovaries in cattle vary from 1 to 20 percent and depend on a variety of factors, such as breed, family, and management practices. Estimates of the proportion of sterile swine vary from 5 to 17 percent of the total female breeding population. The divergence of these figures is undoubtedly due to the fact that the causes of sterility are inherited (compare European and American data for anatomical abnormalities and cystic ovaries in swine).

Largely because the etiology of ovarian cyst formation is not understood, treatment is difficult and results have been inconsistent. In cattle luteinization of cysts can be accomplished by the injection of gonadotrophic hormone, such as LH-rich pituitary gland preparations or chorionic gonadotrophin. Not all cows respond to these treatments, however, and even in those in which normal cycles are reestablished by treatment cysts frequently recur. Moreover, a high proportion of cystic animals (according to some estimates, 50–60 percent) recover spontaneously, complicating the interpretation of the efficacy of treatment. In women a wedge-shaped resection of the cystic ovary has been found effective in reestablishing normal cycles and permitting subsequent conception. Hormone treatments can reestablish normal reproductive performance and ultimate conception in all females (all species) afflicted with cystic ovaries.

Most recently, excellent success has been achieved in the treatment of ovarian cysts in dairy cattle by the injection of synthetic LHRF, which stimulates the animal's pituitary gland to release LH, resulting in either ovulation or luteinization of the cysts. Frequently, normal cycles return and conception follows. It is presently unknown whether cysts recur in females treated with LHRF, as they frequently do in females treated with exogenous LH or (in women) following wedge resection.

The most important question that arises in all discussion of corrective treatments of sterility in domestic animals concerns the wisdom of treating sterility and returning recovered females to the breeding herd or flock. There is a well-founded suspicion that anatomic abnormalities, such as those described for swine, are inherited. For this reason it is fortunate that practically nothing can be done to overcome these defects. Yet the mere elimination of the afflicted animals is not enough. It is necessary to eliminate as many of their relatives as the breeding program of the herd will allow. If that is not done, a gradual increase in sterility will take place (this was seen in several of the herds from which sterile swine were obtained for the University of Illinois study discussed earlier in this chapter).

Similarly, cystic ovaries in cattle are hereditary. In contrast to anatomical abnormalities, nymphomania resulting from cystic ovaries can frequently be corrected by endocrine treatment, or the afflicted females may recover spontaneously for sufficiently long periods to conceive. The available data indicate that it is not advisable to permit nymphomaniacs or related individuals to reproduce, and they should be eliminated from the herd, regardless of the initial investment in them.

An instructive example of what can happen when certain types of sterility are overlooked, and of how a high incidence of sterility can be corrected, is available in Swedish Highland cattle (Eriksson, 1943.). About 1935 it was found that 30 percent of the cattle in these herds had hypoplasia of one ovary (usually the left) and that 5 percent had bilateral hypoplasia. The hypoplastic ovary was significantly smaller than the normal ovary and was completely nonfunctional, thus reducing the fertility of the afflicted animals. A campaign was instigated to detect animals with hypoplastic ovaries and to remove them from the breeding herds, for it was found that the condition was caused by a recessive autosomal gene with incomplete penetrance. The initial prevalence of the condition and the success of the campaign to reduce its frequency in this Swedish breed of cattle can be seen in Table 12-5. The proportion of totally sterile animals with bilateral ovarian hypoplasia was reduced from 5 percent in 1935 to approximately 1 percent in 1948. The same Swedish Highland gene also causes gonadal hypoplasia in the males; as a rule, the left testicle is greatly reduced in size and shows abnormal spermatogenesis. This, or a similar gene, is by no means unknown in other breeds of cattle (Eriksson, 1943), and unilateral testicular hypoplasia has been noted in stallions and men.

DEGREES OF FERTILITY

Impaired Fertility

In addition to cases in which sterility or impaired fertility can be ascribed to definite or plausible causes (discussed in previous sections), there are many cases in which the causes of faulty reproductive perfor-

Table 12–5. Incidence of ovarian hypoplasia in Swedish Highland cattle and the effect of genetic selection against this trait

YEAR	NUMBER OF COWS STUDIED	COWS WITH HYPOPLASTIC OVARIES	
		NUMBER	PERCENT
1936	2,194	384	17.5%
1937–39	1,173	179	15.2
1940–42	1,438	162	11.3
1943–45	1,588	176	11.1
1946–48	1,752	164	9.4

SOURCE: N. Lagerlöf and I. Settergen, *Cornell Veterinarian*, 43:51, 1953.

mance are not immediately apparent. These present an extremely important problem, which can be summarized as follows. Females of most species of domestic animals require, on the average, one and a half services per conception, which is considered a satisfactory performance for either natural matings or artificial insemination. Many individuals, however, require two, three, or many more services before they conceive. They cannot be properly classified as sterile, for if bred often enough they usually conceive and carry young to term. These females are called "hard to settle" or liable to "repeat-breeding." The reasons for such reluctance remain generally unknown, but they seem to be a combination of such factors as failure of the ova to become fertilized and embryonal mortality, the latter being attributable to a variety of causes.

That intrinsic factors in the semen and in the ova may account for a low fertilization rate and high embryonal mortality is demonstrated by studies of cattle. In a University of Wisconsin study bulls were evaluated and then classified into groups of low, medium, and high fertility on the basis of their ability to impregnate cows and to cause them to remain pregnant for 60–90 days after mating. The data from this study are shown in Table 12–6 (the data have been corrected for any abnormalities of cows that may have been responsible for failure to become pregnant). It is obvious that bulls of low fertility fertilize a significantly ($p = 0.01$) smaller number of eggs than bulls rated as highly fertile. The embryonal death rate in cows mated with bulls with a no-return fertility range of 67–79 percent was 25.5 percent; for cows mated with bulls falling in the lower fertility range (no-return rate, 40–66 percent) the estimated death rate was 14.9 percent. In a Cornell University study low-fertility bulls were associated with a higher embryonal mortality (19.6 percent) than high-fertility bulls (10.5 percent).

Just why the semen of some bulls fails to fertilize some ova and why a high proportion of the ova they do fertilize die remains unknown. It is

Table 12-6. Fertilization rates and estimated embryonic death rates in cows bred with bulls having high, medium, and low fertility

FERTILITY CLASSIFICATION	NO. OF BULLS	NONRETURN RATE	ESTIMATED PREGNANCIES	RECOVERED NORMAL OVA FERTILIZED	ESTIMATED FERTILIZED OVA DYING BEFORE 60–90 DAYS
High	22	70–79%	75.8%	100.0%	24.2
Medium	28	60–69	68.3	82.1	16.8
Low	14	40–59	54.1	71.4	24.2
Total:	64		Averages: 65.1%	85.9%	21.0

SOURCE: H. E. Kidder *et al., J. Dairy Science,* 37:691, 1954.

noteworthy, however, that the reproductive performance of the low-fertility bulls was not due to a temporary debility; they had been classi-fied as poor performers for five months preceding the study.

Though some of the difficulties encountered by hard-to-settle females can be laid to defective semen of males with whom they are mated, some of the females themselves have impaired fertility. This becomes apparent from the data shown in Table 12–7. Cows that were reluctant to conceive were bred to bulls that had previously given satisfactory reproductive performance, so that most, if not all, of the reproductive aberrations can be attributed to the cows themselves. Here again it is obvious that two factors are operating: a much lowered fertilization rate (which in highly fertile cows should approach 80 percent) and a high embryonal mortal-ity, about 70 percent. There are no known reasons for this impairment in fertility, which is probably due to a defect of the ova but may also be due to hormonal conditions that are incompatible with normal fertility.

When we consider all factors in a breeding population of cattle com-posed of males and females of all degrees of fertility, it appears that 40–50 percent of all the potential young are lost by the third month after breeding. A Wisconsin study estimates this loss to be 40 percent and apportions it as follows: 3 percent of the cows fail to become pregnant because of genital abnormalities; 9 percent fail because of defective ova; 12 percent fail because ova are not fertilized; 16 percent fail to bear because the embryos die. No data on the heritability of impaired fertility in domestic animals are available, but one study of rats suggests that females in heat unable to conceive after a single exposure to males pro-duced female offspring that required 1.5 heats to conceive; by the fifth generation their offspring required 3.5 heats before becoming pregnant. Also by the fifth generation, only 33 percent of the daughters of the original hard-to-settle females (8 percent of the original population) conceived at the first opportunity. If these data are applicable to domes-tic animals, caution in retaining hard-to-settle females in breeding popu-lations seems indicated.

Table 12-7. **Reproductive performance of hard-to-settle Guernsey and Holstein cows that had no anatomical defects and had previously calved normally**

	3 DAYS AFTER 1ST BREEDING		3 DAYS AFTER 2ND BREEDING		34 DAYS AFTER 1ST BREEDING	
	TOTAL COWS	WITH FERTILIZED OVA	TOTAL COWS	WITH FERTILIZED OVA	TOTAL COWS	WITH FERTILIZED OVA
Guernsey	20	60.0%	11	63.6%	25	20.0%
Holstein	20	70.0	8	75.0	27	25.9

SOURCE: Tanabe and Casida, *J. Dairy Science*, 32:237, 1949.

Table 12-8. **Reproductive performance of ewes in relation to breeding season**

PERFORMANCE	AUG. 1–SEPT. 15	SEPT. 16–OCT. 25
Failure of fertilization due to defective ova	16.7%	2.9%
Failure of fertilization due to poor sperm	47.6	31.4
Embryonal death within 18 days after mating	10.2	6.5
Mating resulting in pregnancies	25.5	59.2

SOURCE: Hulet *et al.*, *J. Animal Science*, 15:607, 1956.

We have noted above that embryonal mortality and failure of fertilization in cattle may be due to hormonal imbalance. That this guess may be correct is suggested by findings on the fertility of sheep in relation to season. Under Illinois seasonal conditions it was found that when ewes were bred only once, and conception rates in August (the beginning of the breeding season), September, and October were compared, the ewes performed as follows: 15 percent conceived in August, 68 percent in September, and 59 percent in October. Similar results were obtained in a University of Wisconsin study in which ewes were bred early and late in the breeding season and the reproductive performances of the two groups were compared (Table 12-8). Only 26 percent of the ewes conceived early in the season, whereas 59 percent conceived after breeding was delayed until later in the season; a large share of this difference is due to failure of the eggs ovulated early in the season to become fertilized. We have already noted that the estrous cycles of sheep are extremely variable in length early and late in the breeding season and deviate significantly from the norm of 16 days (see Table 4-1, p. 124). It was suggested earlier that the length may be variable early in the season

Table 12–9.   Conception and ovulation rates of sheep in relation to season (beginning late in September)

MATING PERIOD	TOTAL EWES MATED	EWES CONCEIVING AT ONE MATING	MATINGS PER CONCEPTION	OVULATIONS PER EWE	EWES WITH QUIET HEATS
Oct. 20–30	22	73%	1.5	1.9	3
Nov. 1–17	20	84	1.3	2.3	1
Nov. 18–30	19	90	1.2	2.4	0
Dec. 15–31	12	67	1.9	2.0	0
Jan. 20–31	22	54	1.9	1.9	2
Feb. 1–28	15	67	1.6	1.9	3
March 1 and later	20	65	1.7	1.8	4
	130	71%	1.5	2.0	13

SOURCE: Averill, *Studies on Fertility*, 7:139, 1955.

because an endocrine imbalance exists before the optimal relation between the ovaries and the pituitary gland has been established and before the rate of hormone flow permits normal reproductive functioning. The length may be variable late in the season because of an increase in the secretion of the gonadotrophic complex, as was postulated in Chapter 3 (p. 89).

A heightened level of the gonadotrophic complex may be incompatible with the cyclic sequence of events normally associated with the estrous cycle, and may therefore result in irregularities in the intervals between heats and ovulations. Now we find that the low fertilization rates and high embryonal mortality observed early and late in the breeding season coincide with the greatest irregularities of the length of the estrous cycle. The extent to which this normal hard-to-settle period of sheep is comparable to the abnormal condition noted for cattle is of course unknown, but the comparability may be real.

Further evidence of the effect of season on the fertility of sheep is provided by data obtained in England that show that the conception and ovulation rates are highest during October, November, and December and then decrease significantly until the end of the season in March (Table 12–9). In this study 3.8 percent of the ewes failed to ovulate, 14.2 percent had unfertilized ova, 7.6 percent had dead ova, and 13.8 percent had lost all their embryos—a total reproductive waste of 39.4 percent.

We have seen ("Bizarre Phenomena Related to Estrous Cycle," p. 134) that heat can be detected in swine in the absence of a boar if pressure is exerted along the back. Females in heat respond by rigidly standing still (lordosis reflex), whereas nonestrous females escape such overtures. Signoret found that in the Large White breed of pigs about 50 percent of the females became immobile in the absence of the male and that the

Table 12-10.   **Relation of conception rate of swine to their ability to show spontaneously the "immobility reflex"**

BEHAVIOR	NULLIPARA		MULTIPARA		ALL FEMALES	
	TOTAL	PERCENT CONCEIVED	TOTAL	PERCENT CONCEIVED	TOTAL	TOTAL CONCEIVED
Very calm	351	56.7%	520	68.6%	871	63.8%
Calm	1,120	49.4	1,507	54.5	2,627	52.3
Agitated	345	31.3	461	36.2	806	34.1
No information	16	68.7	18	55.5	34	61.7
	1.832	47.5%	2,506	54.1%	4,338	51.3%

SOURCE: Du Mesnil Du Buisson, *Ann. Zootech.*, 10:57, 1961.

proportions of females responding was significantly increased by having a male near enough to be smelled or heard.

Du Mesnil du Buisson classified the reflexes of 4,338 sows and gilts as very calm (showed the lordosis reflex after physical contact with man alone), calm, or agitated (could be detected to be in heat only by the use of a boar). All of these animals were artificially inseminated; the numbers conceiving and giving birth to litters are shown in Table 12-10. The data show that the conception rate is 26–33 percent higher in females showing the immobility reflex than in females not showing it. This very important difference implies that attention should be paid to gradations in intensity of estrual behavior and that only those animals with the strongest manifestations of estrual behavior should be used for breeding. Preliminary and incomplete data also suggest that the litter size of females with the weak immobility reflex is lower than that of females with the strong reflex. Nothing is known about the neuroendocrine pathways involved in the manifestation of the reflex of their relation with fertility and fecundity.

Attempts to Increase Fertility

Earlier we made the generalization that fertilization rates in polytocous animals approach 100 percent but are, on the average, significantly lower in monotocous animals. In breeding populations of both types of animals the total loss of potential young due to both failure of fertilization and embryonal mortality is 30–50 percent. Most of this loss occurs very early in pregnancy in all animals studied. The fact that in many individuals, both polytocous and monotocous, there is no evidence of either loss of ova or embryonal mortality raises the question of whether the reproductive efficiency of females that do show reproductive waste can be improved.

The difficulty lies in the fact that it is not known why some females are reproductively so much more efficient than others. Since it is known that implantation does not occur in the absence of progesterone, attempts have been made to improve the reproductive efficiency of rats, pigs, and cattle by the administration of steroid hormones. The results obtained with progesterone alone are not promising; at best, this hormone does not impair the reproductive performance of rats and swine if it is injected in small doses, but it is definitely detrimental to swine in doses of more than 200 mg per day. Combinations of estrogen and progesterone are detrimental to both rats and swine, presumably because the ratio of these two hormones in pregnancy is crucial and the correct ratio has not been found in experiments.

The failure of attempts to preserve a larger proportion of zygotes raises the question whether raising the ovulation rate, by increasing the number of ova fertilized and implanted, would increase the number of them that survive. A classic study of this question indicates that it is indeed possible to increase the fertility of sheep significantly by the injection of PMSG (in this case given on day 14 of the cycle and again after the onset of heat). The dosage of PMSG was intended not to produce superovulation but to increase the proportion of multiple ovulations at the expense of single ovulation (Table 12–11).

The most complete analysis of what happens in ewes in which superovulation is produced by varying doses of PMSG has been made in two excellent studies by T. J. Robinson (1950). Table 12–12 shows that PMSG also increased, is not increased in proportion to the ovulation rate. Nothing in these data precludes the possibility of converting the monotocous ewe into a polytocous animal by the use of hormones.

Much research work has been done on the possibility of bringing about reproduction in sheep during the nonbreeding season. It is easy to cause ovulation or heat in sheep during that season; the major problem, if out-of-season breeding of sheep is ever to become practical, is to synchronize these two events and induce them with only a small number of hormone injections. Treatment of ewes with progesterone or testosterone, followed by an ovulating dose of PMSG, can, for a majority, cause synchronized heat and ovulation, followed by pregnancy, but the number of hormone injections required is at present too great to permit practical application.

Superovulation has been more or less successful in all species in which it has been tried. Cows respond to gonadotrophic hormones by multiple ovulations, but they rapidly become refractory to repeated treatments, probably because antihormones are formed. Even immature females of all laboratory species, as well as immature pigs, sheep, and calves, respond to gonadotrophic hormones by superovulation. In one study involving 17 calves aged 2.5–30 weeks, 262 corpora lutea were produced

Table 12-11.   The effect of PMSG on multiple births in sheep

	TOTAL LAMBING	SINGLES	TWINS	TRIPLETS	QUADRUPLETS
Control	119.1%	68%	29%	1.7%	1.0%
Injected with PMSG	166.8	28	48	20.0	3.6

SOURCE: Lopyrin *et al., Sovetskaia Zootechnika,* 1:82, 1940.

Table 12-12.   Relation between ovulation and implantation rates in super-ovulated ewes

			IMPLANTED OVA	
OVULATIONS PER EWE	NO. OF EWES	EWES CONCEIVING	MEAN NO. PER EWE	PERCENTAGE OF TOTAL OVA
2	5	4	1.4	70%
3	5	5	2.4	80
4	8	8	3.0	75
5	2	2	3.0	60
6	4	4	3.3	54
7	4	4	4.8	68
8–9	3	3	4.0	48
10–12	4	4	5.6	49
13–15	3	2	3.7	26
15+	5	2	4.6	17

SOURCE: T. J. Robinson, *J. Agr. Sci.,* 41:1-63, 1950.

and 70 ova were recovered, 23 of which had been fertilized (Marden, 1953). This is a very low fertilization rate, but it seems to be typical of ova produced by juvenile females under the influence of gonadotrophic hormones.

Of extreme interest is the observation that in rabbits and ewes ova produced when the animal is under the influence of progesterone (pseudopregnancy in rabbits, luteal phase in sheep) show a drastically reduced fertilizability, whereas forced ovulations induced during the follicular phase show a fertilization rate that is about normal. The reason for the difference in fertilizability between ova of the follicular phase and those of the luteal phase remains unknown, but it is not failure of sperm to reach the oviducts. The difficulty appears to lie in the physiological effect of progesterone on the ripening follicle and ovum or in the effect of progesterone on the hormone system controlling follicular maturation.

CONCLUDING REMARKS

If we exclude sterility caused by anatomical defects or infections, the problem of gradations in fertility—from complete fertility through impaired fertility to total infertility—is among the most interesting, important, and difficult problems that the scientist and the practical gynecologist or breeder must face. Its importance lies in the fact that an appreciable number of all animals (including humans) show some degree of impaired fertility, which reduces the breeding potential of the population to which they belong. The difficulty lies in the fact that the problem is multidimensional, being due to a variety of factors acting singly and together. We have seen that females showing reduced fertilizability of eggs also show increased embryonal mortality, and the temptation is strong to ascribe the two to the same cause. There is no evidence either for or against such a conclusion. The problem is further complicated by the fact that there are few if any clues that would permit an intelligent experimental approach. Though it is true, for instance, that the fertilizability of ova in rabbits and in sheep can be reduced to almost zero by injections of progesterone, it is equally true that the fertilizability of ova in sheep is also low very early in the breeding season and that at that time there are no functional corpora lutea in the ovaries. These observations are not necessarily contradictory, for it is possible that, though much progesterone is harmful, a little is essential. It has also become more apparent that we must learn to think of "physiological doses" and of the physiological ebb and flow of hormones before we can learn to imitate the conditions under which reproductive functions are at their optimum.

It is becoming increasingly clearer that not enough attention is paid either by the research worker or the practical breeder to so-called psychogenic factors, which seem to have such a profound effect on the breeding behavior of animals.* The study of reproductive behavior of animals as it is affected by interaction with other animals, by the environment in which the animals live, and by the stresses to which they are exposed, seems a most promising and almost untouched area of research. Those who are less inclined toward reproductive animal psychology, and more toward physiology or endocrinology, can explore the mechanism of action of phenomena that appear to play such an important role in mice and pigs and—who knows—perhaps in man. The chain of neuroendocrine events that can be modified by "odor" (whatever that may be chemically) must indeed be a fascinating area of research. In short, the field of the endocrinology of reproductive behavior—or, if you prefer, of behavioristic endocrinology—is wide open.

Finally, attention is called to another important factor that is too frequently overlooked by clinicians and even by research workers. The

---

*It is again suggested that the reader refer to the discussion "Bizarre Phenomena Related to the Estrous Cycle," Chapter 4.

rate of spontaneous recovery is quite high in subfertile females that show no evidence of anatomical abnormalities. All hard-to-settle females eventually conceive if they are bred often enough, and even females that are sterile because of cystic ovaries recover spontaneously and conceive if enough time is allowed. In view of such spontaneous recovery, it is essential to include in any experimental design an adequate number of control cases from which treatment is withheld. Failure to provide adequate controls is largely responsible for the enormous number of meaningless reports that appear in the clinical literature.

SELECTED READING

Blom, E., and N. O. Christensen. 1947. Studies on pathological conditions in the testes epididymis and accessory sex glands in the bull. *Skand. Veterinartidskrift.*

Eriksson, K. 1943. *Hereditary Forms of Sterility in Cattle.* Ohlsson (Lund, Sweden).

Garm, Otto. 1949. A study on bovine nymphomania. *Acta Endocrinologica,* Suppl. 3.

Kidder, H. E., W. G. Black, J. N. Wiltbank, L. C. Ulberg, and L. E. Casida. 1954. Fertilization rates and embryonic death rates in cows bred to bulls of different levels of fertility. *J. Dairy Sci.,* **37**:691.

Lagerlöf, N., and I. Settergren. 1953. Results of 17 years of control of hereditary ovarian hypoplasia in cattle of the Swedish Highland breed. *Cornell Vet.,* **43**:51.

Laing, J. A. 1957. Female fertility. In *Progress in the Physiology of Farm Animals,* Vol. 3, Butterworth.

Perry, J. S., and R. W. Pomeroy. 1956. Abnormalities of the reproductive tract of the sow. *J. Agr. Sci.,* **47**:238.

Phillips, R. W., and J. H. Zeller. 1941. Some factors affecting fertility in swine. *Am. J. Vet. Res.,* **2**:439.

Index